Emergency Management Threats and Hazards

Emergency Management Threats and Hazards: Water is a resource guidebook, which bridges the work of the emergency management practitioners and academic researchers, specifically for water-related incidents. Practitioners typically follow a disaster phase cycle of preparedness/protection/prevention, response, recovery, and mitigation – all of which have distinct actions and missions to reduce or eliminate adverse impacts from both threats and hazards. Academics will find the connections to allied fields such as meteorology, hydrology, homeland security, healthcare, and more. The book examines many of the distinct differences and variances within the specific scope of water-related incidents, crises, emergencies, and disasters. It provides examples and practical strategies for protection/prevention, response, recovery, and mitigation against adverse impacts to people, property, and organizations. It is also organized in the same construct used by emergency management practitioners (incident command system elements, disaster cycle phases, etc.), which will help align the academic world of emergency management education to both the practice and the training in the emergency management field.

- Takes a global view on threats and hazards, as well as their solutions.
- Provides a single repository of the majority of water-related incidents and provides a "how to" guide for resilience.
- Identifies cascading impacts and provides checklists for resolutions.
- Includes numerous case studies organized by threat and hazard.

Michael Prasad is a Certified Emergency Manager®, by the International Association of Emergency Managers, holding volunteer leadership positions in that organization as well. He is the Executive Director of the Center for Emergency Management Intelligence Research (www.cemir.org), and a national-level expert on mass care specifically on the feeding, sheltering, and caring for children in disasters. He has an MA in emergency and disaster management from American Public University and writes professionally on emergency management policies and procedures. Views expressed are his own, and not necessarily those of any of these organizations.

Emergency Management Threats and Hazards: Water

Michael Prasad

CRC Press is an imprint of the
Taylor & Francis Group, an **informa** business

Designed cover image: Shutterstock

First edition published 2025
by CRC Press
2385 NW Executive Center Drive, Suite 320, Boca Raton FL 33431

and by CRC Press
4 Park Square, Milton Park, Abingdon, Oxon, OX14 4RN

CRC Press is an imprint of Taylor & Francis Group, LLC

Library of Congress Cataloging-in-Publication Data
Names: Prasad, Michael, author.
Title: Emergency management threats and hazards : water / Michael Prasad.
Description: Boca Raton : CRC Press, 2024. | Includes bibliographical references and index. | Summary: "Emergency Management Threats and Hazards: Water is a resource guidebook, which bridges the work of the emergency management practitioners and academic researchers, specifically for water-related incidents. Practitioners typically follow a disaster phase cycle of preparedness/protection/prevention, response, recovery, and mitigation- all of which have distinct actions and missions to reduce or eliminate adverse impacts from both threats and hazards. Academics will find the connections to allied fields such as meteorology, hydrology, homeland security, healthcare, and more"—Provided by publisher.
Identifiers: LCCN 2024011949 | ISBN 9781032755151 (hardback) | ISBN 9781032755892 (paperback) | ISBN 9781003474685 (ebook)
Subjects: LCSH: Flood control. | Water quality management. | Emergency management.
Classification: LCC TC530 .P68 2024 | DDC 363.6/1 dc23/eng/20240415
LC record available at https://lccn.loc.gov/2024011949

ISBN: 978-1-032-75515-1 (hbk)
ISBN: 978-1-032-75589-2 (pbk)
ISBN: 978-1-003-47468-5 (ebk)

DOI: 10.1201/9781003474685

Typeset in Times
by Apex CoVantage, LLC

This book is dedicated to:

My wife, Angela – she is my water: powerful, dangerous, and yet life-sustaining.

My granddaughter, Emma – I wrote this book so that others will make the world better, for you and all the world's children.

> The earth, the air, the land and the water are not an inheritance from our forefathers but on loan from our children. So we have to handover to them at least as it was handed over to us.
>
> **Mahatma Gandhi**

Contents

PART 1 Introduction and Overview

PART 2 Threats around Water

PART 3 Quality Hazards with Water

About the Author

Michael Prasad is a Certified Emergency Manager®, a senior research analyst at Barton Dunant – Emergency Management Training and Consulting (www.barton dunant.com), and the executive director of the Center for Emergency Management Intelligence Research (www.cemir.org). He is also the chair of the Children and Disaster Caucus at the International Association of Emergency Managers, USA, and the vice president of their Region 2 grouping, which covers New Jersey, New York, Puerto Rico, and the U.S. Virgin Islands.

Prasad has held emergency management director-level positions at the State of New Jersey and the American Red Cross, serving in leadership positions on more than 25 disaster response operations, including Superstorm Sandy's response and recovery work. He currently serves as one of the liaison officers for Emergency Support Function #6 – Mass Care for FEMA's Region 2, when they activate their Regional Response Coordination Center. He researches and writes professionally on emergency management policies and procedures from a pracademic perspective, advises non-governmental organizations on their continuity of operations planning, and provides emergency management intelligence analysis for the National Security Policy and Analysis Organization at American Public University. He holds a Bachelor of Business Administration degree from Ohio University and a Master of Arts degree in emergency and disaster management from American Public University. The views expressed do not necessarily represent the official position of any of these organizations.

More details can be found at www.michaelprasad.com

Preface

This book was an idea I had after Superstorm Sandy struck New Jersey in 2012. That hurricane opened my eyes to the concept that water-related threats and hazards can become complex, quite quickly – and that a whole of government, whole-community effort before, during, and after such an incident can be productive. I was with the American Red Cross as an employee in New Jersey back then (I am still a volunteer), and New Jersey saw massive damage across all 21 of its counties. It was very significant on-the-job-training for me, to really understand and become educated on the full extent of federal support to the state – and how all the elements of emergency management can be implemented, even on just one disaster. Much of what I wrote about Superstorm Sandy in the Tropical Storms chapter is from my own first-hand experience. In 2020 – during quite a bit of home time during the pandemic – I started cataloging more and more ideas, links, articles, etc. around the elemental threats and hazards, starting with water. Once the book outline progressed into its parts, I soon learned I knew little about the global impacts to and from water; and even less how water resources are used in the western part of the United States. In many ways, in 2024, I had to stop sending myself website links and ideas for this book, because it is never ending.

Candidly, I have personally taken for granted the clean water supply I have enjoyed all my life. While I have traveled to parts of the world where clean water is not always the norm; back at home, I am still not as good of a steward of water as I should be. And my views on "wet stuff from the sky" and the devastation it can generate, have certainly changed since becoming an emergency manager myself. I never really feared the water – even in the ocean – but I now have much greater respect for water and cherish it, after writing this book. I hope you will, too.

I would like to thank the following people who helped me gain a better understanding of water:

Virginia Hogan, The American Red Cross
Randy D. Kearns, The University of New Orleans
Jun Kinoshita, California Governor's Office of Emergency Services
Kelly McKinney, NYU Langone Hospital
Lenny Layman, Carbon County, WY
Beth McGinnis, Clackamas River Water
David Morris, Albuquerque Bernalillo County Water Utility Authority
Chris Penningroth, WaterBlocks.net
Spencer Pollock, Wyoming Office of Homeland Security
Russell Rains, State of Kentucky
David Reilly, Alaska Division of Homeland Security & Emergency Management
Stuart Reiter, The American Red Cross
Steven Sarinelli, The American Red Cross
Daryl Schaffer, University of Alaska

Part 1

Introduction and Overview

1 Introduction

> Water is life's mater and matrix, mother and medium. There is no life without water.
>
> **Albert Szent-Györgyi, Nobel Prize-winning discoverer of Vitamin C**[1]

Water has always been a source of life on Earth. It can also be the source of death, devastation, and destruction. Leonardo da Vinci said, "Water is the driving force of all nature". And with that force brings both threats and hazards, as well. There are so many facets and infrastructure layers to any nation's oldest utility – the storage and delivery of clean water from somewhere else to where it is needed by people. There is a dam in Syria – the Quatihah Barrage/Lake Homs Dam – which was originally built by Egyptian Pharaoh Sethi during 1319–1304 BC, and which is still in use today.[2] The Roman *Acqua Vergine* was built in 19 BC and is still a functioning aqueduct, bringing clean, drinkable water to certain Roman fountains.[3] New York City's water system, first constructed in the early 1700s using carved out tree trunks for pipes,[4] now extends hundreds of miles into New York State with a watershed area the size of the U.S. State of Delaware, and has aqueduct tunnels underground, some of which are more than 20 ft (7.3 m) in diameter.[5]

And too much water can be a problem, as well. So can too little water – whether it be potable (i.e., drinkable) or non-potable. Quantity and quality issues abound globally when it comes to water. There are many sources of threats by and against forms of water, and there are hazards which can result from them. This book takes an Emergency Management perspective, and from both practitioner and academic viewpoints, as well. The book is designed to provide tangible tactics for all the phases of the disaster cycle (the before, during, and after something bad happens somewhere impacting communities) – to have a higher level of *readiness* to prepare/prevent/protect against the adverse impacts before they happen; to respond to and to recover from them, while they are happening; and finally, to mitigate against adverse impacts from water-related threats and hazards after they happen. The book will also include aspects of water-related incidents where initially there may be a single agency response structure to start, but ones where a more deliberate Whole Community/Whole-of-Government emergency management organizational construct should be put in place, especially to support the response/recovery activities and missions.

DOI: 10.1201/9781003474685-2

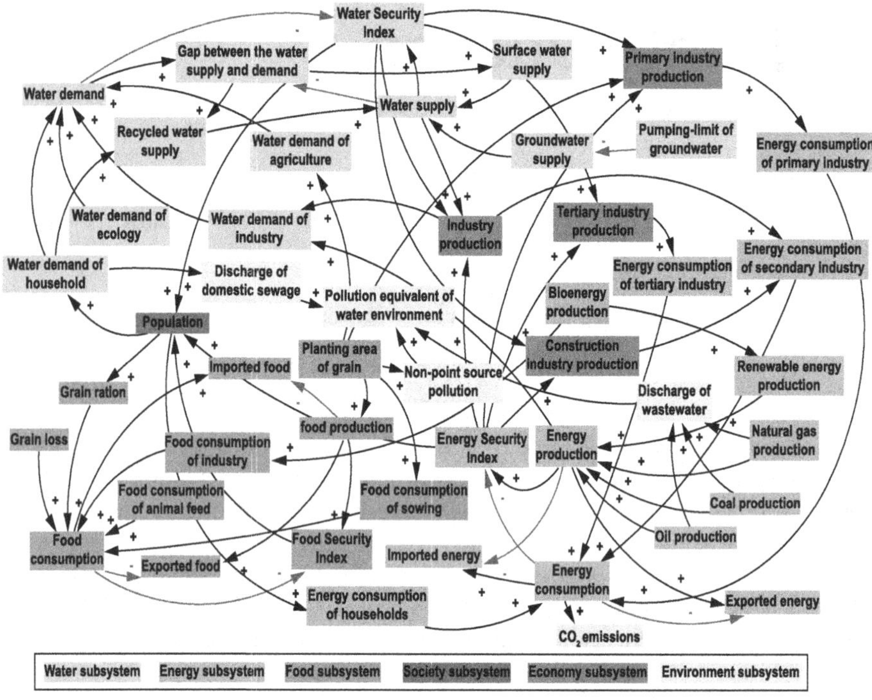

FIGURE 1.1 A system dynamics model of resource-based regions.

Source: Wen, C., Dong, W., Zhang, Q., He, N., & Li, T. (2022). A System Dynamics Model to Simulate the Water-Energy-Food Nexus of Resource-Based Regions: A Case Study in Daqing City, China. *Science of the Total Environment*, *806*, 150497. https://doi.org/10.1016/j.scitotenv.2021.150497. Used with permission.

WHAT DOES PRACADEMIC MEAN?

This book is designed to be "Pracademic". The term was first coined by Volpe and Chandler in 1999, to describe someone who has "dual identities, those of practitioner and academic" (Dickinson and Griffiths, 2023, p. 1).[6] Pracademic fits with Emergency Management quite well, as it is a blend of art, history, and science. There are no bystanders in Emergency Management. Definitively, the phrase "Those who cannot remember the past are condemned to repeat it" (Santayana, 1917, p. 284)[7] is quite applicable to Emergency Management.

This is not a tactical Field Operations Guide (FOG) for emergency operations, nor is it a book with complete, best-practice plans. It is, however, designed as a reference guide for emergency managers in any jurisdiction to move from Strategic into Operational Objectives in building their open Operational Plans, Incident Action Plans, and more – based on some of the unique characteristics of water-related threats and hazards. While this book has recommendations for specific Operational Objectives (i.e., priorities), it cannot be about specific tactics for any given incident. As is said in Emergency Management: If you have seen one disaster, you have seen one disaster. And this book should not be the only document/reference/source of intelligence that an emergency manager should rely on. Many of the historic and past missions described (i.e., tasks, assignments, roles, etc.) are very tactical and operational, but each incident is unique to the geography, staffing, politics, etc.

Water-related threats and hazards can also include those which are chronic, systemic, and endemic. Sea level rise and the corresponding adverse impacts to communities, due to Climate Change[8] is a prime example of both hazards and threats which require a whole-of-community approach to solving them, including non-governmental organizations (NGOs) such as the American Red Cross.[9] The same is true for groundwater supply issues.[10] Emergency Management can help organize boats to rescue families from their flooded homes, but not necessarily change the way people live and work near the water longer term. That falls more along the lines of steady-state urban planning, Whole Community mitigation and other jurisdictional aspects *above and beyond* the level and functionality of Emergency Management. Still, government officials, water system operators, urban planners, and others – including those with non-Emergency Management roles – can use this book for ideas, best practices, etc., when seeking change in their communities in the long term to help become more disaster ready or disaster resilient.

CONNECTING THE DOTS FOR CLIMATE CHANGE TO EMERGENCY MANAGEMENT

Constant increases in greenhouse gas emissions (carbon dioxide, methane, and nitrous oxide) help trap the sun's heat within our atmosphere, and lead to global warming and climate change. Climate Change itself should be considered a systemic disaster (one which occurs over longer periods of time, requiring whole community/whole government *or governments* solutions, and should not be solely managed through Emergency Management).[11] Climate Change has long-term and lasting adverse impacts to populations, economic aspects of every community – including agricultural[12] and tourism – and other sustainable elements of society.

Global warming leads to sea-level rise, which in turn leads to severe swings in both droughts and flooding of populated areas – and extending further into areas which never experienced droughts or flooding previously. Both extremes can adversely impact agriculture and other economic factors, which in turn can create adverse Public Health impacts, such as famine. Climate Change leads to more significant and frequent weather swings – more wildfires and more severe weather storms. These threats and hazards are **massively adversely impactful** when the earth's overall temperature rises by just 1.5°C (2.7°F).[13] And without immediate global intervention, there is research indicating that a 2°C temperature increase will have exponentially higher levels of excessive precipitation, increased change-induced water stress (i.e., droughts), and oceanic expansion (sea-level rise, polar ice melts, and acidity levels impacting ecosystems).[14] More heat waves and more flooding leads to crumbling infrastructure,[15] which can cause water main breaks, dam breeches, etc. Higher tidal levels can lead to "sunny day flooding".[16] All of these fall under the purview of Emergency Management.[17]

1.1 HOW TO USE THIS BOOK

This book is organized into four major parts, as shown in Figure 1.2, after this introduction and a brief review of Emergency Management concepts contained within:

- Threats around Water
- Quality Hazards with Water
- Quantity Hazards with Water
- Complex Incidents, including Both Quantity and Quality Hazards

Each part will have specific chapters covering the many types of unique hazards, described through case examples, as well as potential actions/alternative outcomes for full disaster cycle readiness (before, during, and after). A major goal

FIGURE 1.2 Part and chapter organization of this book.

Source: Barton Dunant. (2024). Part and Chapter Organization Infographic for *Emergency Management Threats and Hazards: Water.*

for this book is to provide insight into both best practices and areas for improvement for "the next time" any incident in and around **water-related threats and hazards** occurs.

The book is designed to fit within a larger written series of referenceable material. With content organized via categories of threats and hazards (for example, the four elemental ones of air, fire, land, and water; **plus**, Terrorism/Violent Extremism, CBRNE, cyber, and others), this reference material will provide an Emergency Management practitioner reference material and knowledge for Response and Recovery operational missions, plus Preparedness and Mitigation work such as threat/hazard/risk assessment analysis, grant applications, and more. Again, while this book is not designed as a FOG, there will be Incident Command System branch recommendations, potential impacts to the Emergency and Recovery Support Functions described, and delineated emergency management intelligence provided via Community Lifelines status reporting.

And in the United States, there are misaligned Federal regulations, laws, etc. between environmental, public health, and emergency management needs. For example, the regulatory environment for water supply organizations, corporations, and local governments must follow both the Clean Water Act[18] and the American Water Infrastructure Act of 2018 (AWIA),[19] and there are major unfunded mandates for water providers. Plus, these regulations do not necessarily align with FEMA guidance, such as their Threat and Hazard Identification and Risk Assessment (THIRA) and Stakeholder Preparedness Review (SPR) Guide Comprehensive Preparedness Guide (CPG 201),[20] or their Hazard Mitigation Plan[21] process for States, Local Jurisdictions, Territories, and Tribal Nations (SLTTs). Emergency managers are best qualified to understand a water utility's place in a community's overall disaster readiness planning and courses of action. Relationships between a water agency's Emergency Management and local government's Emergency Management must be strengthened.

RESOURCE PRIORITIZATION MUST BE COLLABORATIVE

Water providers may be too focused on system restoration that the emergency hydration/community-based missions are not seen as within their scope of operations. The need for meta-leadership[22] and breaking down silos[23] for culture change exists today. Resource prioritization, and also restoration priorities, must be collaborative and focus on life safety first and foremost.

What happens when two utilities are vying for the same resource, for example, a critical machine part? What if one utility knows that the part needed will allow restoration of water flow to a level 1 trauma center. The other utility needs the same part to restore water service to businesses in a prominent retail area. They may be in different counties, but if they coordinate across a multi-jurisdiction MAC System, that water MAC could determine based on ethical principles which utility gets the part. The MAC can also determine the order for who gets the next part when it is shipped. And even though there may be another MAC for healthcare, which wants the part for restoration first, the water MAC has to determine the implementation and determine specialty resources within its specific discipline.

Academics will find historical references and case studies for their use as well. There are additional online web resources, which can be found at www.CEMIR.org, which has an ever-expanding set of academic references, curated along these categories of water-related threats and hazards. And of course, elemental water threats and hazards are global and international – they can exist anywhere and everywhere. So best practices (and worst mistakes) can occur elsewhere – this book provides an opportunity to learn from both.

As our Earth is covered mostly in water, there are, of course, incidents which occur on ships and boats in rivers and the oceans. Unless multiple agencies get involved – and utilize an Emergency Management construct – these types of incidents will not be covered in this book. **There are water hazards even in outer space**. In 2013, an astronaut nearly drowned, during a spacewalk.[24] This book will also not consider these types of incidents as an Emergency Management threat or hazard – one which requires an incident command system to be established, changes in emergency response protocols and procedures, etc. Also, there will be future concerns with water in the stratosphere on airships, in outer space, and eventually on other planets. As people move toward interstellar and interplanetary travel – and work off our planet Earth – there will be new threats and hazards for many elements, including water. This book will not cover any of those interstellar threats or hazards – maybe in a second edition.

As noted above, this book is only focused on water-related incidents here on Earth – those generated by hazards of water. Some examples of Complex or Cascading disasters (the drought exasperated the vegetation's conditions, helping increase the adverse impacts of the wildfire), and those which have water as part of their *causality* (the building collapsed,[25] partially due to water leaks into the foundational structure) will be covered in Part 4, but those will only be examples of this disaster type, in this book. Another example: a mudslide[26] (or landslide) is elementally land-based, not water-based. And while sharks, alligators, and other sharp things can be found in the water, those are not being considered for this book, as water-related threats or hazards. Hopefully, readers will understand that the realm of any elemental threat or hazard, to include every possible water-related incident type, including derivatives, would create an endless book.

And a caution up front – this book should be considered educational guidance only, it is not a legal document, nor is it an authorized set of requirements issued by any governmental entity. While it contains references and segments of some official plan may have actually implemented – and the historic actions or inactions which were taken – those are not necessarily part of your jurisdiction's official plans and doctrine. All material and advice contained within this book must be used at the reader's discretion only. As it is true with any emergency management activity or mission, there are inherent risks to the practitioner, including risk of injury or death. If a suggestion is made to perform an action, such as communicating risks to the public, ordering evacuations of impacted populations, etc., *all* readers should always consider the prioritization of life safety, then incident stabilization, then property/asset protection, then environmental/economic concerns, and then finally recovery operations. Readers must have the **foresight** to know their own limitations, capabilities, and maintain their own personal safety and security. This book contains no material which is inaccurate, nor does it contain any statement, instruction material, or formula that involves the ***foreseeable*** risk of injury to readers or users of this book.

NOTES

1 Szent-Györgyi, A. (1971). Biology and Pathology of Water. *Perspectives in Biology and Medicine, 14*(2), 239–249. https://doi.org/10.1353/pbm.1971.0014
2 Water-Technology.net. (2013). *The World's Oldest Dams Still in Use.* Verdict Media Limited. Accessed November 10, 2023. See www.water-technology.net/features/feature-the-worlds-oldest-dams-still-in-use/
3 Water Science School. (2018). *Aqueducts Move Water in the Past and Today.* United States Geological Survey. Accessed November 10, 2023. See www.usgs.gov/special-topics/water-science-school/science/aqueducts-move-water-past-and-today
4 New York City. (n.d.). Water Supply History of New York City Drinking Water. *NYC Environmental Protection.* Accessed November 10, 2023. See www.nyc.gov/site/dep/water/history-of-new-york-citys-drinking-water.page
5 Water Technology. (n.d.). *New York City Water Tunnel 3.* Verdict Media Limited. Accessed November 10, 2023. See www.water-technology.net/projects/new-york-tunnel-3
6 Dickinson, J., & Griffiths, T.-L. (2023). Introduction. In J. Dickinson & T.-L. Griffiths (Eds.), *Professional Development for Practitioners in Academia: Pracademia* (pp. 1–8). Springer International Publishing. https://doi.org/10.1007/978-3-031-33746-8_1
7 Santayana, G. (1917). *Introduction, and Reason in Common Sense.* C. Scribner's Sons. See www.google.com/books/edition/Introduction_and_Reason_in_common_sense/_qL60vE_nKYC?hl=en&gbpv=0
8 See https://theconversation.com/faster-disaster-climate-change-fuels-flash-droughts-intense-downpours-and-storms-213242
9 See www.redcross.org/content/dam/redcross/about-us/publications/2023-publications/2023_ESG_Report_FINAL.pdf
10 See https://thehill.com/policy/energy-environment/4176511-four-in-10-groundwater-wells-hit-all-time-lows-in-last-decade-nyt/
11 See www.hsaj.org/articles/22285
12 Jiri, O., Phophi, M. M., Mafongoya, P. L., & Mudaniso, B. (2022). Climate Change Adaptation and Resilience on Small-Scale Farmers. In S. Eslamian & F. Eslamian (Eds.), *Disaster Risk Reduction for Resilience: Disaster and Social Aspects* (pp. 451–462). Springer International Publishing. https://doi.org/10.1007/978-3-030-99063-3_20
13 See https://news.un.org/en/story/2023/11/1143607
14 See https://climate.nasa.gov/news/2865/a-degree-of-concern-why-global-temperatures-matter/
15 See www.preventionweb.net/news/intense-heat-waves-and-flooding-are-battering-electricity-and-water-systems-americas-aging

16 Also known as nuisance flooding and King Tide flooding. Example at https://oceantoday.noaa.gov/flooding-sunny-day/
17 See www.preventionweb.net/understanding-disaster-risk/risk-drivers/climate-change
18 See www.epa.gov/laws-regulations/history-clean-water-act
19 See www.epa.gov/ground-water-and-drinking-water/americas-water-infrastructure-act-2018-awia
20 See www.fema.gov/sites/default/files/2020-07/threat-hazard-identification-risk-assessment-stakeholder-preparedness-review-guide.pdf
21 See www.fema.gov/emergency-managers/risk-management/hazard-mitigation-planning/create-hazard-plan/mitigation-planning-training
22 See https://domesticpreparedness.com/emergency-management/meta-leadership-2-0-more-critical-than-ever
23 Wolf-Fordham, S. (2020). Integrating Government Silos: Local Emergency Management and Public Health Department Collaboration for Emergency Planning and Response. *The American Review of Public Administration*, *50*(6–7), 560–567. https://doi.org/10.1177/0275074020943706
24 See https://abcnews.go.com/US/astronaut-drowned-space-due-nasas-poor-communication-report/story?id=22687977
25 See www.nist.gov/disaster-failure-studies/champlain-towers-south-collapse-ncst-investigation/background
26 See www.nytimes.com/2023/01/17/us/montecito-mudslide-2018-california-storms.html

BIBLIOGRAPHY

Dickinson, J., & Griffiths, T.-L. (2023). Introduction. In J. Dickinson & T.-L. Griffiths (Eds.), *Professional Development for Practitioners in Academia: Pracademia* (pp. 1–8). Springer International Publishing. https://doi.org/10.1007/978-3-031-33746-8_1

Jiri, O., Phophi, M. M., Mafongoya, P. L., & Mudaniso, B. (2022). Climate Change Adaptation and Resilience on Small-Scale Farmers. In S. Eslamian & F. Eslamian (Eds.), *Disaster Risk Reduction for Resilience: Disaster and Social Aspects* (pp. 451–462). Springer International Publishing. https://doi.org/10.1007/978-3-030-99063-3_20

New York City. (n.d.). Water Supply History of New York City Drinking Water. *NYC Environmental Protection*. Accessed November 10, 2023. See www.nyc.gov/site/dep/water/history-of-new-york-citys-drinking-water.page

Santayana, G. (1917). *Introduction, and Reason in Common Sense*. C. Scribner's Sons. See www.google.com/books/edition/Introduction_and_Reason_in_common_sense/_qL60vE_nKYC?hl=en&gbpv=0

Szent-Györgyi, A. (1971). Biology and Pathology of Water. *Perspectives in Biology and Medicine*, *14*(2), 239–249. https://doi.org/10.1353/pbm.1971.0014

Water Science School. (2018). *Aqueducts Move Water in the Past and Today*. United States Geological Survey. Accessed November 10, 2023. See www.usgs.gov/special-topics/water-science-school/science/aqueducts-move-water-past-and-today

Water Technology. (n.d.). *New York City Water Tunnel 3*. Verdict Media Limited. Accessed November 10, 2023. See www.water-technology.net/projects/new-york-tunnel-3

Water-Technology.net. (2013). *The World's Oldest Dams Still in Use*. Verdict Media Limited. Accessed November 10, 2023. See www.water-technology.net/features/feature-the-worlds-oldest-dams-still-in-use/

Wen, C., Dong, W., Zhang, Q., He, N., & Li, T. (2022). A System Dynamics Model to Simulate the Water-Energy-Food Nexus of Resource-Based Regions: A Case Study in Daqing City, China. *Science of The Total Environment*, *806*, 150497. https://doi.org/10.1016/j.scitotenv.2021.150497

Wolf-Fordham, S. (2020). Integrating Government Silos: Local Emergency Management and Public Health Department Collaboration for Emergency Planning and Response. *The American Review of Public Administration*, *50*(6–7), 560–567. https://doi.org/10.1177/0275074020943706

2 Emergency Management 101

Thousands have lived without love, not one without water.

W. H. Auden, poet, from "First Things First"[1]

Experienced emergency managers can probably skip reading this entire chapter. However, reading this chapter when a threat is looming on the horizon – or to be able to refer to this section when there is a need to explain to other non-emergency management professionals the "science" behind some of the work that Emergency Management does may be worthwhile. There are international concepts in the book as well, so there may be some missions or courses of action which have been performed more effectively in the past, in other jurisdictions elsewhere. This part of the book is where the level-setting, overuse of acronyms, and emergency management mantras and dictums will be shared. And as the book was written in 2023–2024, some of these will probably become outdated in the future. There may also be undiscovered good ideas for writing grant applications,[2] meeting compliance requirements for risk analysis or hazard mitigation planning,[3] or fact-based explanations to leadership as to why the Emergency Management team needs more staff,[4] due to climate change or other new threats. The overall goal for this book is to help communities become more resilient against threats and hazards to and from water, through the effective use of full cycle, all hazards Emergency Management.

THREATS VERSUS HAZARDS

Think of threats and hazards, like causes and effects. The threat is what has happened (or could happen), and the hazard is the adverse impact to people, places, and things that are important to us. People determine if it is a hazard. Some threats have some obvious-to-the-public hazards, such as tornadoes, wildfires, or tropical storms. Other threats are not as clear, such as an atmospheric river or global terrorism. Hurricanes, wildfires, earthquakes, and many other threats can become complex in their hazards, very quickly.

Threats are manifestations of risks. Many of them can be managed, mitigated against, etc. before they become hazards to populations. Risk factors covered in this book, originated by Harold D. Lasswell in 1936, expanded on by Peter M. Sandman to add outrage[5] as a factor, form a risk matrix, which help guide jurisdictions on allocating resources and problem-solving activities.

DOI: 10.1201/9781003474685-3

> Extreme weather events, critical change to Earth systems (a new entrant this year), and biodiversity loss and ecosystem collapse are the top three long-term risks featured in the (World Economic Forum's) Global Risks Report[6] in 2024. They are interrelated and mutually reinforcing. Abrupt and irreversible changes to Earth systems lead to more extreme weather events and risk collapses in ecosystems that are not well-adapted to new climates. The priority solution is faster emissions reduction and credible steps by all actors in our economic system to accelerate the speed and scale of a clean transition.
>
> Reducing human emissions is the swiftest lever to postpone or avoid critical changes to Earth systems. Once a climate tipping points has been reached, Earth's natural systems reinforce changes and so delaying these for as long as possible will give our civilization time to develop appropriate adaptation and resilience strategies.
>
> Here then lies the second priority for addressing systemic collapse from environmental risk – effectively adapting to coming changes. As sea levels rise, an ecosystem of interconnected solutions is needed to address threats to human life, landscapes and property.
>
> **(World Economic Forum, 2024, p. 1)[7]**

This book is organized with the threats first and then the possible hazards next. This part covers a number of basic concepts of Emergency Management: The 5W1H methodology[8] of "Who", "What", "When", "Why", and "How". The last "W" the "Where" is fluid (sorry for the pun), it is wherever the hazards occur or will occur. The book will also briefly highlight critical aspects of Emergency Management which needs whole community and whole of government attention. These are the threats and hazards – and adverse impacts – for which emergency managers are typically *responsible* for, in most cases, but higher levels of government and society itself, needs to be held *accountable*.

If Emergency Management is a new concept, this is a good chapter to start reading this book. For both academics (and those students who are studying Emergency Management, Disaster Management, Crisis Management, etc.) and practitioners who are new to the profession, this chapter will provide the U.S. models, policies, practices, etc. in effect in 2024. If located in a different country, the Emergency Management construct will be different. In the book, there will be case examples from around the world and recommendations but based primarily on the U.S. models.

Emergency Management is the operationalization of Risk Management, Consequence Management, Crisis Management, etc. It is performed before, during, and after incidents – generated by hazards from threats – occur. Emergency Management is further organized into branches or sections, which while in use during the Response phase, should also be staffed and allocated resources in the disaster phases which occur before and afterward. The problems which are solved can be modeled using an agile project management structure, which is also one which

FIGURE 2.1 Key emergency management acronyms and mantras.

Source: © Barton Dunant. Used with permission.

continuously improves itself through planning, organizing, equipping, training, and exercising.

Emergency Management in the United States has a number of five letter mantras, which are the "who, what, when, how, and why" of what needs to be performed, associated with threats and hazards. Figure 2.1 has them all in one infographic, and this chapter will detail each row. In this chapter, both commonly used acronyms and commonly used protocols, procedures, U.S. Federal doctrine, etc. will be introduced. Those acronyms can also be found in the Index.

2.1 WHO

Response staffing can come from within an Emergency Management organization, other local (same or above) jurisdictions as part of a memorandum of agreement/understanding, or even from other jurisdictions via an Emergency Management Assistance Compact (EMAC)[9] request from one governor to another. See below for more information on EMAC.

And note that Liaison Officers or LNOs can be from other government agencies, corporations (such as commercial water treatment facilities), and even non-governmental organizations (NGOs), such as the American Red Cross. More on the U.S.-based ICS system, below.

How many people does it take to provide emergency management services? That depends on whether it is considered steady-state work (Preparedness/Protection/Prevention and Mitigation) or disaster state work (Response and Recovery). For steady-state work, most jurisdictions have a small cadre of staff – and many have additional volunteers and part-time designated staff – who perform Emergency Management work. The disaster state work is primarily considered the purview of Emergency Services (police, fire, medical, public health, public works, etc.) but in larger scale incidents is supported and supplemented with additional staff – especially to fulfill the non-operational branches/sections of the incident command system constructed for this incident. This can be from mutual aid from other jurisdictions, partners who provide additional staff, national guard units, etc. The size and scope of how much staff is "deployed" or dedicated to an incident has been defined to five types by FEMA, as shown in Table 2.1.

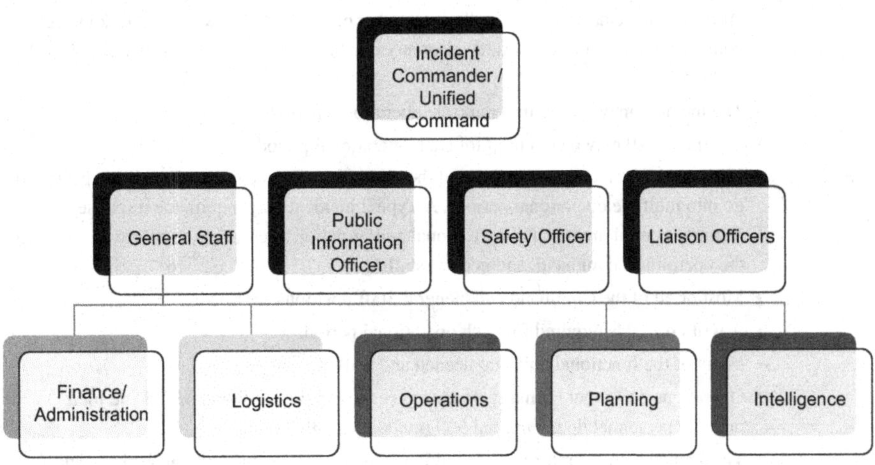

FIGURE 2.2 U.S. National Incident Management System/Incident Command System chart, including Intelligence.

Source: © Barton Dunant. Used with permission.

TABLE 2.1
U.S. Federal Incident Types

Incident Type	Description
Type 5	• The incident can be handled with one or two single resources with up to six personnel. • Command and General Staff positions (other than the Incident Commander) are not activated. • No written Incident Action Plan (IAP) is required. • The incident is contained within the first operational period and often within an hour to a few hours after resources arrive on scene. • Examples include a vehicle fire, an injured person, or a police traffic stop.
Type 4	• Command staff and general staff functions are activated only if needed. • Several resources are required to mitigate the incident, including a Task Force or Strike Team. • The incident is usually limited to one operational period in the control phase. • The agency administrator may have briefings, and ensure the complexity analysis and delegation of authority are updated. • No written Incident Action Plan (IAP) is required but a documented operational briefing will be completed for all incoming resources. • The role of the agency administrator includes operational plans, including objectives and priorities.
Type 3	• When capabilities exceed initial attack, the appropriate ICS positions should be added to match the complexity of the incident. • Some or all of the Command and General Staff positions may be activated, as well as the Division/Group Supervisor and/or Unit Leader level positions. • A Type 3 Incident Management Team (IMT) or incident command organization manages initial action incidents with a significant number of resources, an extended attack incident until containment/control is achieved, or an expanding incident until transition to a Type 1 or 2 IMT. • The incident may extend into multiple operational periods. • A written IAP may be required for each operational period.
Type 2	• This type of incident extends beyond the capabilities for local control and is expected to go into multiple operational periods. A Type 2 incident may require the response of resources out of area, including regional and/or national resources, to effectively manage the operations, command, and general staffing. • Most or all of the Command and General Staff positions are filled. • A written IAP is required for each operational period. • Many of the functional units are needed and staffed. • Operations personnel normally do not exceed 200 per operational period and total incident personnel do not exceed 500 (guidelines only). • The agency administrator is responsible for the incident complexity analysis, agency administrator briefings, and the written delegation of authority.

TABLE 2.1 *(Continued)*
U.S. Federal Incident Types

Incident Type	Description
Type 1	• This type of incident is the most complex, requiring national resources to safely and effectively manage and operate. • All Command and General Staff positions are activated. • Operations personnel often exceed 500 per operational period and total personnel will usually exceed 1,000. • Branches need to be established. • The agency administrator will have briefings, and ensure that the complexity analysis and delegation of authority are updated. • Use of resource advisors at the incident base is recommended. • There is a high impact on the local jurisdiction, requiring additional staff for office administrative and support functions.

Source: FEMA. (2019). *IS-0200.c Basic Incident Command System for Initial Response*. FEMA. See https://emilms.fema.gov/is_0200c/groups/518.html

WHEN TO USE AN ICS: BENEATH AND BEYOND

It's been said that "all disasters start and end locally" and this is also applicable to staffing for incidents of any scale. The more the resources needed – especially from beyond a single jurisdiction – the higher the level of typing for an incident. FEMA starts their credentialing, training, and experience monitoring/evaluating at the Type 3 level, but that does not mean that local responders to Type 4- and Type 5-level incidents should not have the credentialing training and experience monitoring for a higher level. An ICS can be used on a Type 4 or Type 5 incident – in some jurisdictions, it is required. In all cases, the benefits of using an ICS – including Incident Management System benefits such as span of control, unity of effort,[10] unified command, flexibility, and scalability – can benefit any single jurisdiction. Emergency managers should always consider the "what-if" the incident escalates – and how they need staffing to support additional needs and courses of action. Some incidents appear to be only responded to by a single emergency services response group or type, but the reality is that all incidents have emergency management components to them such as additional resource request processes, common operating picture, interoperability, planning, operational coordination, public information and warning, and emergency management intelligence. It is very apparent to a professional emergency manager when an incident is managed by a single emergency services group who does not have an Incident Command System in place. Any group who believes that ICS is "beneath them" on incidents usually results in ineffective protection of life safety, maintenance of incident stability, protection of assets and property, care for the environment, and/or the successful transition to Recovery.

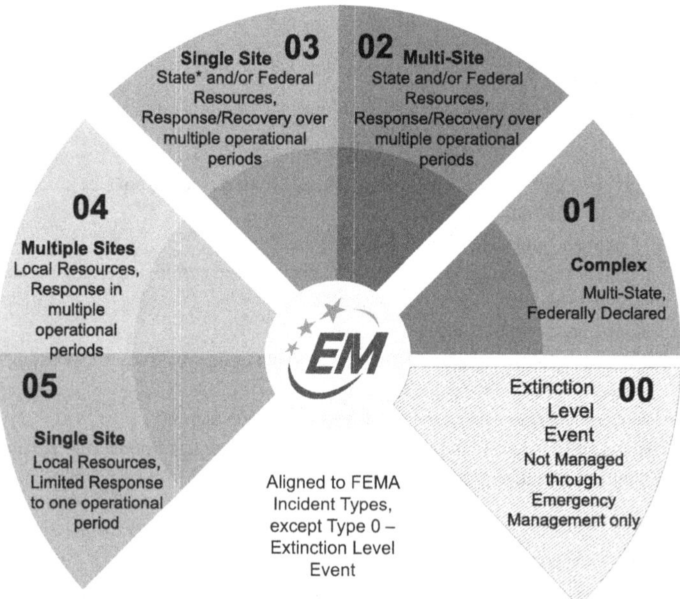

FIGURE 2.3 Incident Types.

Source: © Barton Dunant. Used with permission.

And there should be one more level – a Type 0 for extinction level event. Emergency managers need to understand there is a limit to the extremes of what types of incidents can be managed solely through Emergency Management. See Figure 2.3 for a more complete version of Incident Types.

> Emergency management has as one of its standards to "ratchet up and down" the resource support needed, based on the incident type. For example, when a Type 4 incident is still scaling up – meaning not yet under control or expanding – additional resources should be requested as if the incident could grow to a Type 3. While COVID-19 may have started in the United States as a small outbreak in Washington state on January 20, 2020, it expanded exponentially across the entire country in a matter of weeks. And at the same time, the virus spread worldwide. The size and scope moved the incident typing off the scale; there were no additional resources available anywhere, nor was there proper planning in place for this level of incident. There are hazards and threats for which emergency management cannot plan, organize, equip, train, and exercise because they are too complex. One way to quantify them is to describe them as Type 0 – Extinction Level Events.

> When the phrase "Extinction-Level Event" is mentioned, thoughts turn towards world-changing events – such as asteroid strikes, nuclear war, and even climate change/global warming. None of those tragedies have their response and recovery missions coordinated through their national emergency management process. There is a cap to the maximum of maximums of the capabilities and capacities for national and jurisdictional emergency management agencies and departments – as well as the concepts of the Incident Command System within the field of emergency management itself. When a disaster expands beyond the capability of the internal sub-jurisdictions within a nation, that jurisdiction usually requests assistance upward, all the way to the national level for support. When the nation itself needs support beyond its own capabilities, it can choose to reach out to partner nations, intergovernmental organizations, or non-governmental organizations for additional support (i.e., NATO, the United Nations, the Red Cross/Red Crescent National Societies across the globe, etc.). When all the nations are impacted at the same time by the same incident – and there is no one unimpacted left to help – that constitutes a worldwide catastrophe. Can any such incident be managed within a single nation's borders? Maybe, but not by or through emergency management, since the decisions about all aspects of the disaster phase cycle missions of Preparedness/Prevention/Protection, Response, Recovery, and Mitigation are a matter of national security and economic development.
>
> **(Prasad, 2023, p. 1)**[11]

2.2 WHAT

A full-cycle emergency manager must look at the totality of the incident, not just the Response aspects. Emergency Services[12] can (and should) focus on the first three letters of the acronym "LIP", which represent **L**ife Safety, **I**ncident Stabilization, and **P**roperty/Asset Protection (or conservation). Holistic Response (covering more than just what impacts people), Recovery and Mitigation work adds the last two letters of '**E**nvironmental/Economic Resilience' and transition to **R**ecovery, to round out the full acronym of LIPER: life safety, incident stabilization, property/asset protection, environmental/economic resilience, and recovery initiatives.

FIGURE 2.4 LIPER

THE LIPER, WATER UTILITIES, AND HOSPITALS AROUND PORTLAND, OR

The Multi-Agency Coordination System (MACS) is where this work happens in the Portland Metro Area. The MACS allows every agency to maintain their individual command and control, based on their legal authorities, but allows for coordination and prioritization as needed to ensure common good. The water system pumps water up a hill – under a lake – to three of the major medical facilities (including a Level 1 Trauma Center) serving Portland. Portland Water Bureau also serves many other small communities from the Bull Run Reservoir (on best days).

The MACS have a broader picture of the needs of multiple communities – and multiple sectors within each community. Utilizing the LIPER and the Community Lifelines, the Portland Metro Area gets everyone on the same page. For example, the MACS system has been extremely helpful in healthcare coordination through their healthcare coalition.[13] This allows healthcare facilities to understand the reality of restoration time frames and what it means for hospital services, after a water-related (or any other) disaster or incident. Restoration of water service to hospitals is more encompassing than just drinking water. It includes clinical water for steam sterilization, medication management and compounding, ultrafiltrations, etc. The water use audit tool from the CDC[14] is one aspect of Preparedness which hospitals can use to advise local Emergency Management groups of their potential water consumption rates (well beyond the Joint Commission's 72-hour water storage rule).[15] Until a hospital can articulate what water is needed for each clinical service line it has, it cannot self-prioritize how to discontinue services. Thus, the work of the healthcare coalition is to prioritize its own facilities for water restoration or temporary service to ensure equity across the healthcare system in a large urban metropolitan area, such as Portland. Water providers should not be performing that service prioritization, they should be determining their capacity to process the restorations based on what the healthcare community – through the MACS – tells them. And the MACS can help where ethical decision-making may come in between water providers. For example, eliminating competition for scarce resources (like staffing, specialized filters, pipe fittings, valves, etc.).

B. McGinnis, personal communication, December 21, 2023

These strategic objectives are codified in the U.S. National Preparedness Goal,[16] as mission areas and core capabilities. The five mission U.S. national areas are Prevention, Protection, Mitigation, Response, and Recovery. The 32 core capabilities can be found in Figure 2.5.

Prevention	Protection	Mitigation	Response	Recovery
Planning				
Public Information and Warning				
Operational Coordination				
Intelligence and Information Sharing		Community Resilience	Infrastructure Systems	
Interdiction and Disruption		Long-term Vulnerability Reduction	Critical Transportation	Economic Recovery
Screening, Search, and Detection		Risk and Disaster Resilience Assessment	Environmental Response/Health and Safety	Health and Social Services
Forensics and Attribution	Access Control and Identity Verification	Threats and Hazards Identification	Fatality Management Services	Housing
	Cybersecurity		Fire Management and Suppression	Natural and Cultural Resources
	Physical Protective Measures		Logistics and Supply Chain Management	
	Risk Management for Protection Programs and Activities		Mass Care Services	
	Supply Chain Integrity and Security		Mass Search and Rescue Operations	
			On-scene Security, Protection, and Law Enforcement	
			Operational Communications	
			Public Health, Healthcare, and Emergency Medical Services	
			Situational Assessment	

FIGURE 2.5 Core capabilities and mission areas infographic.

Source: FEMA. (2015). *National Preparedness Goal Second Edition*. U.S. Department of Homeland Security. See www.fema.gov/sites/default/files/2020-06/national_preparedness_goal_2nd_edition.pdf, p. 3.

While any threat or hazard can impact all or some of the core capabilities of a jurisdiction, for **water-related threats and hazards**, the following will most likely be adversely affected:

Mitigation: Community Resilience.
Response: Environmental Response/Health and Safety, Fire Management and Suppression, Public Health, Healthcare, and Emergency Medical Services.
Recovery: Natural and Cultural Resources.

This can be one major difference between Emergency Services/Emergency Responders and emergency managers. Emergency managers facilitate the prioritization of the LIPER – as shown in Figure 2.4 – against the mission areas and the core capabilities, in support of partners – including Emergency Services/Emergency Responders – and other groups. Emergency Management provides the full disaster phase cycle wrap-around support for all of the groups supporting Core Capabilities, across the Mission areas.

2.2.1 Chronic, Systemic, and Endemic Disasters

There are disasters which are beyond the scope of Emergency Management. As noted above, Extinction Level Events (Type 0) certainly require more ownership and collaboration above what only Emergency Management can provide. The same is true for continual, long-term disasters which are defined as chronic, systemic, and endemic – even when coupled[17] with disasters for which Emergency Management is in command. Disasters such as the opioid crisis in the United States and elsewhere, poverty, racism, refugee migration, etc. all fit into this grouping of disasters for which an Emergency Management command and control will not work effectively and may even harm critical aspects the LIPER. In the same way, Prasad (2023) noted that the global COVID-19 pandemic was beyond the capabilities of national Emergency Management's control; chronic, systemic, and endemic disasters have the same reasoning:

- Emergency management practice is jurisdictionally bound and generally follows a "bottom-up" approach, with resources for unmet needs coming from a higher level.
- The size and scope of the management system for any disaster response and recovery efforts are limited. At some point, the response efforts must become a whole-of-government approach, and therefore change management systems, because whole governments operate under a political management system instead of the ad hoc temporary structure of an emergency management system.
- Emergency management applies a straight-line approach to disasters, in a cyclical pattern. Even if there is an overlap between adjacent disaster cycle phases, they generally occur in order.
- Emergency management follows a unity of effort model; everyone in the response and recovery Incident Command System (ICS) is working toward the same goals and the same end-state.
- Emergency management – through any ICS in any country – is organized differently than the steady-state political-oriented governmental day-to-day operations. COVID-19, like any worldwide impacting incident, turned those systems upside-down. The ICS organizational branches of Command, Intelligence, Finance/Administration, Logistics, Operations, and Planning for every level of government were significantly impeded during COVID-19 (Prasad, 2023, p. 1).[18]

WHEN THREATS DO NOT GENERATE HAZARDS

And just because something is a threat does not automatically mean that it has hazards. There are causal and correlative connections between threats and hazards, but not always certainty. Wildfires occur all over the world in places where there are no people, no infrastructure, no real adverse impacts. That does not turn those incidents into disasters. Tropical storms spin up and then fizzle out, without ever making landfall or even impacting shipping lanes. Even earthquakes can have this lack of hazard impact, as shown in Figure 2.6.

A 7.0 magnitude earthquake is considered "major".[19] This one was in the ocean, off the coast of Indonesia. Open-source media indicated it was *actually* two earthquakes (averaging about 7.0 each), in an area approximately 255 km (158 miles) from the nearest island. No tsunami warning and no apparent damage to any infrastructure.[20] This same level of earthquake devastated Haiti in 2010. It is not the earthquake threat which generates a disaster, but rather the ways communities are adversely impacted by its hazards.

And while Emergency Management in the United States, through the National Preparedness Goal, has major mission areas of Protection and

Subject: [EXTERNAL] 2023-11-08 04:53:51 (M7.0) Banda Sea -6.5 129.4 (86c07)

M7.0 Earthquake - Banda Sea

Preliminary Report	
Magnitude	7.0
Date-Time	• Universal Time (UTC): 8 Nov 2023 04:53:51 • Time near the Epicenter (1): 8 Nov 2023 13:53:51 • Time in your area (1): 7 Nov 2023 23:53:51
Location	6.515S 129.366E
Depth	10 km
Distances	• 338.2 km (209.7 mi) SSE of Ambon, Indonesia • 342.8 km (212.6 mi) NE of Lospalos, Timor Leste • 354.6 km (219.9 mi) S of Amahai, Indonesia • 358.5 km (222.2 mi) S of Masohi, Indonesia • 387.4 km (240.2 mi) WSW of Tual, Indonesia
Location Uncertainty	Horizontal: 10.6 km; Vertical 1.8 km
Parameters	Nph = 41; Dmin = 835.0 km; Rmss = 1.14 seconds; Gp = 62° Version =
Event ID	us 7000l9h4

(1) Nominal Standard Time; local and daylight savings time zone may differ.
For updates, maps, and technical information
see: Event Page or USGS Earthquake Hazards Program
National Earthquake Information Center
U.S. Geological Survey

FIGURE 2.6 Screenshot of an alert sent by the U.S. National Geological Survey, for a 7.0 magnitude earthquake in the Banda Sea.

Source: Accessed August 11, 2023, from the author's e-mail account.

Prevention (typically utilized against human-made threats, such as terrorism), the causality associated with the Threats from water may not be capable of elimination through Mitigation (FEMA, 2015). On the other hand, Hazards can be mitigated against.

2.3 WHEN

In the United States, the current Federal model for the Disaster Phase Cycle is described as a linear one, from the Federal Emergency Management Agency's (FEMA's) National Disaster Recovery Framework (NDRF), as shown in Figure 2.7.

The worldwide SARS-CoV-2 coronavirus pandemic which started in late 2019 has turned this model "inside out" – no longer will phases only follow one another, but incidents themselves (or other incidents in the same jurisdiction, impacting the same Emergency Management teams) can occur over longer timespans and even at the same time, in more of a spiral, three-dimensional model, shown in Figure 2.8.

This book will provide an overall disaster readiness (or resiliency) view, in order to identify actions and activities necessary to protect and preserve life safety (and other elements of the LIPER), regardless of where a community is in the disaster phase cycle. Rather than looking at only a piece of the phases, it might be better to have a lens which can view more than one mission at a time, as in Figure 2.9.

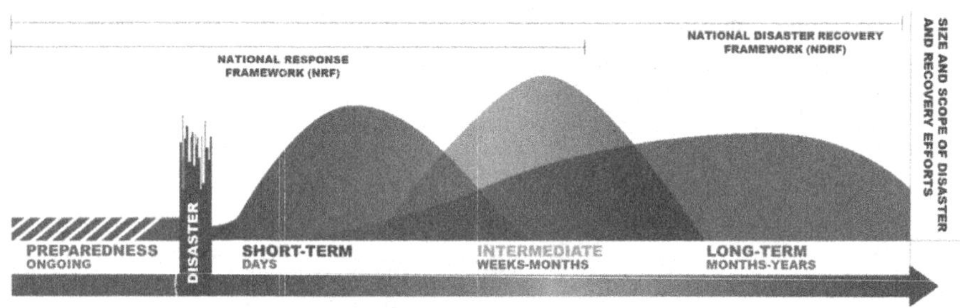

SHORT-TERM

- Health and safety needs beyond rescue
- Assessment of damages
- Restoration of basic infrastructure
- Mobilization of recovery organizations and resources
- Restarting and restoring essential services for recovery decision-making

INTERMEDIATE

- Return of individuals, families, critical infrastructure, and essential government or commercial services to a functional temporary or pre-disaster state
- Regional priorities, strategies, goals, and decision-making
- Risk assessment for remediation decision-making, including considerations for long-term human and environmental health monitoring
- Regional prioritization of area and facility recovery
- Characterization of extent of contamination
- Decontamination of contaminated items, areas, and facilities
- Site-specific remediation and restoration (characterization, decontamination, clearance), and reoccupation
- Waste management

LONG-TERM

- Continue remediation and restoration to meet clearance, restoration, and reoccupancy goals
- Complete redevelopment and revitalization of the impacted area
- Rebuild or relocate damaged or destroyed social, economic, natural, and built environments
- Move to self-sufficiency, sustainability, and resilience
- Continue waste management

FIGURE 2.7 NDRF Disaster Continuum.

Source: FEMA. (2016). *National Disaster Recovery Framework*. FEMA. See www.fema.gov/emergency-managers/practitioners/national-disaster-recovery-framework

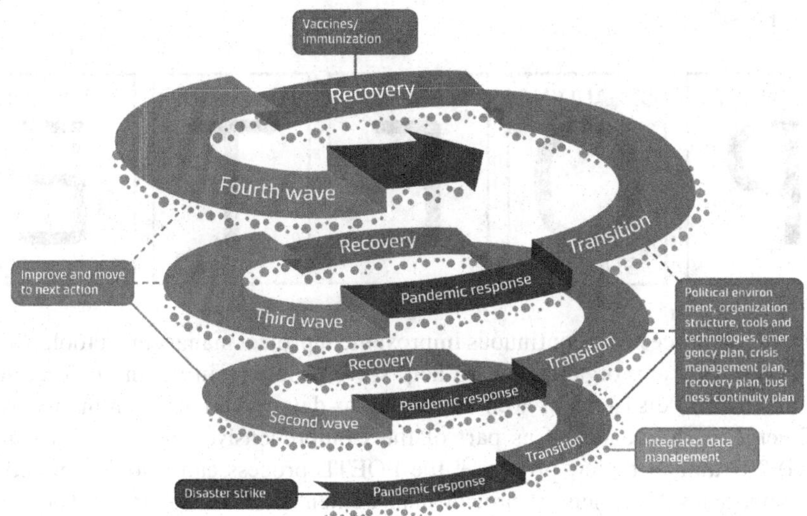

FIGURE 2.8 Spiral view of the COVID-19 disaster phase cycle.

Source: Fakhruddin, B., Blanchard, K., & Ragupathy, D. (2020). Are We There Yet? The Transition from Response to Recovery for the Covid-19 Pandemic. *Progress in Disaster Science*, *7*, 100102. https://doi.org/10.1016/j.pdisas.2020.100102. See www.sciencedirect.com/science/article/pii/S2590061720300399. CC BY-NC.

FIGURE 2.9 A representation of the four disaster phases aligned to the five U.S. mission areas.

Source: © Barton Dunant. Used with permission.

2.4 HOW

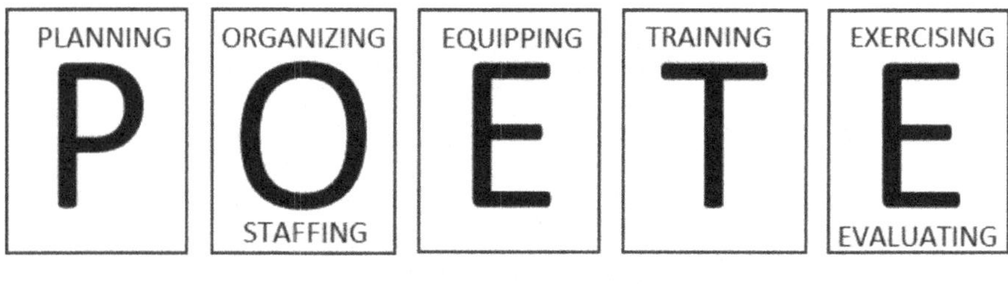

The POETE process is a continuous improvement project management tool, which can be used by any organization to help solve problems in (as in "before, during, and after") disasters, as well as for day-to-day operations. Commonly used in Emergency Management as part of the Comprehensive Preparedness Guide (CPG) 201 annual review process,[21] the POETE process can also be applied in non-Emergency Management project management/problem-solving.[22] The acronym represents Planning, Organizing (aka staffing), Equipping, Training, and Exercising – and can be applied on an all-hazards/all-threats basis for emergencies, disasters, and more:

Planning: It is the process of developing and implementing plans, in any organization – including continuity plans, contingency plans, emergency action plans, etc. To paraphrase U.S. President Dwight Eisenhower,[23] "Plans are not as valuable as is the Planning process". And of course, there is the classic:[24] "If you fail to plan, you plan to fail". This collaborative process of planning should result in new or modified plans, but there is value in the networking, the understanding of the capabilities and capacities of others, and your own internal organization's needs for cross-walking one set of plans with others.

Organizing: It is defining who does what in your plans, as well as who designed what in your plans. And who their backups are. This is usually more position-driven than name-driven (i.e., the press secretary will become the public information officer during the response operations, rather than "Sam is the public information officer"). This way the plans do not need to be updated as frequently as staff may change. This step also helps "right-size" the various roles needed for supporting steady-state and disaster state operations and needed continuity roles. In other words, it helps identify when some is assigned more than one hat to wear at the same time.

Equipping: It is the logistical tools (sites, systems, supplies, stuff) needed by the staff to support the work in the plans. This should also include contingency equipment (or upgrades to existing tools) needed at continuity sites when the threat dictates an evacuation from the primary site(s). In other words, if you use a fax machine for your day-to-day work, and must relocate to a hotel conference room to work, can you get a fax machine there? Or maybe your work cell phone does not have its "hotspot" activated for budgetary reasons, but during a disaster that feature could be turned on to allow you to work from home or on the road.

CRIME PREVENTION THROUGH ENVIRONMENTAL DESIGN

Emergency managers – especially those with homeland security or law-enforcement backgrounds/education – may have heard of the term Crime Prevention through Environmental Design, or CPTED. Much[25] of CPTED is the physical elements used to "harden" a target, such as building or critical infrastructure site.[26] What can be put in place to help protect against the adverse impacts of a terrorist attack, will also work against accidental mishaps as well. If emergency managers take an all-hazards approach (rather than just focusing on terrorism, per se), the elements of CPTED can be applied more holistically and collaboratively.

For example, adding bollards and high-earth berms around a water pump facility will help thwart a vehicle ramming attack on that site. It will also prevent someone from having a medical emergency in their vehicle and accidentally driving into the water pump facility. On the other hand, if the CPTED elements are not implemented collaboratively with other elements of public safety (such as fire suppression), then the all-hazards approach will not be met. If the water pump facility – or the water system's building next to it – is inaccessible to the fire department because of excessive CPTED elements – and there is a fire – it may not be suppressed.

Training: It is just that the training needed by the staff – using the equipment – to support the work is defined in the plans. If you are the press secretary, have you had crisis communications training? Do you know how to operate in a Joint Information Center (JIC), where you may not be the biggest fish in the pond? Training is also where emergency managers and others get to collaborate, coordinate, cooperate, and even communicate with our partners, suppliers, stakeholders, etc. Training together is a great opportunity to network and learn the capabilities and capacities of others, as well. "Train like you fight, fight like you train".[27]

Emergency managers have ample opportunity to train and educate themselves – both before they start their careers and through continuing educational opportunities. Regardless of whether one is a full-time paid or part-time volunteer, the needs of the profession demand constant and cyclical training. In the United States, water-related threats and hazards training[28] – along with overall all-hazards, full disaster phase cycle training – is available through the states (including free course offerings, such as MGT-343 Disaster Management for Water and Wastewater Utilities[29] from the National Domestic Preparedness Consortium – ndpc.us) and at the federal level (FEMA – training.fema.gov). Emergency Management training is also available from non-governmental organizations, such as the American Red Cross. Worldwide, the World Health Organization has online training classes and curriculums available at openwho.org.

Exercising: This is the step in the process which puts all of the other steps together – practicing what is in the plan, by the people designated in the plan (and their backups), using both primary and secondary equipment, and the training they have already received beforehand. Another key element of Emergency Management exercising is the after-action review/improvement plan process, which is the feedback loop for changes needed (to the plans, organization, equipment, training, and even future exercises, as determined by those who participated in the exercise(s) themselves).

In the United States, the U.S. Department of Homeland Security (USDHS) created a common platform and training for developing and implementing exercises, called the Homeland Security Exercise and Evaluation Program (HSEEP).[30] Emergency Management groups across all organizations, organization types, etc. should utilize the HSEEP modeling for collaborative and coordinated exercise development and execution.

WHAT EMERGENCY MANAGEMENT NEEDS THE PUBLIC TO DO, IN THE RESPONSE PHASE

1. Follow the directions given by local emergency management officials.
2. Help themselves and their neighbors, as much as they can.
3. Think Life Safety First.
    a. *Shelter-in-Place:* This means staying where you are and not evacuating. And the "you" described can mean you and your family, your coworkers, your pets, your personal care assistant, etc. You still get to be you – and please take into consideration all the unique aspects of you not mentioned specifically. Sheltering-in-Place could mean moving to a safer room in the building or a safer location in the complex or on the campus – but generally this is an easier decision where and when to go, when you have your own existing POETE process for this action. For evacuating, Planning means having a plan as to where you would go to shelter-in-place, knowing what you would need there for an extended stay perhaps, and how you would communicate with others if circumstances of the emergency change (for the better or worse). Organization means knowing who is doing what – who may be needed to help you shelter-in-place and then get you back to "normal". Equipment is the stuff you need to shelter-in-place. Emergency managers call this a "Stay Box". This is a container which can hold everything from shelf-stable food/pet food, medicine, battery-powered cell chargers, a flashlight, potable water, a change of clothes, sleeping bag, whatever. And this is stuff you need to have everywhere you might shelter-in-place, not just at home. So, you will need multiple stay boxes. Training is the process of reviewing the plan, asking questions, taking classes, and attending seminars on personal preparedness and asking questions of your leadership at work about emergency preparedness and

response. And finally, Exercising is the practice, practice, practice that everyone needs to do for emergency action planning. Practice your plan, practice using your equipment, practice with others, and practice some "what-ifs" from your training.

b. *Evacuate:* Evacuation means leaving where you are now, and not sheltering-in-place. It could mean evacuating the entire building, complex, or campus – or just evacuating to a safer place within the site where you reside. For that high-rise building fire, residents on the upper floors may evacuate to a different floor which has more fire protection and/or access to a fire stairwell, for example. Residents with mobility concerns may evacuate to what's called an Area of Refuge, which is a meeting spot for residents to be further evacuated by the fire department. The building may have a designated Emergency Assembly Point outside, for people to meet for accountability (where officials help make sure everyone has safely evacuated the building). And there is a POETE process for Evacuations. It is going to sound somewhat familiar to Sheltering-in-Place. Planning means having a plan as to where you would go now and the "what-ifs" for *if* you are able to return or not, knowing what you might need while you wait at the Emergency Assembly Point, and how you would communicate with others if circumstances of the emergency change (for better or worse). Organization means knowing who is doing what – who may be needed to help you evacuate and then get you back to "normal". Equipment, as mentioned before, is the stuff you need to evacuate. "Need" is the operative word. Emergency managers call this a "Go Bag". This can hold everything from food, medicine, battery-powered cell charger, electric wheelchair charging devices, flashlight, small amount of drinking water, glow stick, whatever. And as with the Stay Box, you need a Go Bag in different places, too. A handbag or backpack you carry with you everywhere will do. Training is the process of reviewing the plan, asking questions, taking classes, and attending seminars on personal preparedness and asking questions of your leadership at work about emergency preparedness and response. And finally Exercising is the practice, practice, practice that everyone needs to do for emergency action planning. Practice your plan, practice using your equipment, practice with others, and practice some "what-ifs" from your training. Note that the steps and actions for Evacuations should sound familiar to those for Sheltering-In-Place. Emergency managers designed this to be similar, for muscle-memory and consistency. There is a lot of preparation work involved for both, but it will be worth it. The point is to act on your plan, and not to panic.

c. *Both or One, Then the Other:* This is the case where different sets of directions from Emergency Management apply to different

> geographic areas of a jurisdiction. For example, it may be a storm which requires coastal evacuation, but sheltering-in-place, elsewhere. It may also mean for your family to evacuate first, then to shelter-in-place elsewhere, such as what is typically the directions from Emergency Management for tropical storm impacts. It can also be a situation where you shelter-in-place, and then evacuate when it is safer to do so (Prasad, n.d., p. 1).[31]

2.4.1 Emergency Support Functions

At the U.S. Federal level (and for most U.S. states/territories), the standardized Emergency Support Functions (ESFs) are the major components of the National Response Framework,[32] and are shown in Table 2.2. Governmental organizations are aligned as leads to these various ESFs. For example, the U.S. Department of Health and Human Services is the lead for ESF#8, as is most state departments of health. Table 2.2 also shows some examples of supporting actions or capabilities related to Public Health. In Chapter 3, the ESF specifics associated with water-related threats will be highlighted.

While not an ESF, educational and childcare facilities will be adversely affected as well, by any threats or hazards, not just water-related ones. A shortage of childcare facilities may mean emergency responders are impacted as well. If those responders

Federal Emergency Support Functions (ESFs)				
#1 - Transportation	#2 - Communications	#3 – Public Works and Engineering	#4 – Firefighting	#5 – Information and Planning
#6 – Mass Care, Emergency Assistance, Temporary Housing, and Human Services	#7 – Logistics	#8 – Public Health and Medical Services	#9 – Search and Rescue	#10 – Oil and Hazardous Materials Response
#11 – Agriculture and Natural Resources	#12 – Energy	#13 – Public Safety and Security	#14 – Cross-Sector Business and Infrastructure	#15 – External Affairs

Federal Recovery Support Functions (RSFs)		
Economic	Health & Social Services	Community Assistance
Infrastructure Systems	Housing Recovery	Natural & Cultural Resources

FIGURE 2.10 The U.S. Federal-level Emergency Support Functions and the Recovery Support Functions.

Source: Barton Dunant.

TABLE 2.2
U.S. Federal-Level Emergency Support Functions

Emergency Support Function (ESF)	Example Supporting Actions or Capabilities Related to Public Health and Healthcare
#1 – Transportation	Coordinate the opening of roads and manage aviation airspace for access to health and medical facilities or services.
#2 – Communications	Provide and enable contingency communications required at health and medical facilities.
#3 – Public Works and Engineering	Install generators and provide other temporary emergency power sources for health and medical facilities.
#4 – Firefighting	Coordinates federal firefighting activities and supports resource requests for public health and medical facilities and teams.
#5 – Information and Planning	Develop coordinated interagency crisis action plans addressing health and medical issues.
#6 – Mass Care, Emergency Assistance, Temporary Housing, and Human Services	Integrate voluntary agency and other partner support, including other federal agencies and the private sector, to resource health and medical services and supplies.
#7 – Logistics	Provide logistics support for moving meals, water, or other commodities.
#8 – Public Health and Medical Services	Provide health and medical support to communities, and coordinate across capabilities of partner agencies.
#9 – Search and Rescue	Conduct initial health and medical needs assessments.
#10 – Oil and Hazardous Materials Response	Monitor air quality near health and medical facilities in close proximity to the incident area.
#11 – Agriculture and Natural Resources	Coordinate with health and medical entities to address incidents of zoonotic disease.
#12 – Energy	Coordinate power restoration efforts for health and medical facilities or power-dependent medical populations.
#13 – Public Safety and Security	Provide public safety needed security at health and medical facilities or mobile teams delivering services.
#14 – Cross-Sector Business and Infrastructure	Be informed of and assess cascading impacts of health or medical infrastructure or service disruptions, and deconflict or prioritize cross-sector requirements.
#15 – External Affairs	Conduct public messaging on the status of available health and medical services or public health risks.

Source: FEMA (2019b, pp. 21–22). FEMA. (2019a). *National Response Framework.* U.S. Department of Homeland Security. See www.fema.gov/sites/default/files/2020-04/NRF_FINALApproved_2011028.pdf

have unmet childcare needs, they may not deploy, call out, etc. to take care of their families. This can be the case for all types of threats and hazards, regardless of whether there are evacuations or sheltering-in-place. Many jurisdictions utilize educational facilities for Response missions – if those sites do not have their own – or alternative methods – for supporting themselves for water quality and quantity threats, they may become unavailable/unusable for any other ESF's use – especially ESF #6 – Mass Care.

2.4.2 Recovery Support Functions

Recovery Support Functions (RSFs) are similar in structure and guidance modules, as the ESFs. The U.S. Federal Government first published the *National Disaster Recovery Framework*[33] in 2011, after more than two years of building the framework through a national stakeholder process:

This framework provides the following guidance to recovery leaders and stakeholders:

- Identifying guiding principles for achieving successful recovery.
- Outlining pre- and post-disaster roles and responsibilities for recovery stakeholders and recommending leadership roles across all levels of government.
- Describing how the whole community will build, sustain, and coordinate the delivery of the Recovery core capabilities.
- Explaining the relationship between Recovery and the other mission areas (Prevention, Protection, Mitigation, and Response).
- Promoting inclusive and equitable coordination, planning, and information-sharing processes.
- Encouraging the whole community to leverage opportunities to increase resilience and incorporate climate adaptation and mitigation measures pre- and post-disaster, such as continuity planning and land use and environmental regulations.
- Identifying scalable and adaptable organizations for coordinating recovery.
- Describing key factors, activities, and considerations for pre- and post-disaster recovery planning.
- Ensuring recovery resources are sourced from a wide range of whole community partners, including individuals and voluntary, nonprofit, philanthropic, and private sector and governmental agencies and organizations.

(USDHS, 2016, p. 2)[34]

The six U.S. Federal level RSFs are shown in Table 2.3. As with the ESFs, each U.S. state or territory can align their Recovery work differently, if they want to. Chapter 3 will have more specifics on the impacts of water-related threats on the RSFs. The chapters in Parts 3 and 4 will also have specifics for each hazard type.

TABLE 2.3
Federal Recovery Support Functions (RSFs)

Economic	Health and Social Services	Community Assistance
Infrastructure Systems	Housing	Natural and Cultural Resources

Since the Recovery Support Functions have coordinating, primary, and supporting agencies both inside and outside of the U.S. Federal Government, FEMA created a Recovery Support Function Leadership Group:

> The Recovery Support Function Leadership Group (RSFLG) is made up of multiple departments and agencies across the federal government that work together to help communities recover from a disaster. The RSFLG allows federal agencies to coordinate disaster recovery work under the National Disaster Recovery Framework (NDRF) across the six Recovery Support Functions in order to provide communities with unified federal assistance as quickly and effectively as possible.
>
> **(FEMA, 2022b, p. 1)**[35]

The Recovery Support Functions directly align to U.S. Federal-Level Recovery and Other Core Capabilities of Planning, Public Information and Warning, Operational Coordination, **Economic Recovery**, **Health and Social Services**, **Housing**, **Infrastructure Systems**, and **Natural and Cultural Resources**.[36]

2.4.3 Community Lifelines

In the United States, FEMA established a collection of essential elements of information, to enable the continuous operation of critical government and business functions, and which are essential to human health and safety, economic security, or both. In August 2023, FEMA updated their list of Community Lifelines (CLs) to break out Potable Water Systems and Wastewater Systems into its own Lifeline.[37]

FIGURE 2.11 FEMA's Community Lifelines, Version 2.1.

Source: © Barton Dunant. Used with permission.

Lifelines are the most fundamental services in the community that, when stabilized, enable all other aspects of society to function.

- FEMA has developed a construct[38] for objectives-based response that prioritizes the rapid stabilization of Community Lifelines after a disaster.
- The integrated network of assets, services, and capabilities that provide lifeline services are used day-to-day to support the recurring needs of the community and enable all other aspects of society to function.
- When disrupted, decisive intervention (e.g., rapid re-establishment or employment of contingency response solutions) is required to stabilize the incident.

Safety and Security (Law Enforcement/Security, Fire Service, Search and Rescue, Government Service, Community Safety): Human-made incidents will always have an intelligence/investigation aspect. There should always be a concern for a nexus to terrorism[39] and this can have multinational impacts as well. Law Enforcement and Security will also be tasked with protective missions during the Response phase.

Food, Hydration, Shelter (Food, Hydration, Shelter, Agriculture): People may evacuate their homes, even days after the incident. They may also need assistance with Food and Hydration as they shelter-in-place (i.e., stay at home or work).

Health and Medical (Medical Care, Public Health, Patient Movement, Medical Supply Chain, Fatality Management): Every incident, emergency, or disaster has a public health impact. Restoring the full- or near-functional capability of health and medical care is paramount to a community.

Energy (Power Grid, Fuel): All of the other Community Lifelines are in some way connected to this one. Energy produces electricity for consumer, industrial, and other organizational use. We are a global set of societies very dependent on continuous electricity supply.

Communications (Infrastructure, Responder Communications, Alerts Warnings and Messages, Finance, 911 and Dispatch): The systems, staffing, and crisis communications messaging are all part of this CL.

Transportation (Highway/Roadway/Motor Vehicle, Mass Transit, Railway, Aviation, Maritime): The ability to move goods, services, people, etc. needs to be restored after a disaster.

Hazardous Material (Facilities, HAZMAT, Pollutants, Contaminants): The cleanup of any hazardous materials is a critical component to the LIPER for an impacted community.

Water Systems (Potable Water Infrastructure, Wastewater Management): This new CL is also one which applies to all of the others. No potable water, no community. This book will cover this CL in much more detail and focus.

And the ESFs, RSFs, and CLs are all interconnected. Adverse impacts to one can and will impact others. These are all complex adaptive systems,[40] which require emergency managers **consistently and constantly review** their plans, organization, equipment, training, and exercises (the POETE), for all hazards and threats – including water-related ones.

WHOLE COMMUNITY AND WHOLE-OF-GOVERNMENT FUNDING

As noted previously, the phrase "Whole Community" appears more and more in Emergency Management doctrine, as it is one of the U.S. Federal level's guiding principles. It can represent two things, as the U.S. Federal level:

1. Involving different people in the development of the national Emergency Management doctrine
2. Ensuring their roles and responsibilities are reflected in the content of the materials produced at the national level

Whole Community and Whole-of-Government includes

- Individuals and families, including those with disabilities and access/functional needs (DAFN)
- Businesses
- Faith-based and community organizations
- Nonprofit groups
- Schools and academia
- Media outlets
- All levels of government, including state, local, tribal, territorial, and federal partners

While many human-made hazards – such as water contamination – will not qualify for federal financial assistance through FEMA and the NDRF, there are other aspects of U.S. Federal laws which may provide financial help to states, local governments, tribal entities,[41] and territories:

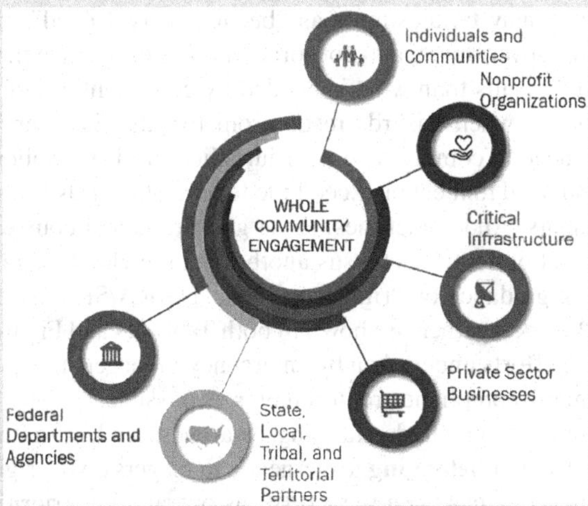

FIGURE 2.12 Whole community relationships.

Source: FEMA. (2022a). *E0237: Planning Process Theory and Application Student Manual*. FEMA.

> This Framework is not intended to alter or impede the ability of any local, regional/metropolitan, state, tribal, territorial, insular area, or Federal government department or agency to carry out its authorities or to comply with applicable laws, executive orders, and directives. Instead, it requires the whole community to coordinate or integrate individual authorities and missions. As the NDRF applies to all incidents, its structures and procedures apply equally to incidents where Federal support to local, regional/metropolitan, state, tribal, territorial, and insular area governments is coordinated under the Robert T. Stafford Disaster Relief and Emergency Assistance Act (Stafford Act), and to incidents where Federal departments and agencies exercise other authorities and responsibilities outside the Stafford Act. After the 2010 Deepwater Horizon oil spill, for example, Federal response and support was managed pursuant to the Oil Pollution Act. Other statutes such as the Homeland Security Act, Small Business Act, the Farm Bill, and the Public Health Service Act authorize substantive Federal assistance in response to certain types of incidents. The costs of direct Federal recovery support will continue to be borne by agencies using appropriations made for such purposes, except for those expenses authorized for reimbursement under the Stafford Act or as otherwise provided by law. When recovery requirements extend over long periods of time, steady state programs may be leveraged to support recovery efforts (FEMA, 2016a, p. 3).[42]

2.5 WHY

Why do emergency managers and Emergency Management exist? The long answer to these questions can be found in other books[43] and higher education programs.[44] The shorter answer may be as simple as "because it is needed". Our steady-state governmental, non-government, and corporate models of management systems have different goals and results than what is needed by communities before, during, and after a disaster state, when hazards result from threats. Emergency Management from SLTTs will need to comply with their jurisdiction's laws, policies, procedures, etc. They will also need to meet or exceed the federal standards from multiple agencies and departments. And sometimes, these guardrails and courses of action will conflict with one set of doctrines versus another. At the U.S. Federal Level, FEMA has provided some guidance on "rightsizing" the THIRA/SPR and Natural Hazard Mitigation Plan[45] work together, as shown in both Table 2.4 and Figure 2.11. The goal is to maximize the efforts undertaken by emergency managers toward Mitigation, so that it covers as many compliance requirements as possible.

Key to all of this is a better understanding, awareness, and, most importantly, **integration** of the concept of **belonging**. Emergency managers do the right things for the communities they serve, based on the LIPER as overarching priorities for missions and core capabilities. The best professional emergency managers and Emergency Management teams know and are already engaged with the varying demographics of their communities. This is a Whole Community approach.

TABLE 2.4
Aligning Mitigation Planning and Thira/Spr Processes.

Mitigation Planning Process	Unified Approach	THIRA-SPR Process
Mitigation Planning Step 1: Organize the Planning Process and Resources	Step 1: Involvement Across the Planning Area	Involve the Whole Community throughout each step of the THIRA/SPR process
Mitigation Planning Step 2: Assess Risks	Step 2: Threat and Hazard Identification	THIRA Step 1: Identify the Threats and Hazards of Concern
	Step 3: Risk Assessment	THIRA Step 2: Give the Threats and Hazards Context
Mitigation Planning Step 3: Develop a Mitigation Strategy	Step 4: Develop Capability Targets	THIRA Step 3: Establish Capability Targets
	Step 5: Identify Gaps	THIRA Step 3: Establish Capability Targets
	Step 6: Develop, Prioritize, and Operationalize Strategies	SPR Step 2: Identify and Address Capability Gaps
Mitigation Planning Step 4: Adopt and Implement the Plan	Step 7: Monitor and Adjust	THIRA/SPR Process

Source: FEMA. (2020). *FEMA Job Aid Increasing Resilience Using THIRA/SPR and Mitigation Planning*. FEMA. See www.fema.gov/sites/default/files/2020-09/fema_thira-hmp_ jobaid.pdf

WHAT DOES WHOLE COMMUNITY REALLY MEAN?

As of 2011, FEMA has taken a Whole Community approach to their POETE for major incidents and disasters.

> A Whole Community approach attempts to engage the full capacity of the private and nonprofit sectors, including businesses, faith-based and disability organizations, and the general public, in conjunction with the participation of local, tribal, state, territorial, and Federal governmental partners. This engagement means different things to different groups. In an all-hazards environment, individuals and institutions will make different decisions on how to prepare for and respond to threats and hazards; therefore, a community's level of preparedness will vary. The challenge for those engaged in emergency management is to understand how to work with the diversity of groups and organizations and the policies and practices that emerge from them in an effort to improve the ability of local residents to prevent, protect against, mitigate, respond to, and recover from any type of threat or hazard effectively.
>
> **Whole Community is a philosophical approach in how to conduct the business of emergency management.**
>
> Benefits include:

- Shared understanding of community needs and capabilities
- Greater empowerment and integration of resources from across the community

- Stronger social infrastructure
- Establishment of relationships that facilitate more effective prevention, protection, mitigation, response, and recovery activities
- Increased individual and collective preparedness
- Greater resiliency at both the community and national levels (FEMA, 2011, p. 3).[46]

The Whole Community approach is not infallible. If performed in an incomplete or discriminatory manner (for example, specifically – if not overtly and deliberately – excluding one group or another in the process), or if governmental officials believe that disaster Readiness in their jurisdictions can *organically be organized* by these various community groups themselves, **it will fail**. McKinney (2018) describes this mistake *by government* to *assume* Whole Community Readiness as a myth:

> And, like all myths, the Whole Community myth contains a grain of truth, because there are plenty of people working to be ready for disasters. It's just that the idea that it is happening spontaneously everywhere, in an organized way, to increase our collective resilience is a fiction, It is classic muddled thinking to say that everybody is doing something, since it is the same thing as saying that nobody is.
>
> **(McKinney, 2018, p. 39)**[47]

2.5.1 Inclusion, Diversity, Equity, and Accessibility

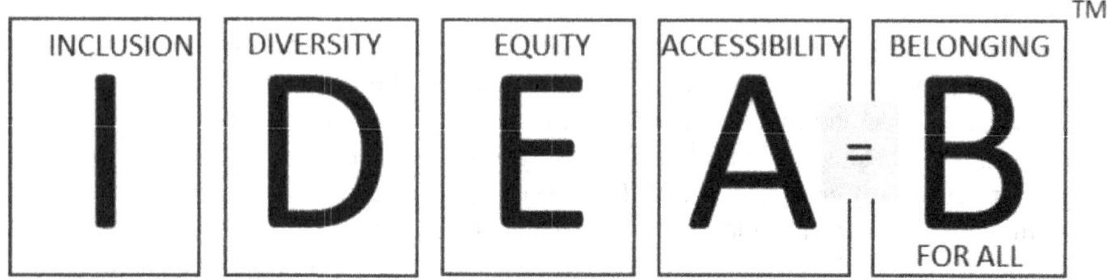

Source: Barton Dunant. © 2024. Used with permission.

In fact, in a best-case scenario, they themselves represent the communities they serve. They understand the different needs of their stakeholders, including those who are very dissimilar from any preconceptions of "the norm". This can all be performed through IDEA=B™: Inclusion, Diversity, Equity, and Accessibility equals a strong state of Belonging. In 2021, U.S. President Joseph Biden issued an *Executive Order*,[48] applicable to all Federal executive offices and departments, which had easy and yet strong definitions for the following:

Inclusion: The recognition, appreciation, and use of the talents and skills of employees of all backgrounds.

Diversity: Means the practice of including the many communities, identities, races, ethnicities, backgrounds, abilities, cultures, and beliefs of the American people, including underserved communities.

Equity: The consistent and systematic fair, just, and impartial treatment of all individuals, including individuals who belong to underserved communities that have been denied such treatment.

Accessibility: The design, construction, development, and maintenance of facilities, information and communication technology, programs, and services so that all people, including people with disabilities, can fully and independently use them. Accessibility includes the provision of accommodations and modifications to ensure equal access to employment and participation in activities for people with disabilities, the reduction or elimination of physical and attitudinal barriers to equitable opportunities, a commitment to ensuring that people with disabilities can independently access every outward-facing and internal activity or electronic space, and the pursuit of best practices such as universal design (FEMA, 2023, p. 3).[49]

Putting all four of these aspects together equals a greater sense of **Belonging**, when staff, partners, stakeholders, and the public are equitably treated, respected, and provided services fairly.

Belonging: Emergency Management has a distinct operational opportunity and mandate to ***be kind***. Kind to our coworkers, partners, and the public served. Kind to animals and the environment. Kind to a community's needs, as they deem as priorities. This is not only because in many cases it is required by law, but it is always the right thing to do.

Emergency Management should be viewed as one of many types of management techniques, styles, and/or systems. From a business school perspective, think of the various models such as "Management by Walking Around" and "Total Quality Management", as well as process improvement programs[50] such as "Six Sigma" and "Agile". Emergency Management is a blend of art and science and requires ***meta-leadership*** skills – the ability to drive results from those whom you have no power over, or even a presumed leadership position.[51] Organizations who have an emergency management component – such as governments – typically operate "day-to-day" using one of the traditional management techniques, styles, and/or systems. When they go into "disaster mode" (in government-speak, when changes in the command and control in a jurisdiction can be legally modified, based on emergency management laws and usually an emergency declaration by an authority having jurisdiction), they need to switch management and leadership roles (and many times rules) and may even delegate roles and responsibilities to others. The need to break the siloes of operations that may exist, even if they are "silos of excellence".[52] The CEO (or elected municipal leader) is rarely the Incident Commander. And emergency managers become resource wizards – of what has been described as the 4 S's of Logistics: Space (or Sites), Staff, Stuff, and Systems.[53]

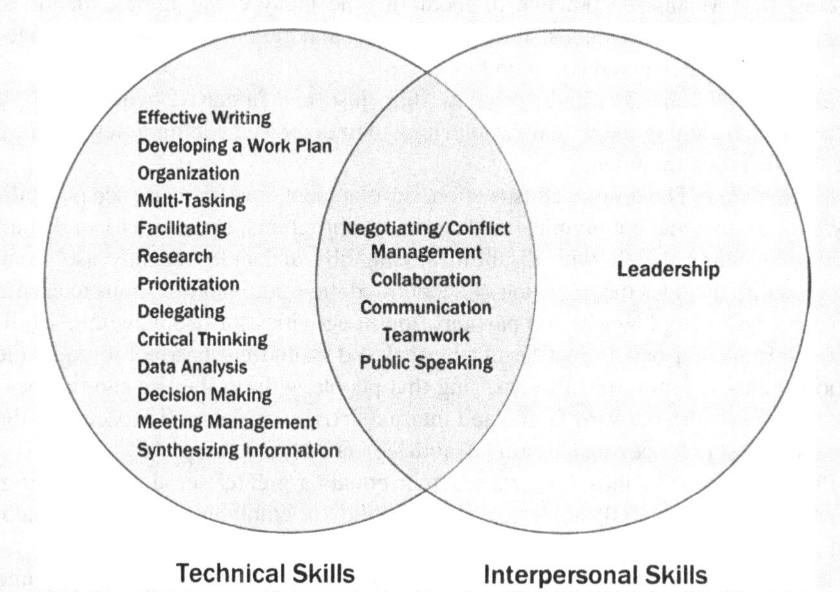

FIGURE 2.13 Emergency management skill sets.

Source: FEMA (2022). *E0237: Planning Process Theory and Application Student Manual*. FEMA.

See Figure 2.13 for a graphical depiction of *some* of the skills which an emergency manager needs to be successful.

Emergency managers need to be a leader and also a team player, kind and yet direct, ambitious as well as centered, and most importantly an expert while still learning.

2.5.2 International Aspects of Emergency Management

2.5.2.1 Incident Command Systems

In the United Kingdom, these Emergency Management actions may be performed by a Strategic Co-ordinating Group (SCG) under the Joint Doctrine's Interoperability Framework.[54] In Qatar, various elements may be distributed through departments of the National Command Centre,[55] such as finance/logistics elements through the Administrative Affairs Department and others through the Central Operations Department. Emergency Management Intelligence may be curated and distributed through the Technical Affairs Department.

New Zealand's Coordinated Incident Management System[56] and the Australasian Inter-service Incident Management System (AIIMS)[57] both are similar to the US' NIMS. In New Zealand, there are branches for Control, Safety, Intelligence, Planning, Operations, Logistics, Public Information Management, Welfare and Recovery (in Response). In Australia, it may be very dependent on which agency is the primary responder, as to how the incident command system is organized.

Regardless of **who** is performing the missions, they need to be performed. Regardless of who reports to whom throughout the disaster phase cycle, emergency managers recognize the important concepts of adaptability, scalability, and uniform technology. They also know the importance of *Management by Objectives*, ***Functional Management Structures***, and the concept of *Span of Control*. It should be straightforward for a seasoned emergency management practitioner in any country, to find the right place for the right role to accomplish the missions needed.

2.5.2.2 Tropical Storm Naming Conventions

Where a severe tropical storm is located will determine what it is called. It is a "hurricane" if it is in the North Atlantic Ocean or the Northeast Pacific Ocean (i.e., both coasts of the United States). It is called a "cyclone" if it is found in the South Pacific Ocean and the Indian Ocean. It is also called a "typhoon" in the Northwest Pacific Ocean. See Figure 2.14. All these terms mean the same thing.

The World Meteorological Organization names tropical cyclones (storms) internationally. They produce lists of names for each of the ocean areas where tropical storms can occur. Their names are designed to be short in character length, easy to pronounce and communicate, be culturally appropriate in different languages, and unique in that the same names are not used in multiple regions in the same year.[58]

2.5.3 Essential Elements of Intelligence

While most emergency managers will learn of the term "Essential Elements of Information", this book will replace the word "Information" with the word "Intelligence". It connotates a much deeper and broader understanding of the needs for Emergency Management. Knowing the name and telephone number of the Water Supply company is information. Knowing whether they are interconnected to other water systems now is intelligence. The Essential Elements of Intelligence (EEIs) can be directly correlated to the Community Lifelines, plus some communities may have additional ones such as political boundaries/cross-border relationships. How a jurisdiction interacts with a sovereign tribal nation which is geographically located in that jurisdiction should be an EEI for both groups' emergency managers.

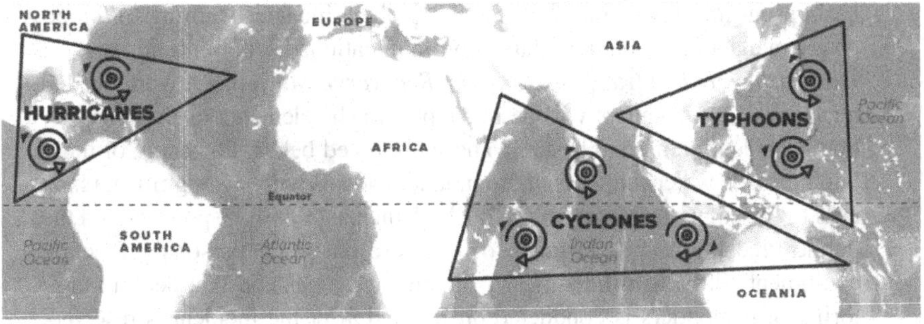

FIGURE 2.14 Tropical storms 101.

Source: Humanitarian Coalition of Canada. (2023). *Crash Course: Tropical Storms 101*. Humanitarian Coalition. Accessed January 3, 2024. See www.humanitariancoalition.ca/crash-course-tropical-storms-101. Used with permission.

2.5.4 Impacts to the Disaster Phase Cycles and Mission Areas

Again, as noted previously, threats generate hazards. So, the reverse is true as well – if you can mitigate against the adverse impact of hazards consistently, you will also mitigate against a number of different threats, which generate those same or similar hazards. For example, creating alternative water treatment plants helps mitigate against both accidental and intentional threats to the community's water supply systems. So does interconnecting one community's water supply system to another's. Will these actions solve all possible threats? Probably not – but the more layers, alternatives, backups, etc. a community has, the better.

For human-made water supply contamination, a significant water quality hazard, there are adverse impacts to all of the disaster phase cycles: before, during, and after. This is a way to describe overall community (and therefore individual) disaster readiness or resiliency. When communities are protecting against, preventing, preparing to respond to, recovering from, and mitigating against the adverse impacts of water-related threats and hazards, then the response to those adverse impacts is greatly reduced and/or more effective.

1. *Before the (next) Crisis Occurs (Preparedness/Protection/Prevention, Mitigation work):* Much of the work here by emergency managers is informative, through communications with industry, other governmental entities, and the public. The same type of messages to protect watersheds and waterways from pollution and abuse can be applied to multiple groups. For example, in the United States, their federal Environmental Protection Agency has a program on watershed protection[59] which includes training for emergency management and templated communications for communicators.
2. *During the incident (Response, interim Recovery):* Much of the work needed by emergency managers will be in support of the elements of that jurisdiction's incident command system. Those are noted below. For many of these human-made water supply contamination, incidents will have significant hazards to life safety (and the other LIPER elements) of both responders and the public. Precautions must be taken to make sure that proper personal protection equipment (PPE) is utilized near contaminated areas and that health monitoring of responders is conducted during and after the incident is resolved. All incidents have a public health (Emergency Support Function #8) impact, including all of the water-related threats and hazards. A full list of example impacts of human-made water supply contamination on the ESFs and RSFs

are noted below. There may also be parallel or collaborative criminal investigation missions being conducted, which may be part of response and recovery work. Those will be noted in the Impacts to the Incident Management System(s) used in the impacted jurisdiction as well.

3. *After the incident Response has concluded or reverted back to steady-state management organization and away from emergency management (long-term Recovery, initial Mitigation planning):* This area of work for emergency managers can be the most arduous: The community needs are still excessive, but the spotlight and "attraction" by the non-impacted public, the media, politicians, etc. is significantly diminished. Writing grant applications for hazard mitigation grant funding,[60] balancing urban planning needs with environmental ones, coordinating with non-governmental organizations for individual assistance, and continued public health surveillance monitoring[61] any potential health concerns are just a few examples of the emergency management work needed in this disaster cycle phase. In the United States, work with federal agencies, such as FEMA and the U.S. Army Corps of Engineers (USACE).[62]

2.5.5 Incident Command System

This book will use the U.S. NIMS Incident Command System model, shown in Figure 2.15 – with the formal addition of Emergency Management Intelligence – as the major groupings or categorization of elements of the incident command system:

a. Incident/Unified Command: Commander's Intent must incorporate the long-term view of problem-solving. And that view for Response missions must take into account environmental concerns along all of the Recovery Support Functions. For example, solutioning a bypass to a section of contaminated land in a riverbed by creating a new channel for the river water to flow around the contaminated area will require multiple entities and agencies to evaluate existing environmental concerns, economic impacts to the local jurisdiction of these changes, long-term testing of water quality, and more. Those decisions will undoubtably have financial, staffing, and other impacts beyond those involved in the incident command system for this incident.

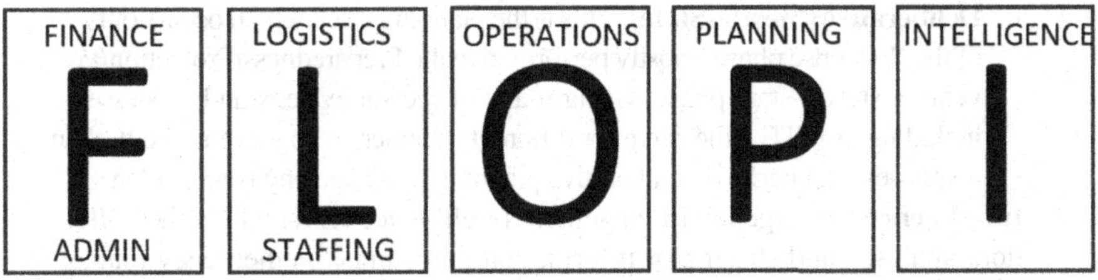

FIGURE 2.15 Major elements of the U.S. NIMS ICS model.

Source: Barton Dunant. Used with permission.

   b. **Safety Officers:** As previously noted, the role of Safety Officer is critical to ensure responder safety, especially in geographic areas where there is hazardous material contamination.
   c. **Liaison Officers:** Can be from other government agencies, corporations (such as commercial water treatment facilities), and even non-governmental organizations (NGOs), such as the American Red Cross. They represent those organizations to Command and can help connect disparate Incident Command Systems together.
   d. **Public Information Officers:** Are the communicators to the public – and usually internally within the operation – on behalf of Command. Their job is also to collect Emergency Management Intelligence from the media and the public directly. This feedback loop is critical for a successful operation.
4. **Finance/Administration:** Provides the support for funding and documenting the resources utilized throughout the ICS. This group also contains any compliance or regulatory reporting requirements.
5. **Logistics:** Provides the staffing, systems, sites, support items and materials – the "stuff" for all elements of the ICS.
6. **Operations:** This is the public facing group of the ICS, besides the PIO. Missions, courses of action, etc. via the priorities of the LIPER are conducted through the Operations branch.
7. **Planning:** Can be divided into three parts, as part of the ICS – and there is also a steady-state function of deliberative planning:
   a. **Strategic:** The highest level, usually "marching orders" from non-emergency management leadership (i.e., elected/appointed officials) and should always meet or exceed the LIPER priorities. "Make sure that impacted residents are feed and hydrated" is one example of a strategic priority, as part of the planning process.
   b. **Operational:** What is planned for this operational period to meet the strategic planning. "Big Ticket" items such as "Provide potable water to 10,000 residents today" are an operational level interpretation of the strategic priority above.
   c. **Tactical:** The detailed planning efforts involved in the "how" the operational objective will be accomplished, and in this operational period. "Relocate tractor-trailers with sixteen pallets of bottled water (30,000 16-oz individual bottles) to Point-of-Distribution #3 by 0800 hours" is but one line in a tactical plan.
   d. **Deliberative (Steady-State):** This is the planning work which occurs outside of the Response phase, mostly performed in the Preparedness/Protection/Prevention steady-state phase. Using many of the same steps and processes – including POETE – the same (or different) planners who serve in the ICS on responses, can perform deliberative planning to update any type of plan.
8. **Intelligence:** Emergency Management Intelligence (EMINT) is the collection, analysis, and sharing of information that can help emergency managers make informed decisions before, during, and after disasters. EMI can come from a variety of sources, including public records, social media, and eyewitness accounts. It can be used to identify potential threats, assess damage, and plan for recovery. EMI is an important part of emergency

management because it can help to reduce the adverse impacts of disasters. By understanding what is happening and what is likely to happen, emergency managers can make better decisions about how to allocate resources and how to better protect people, places, and things.[63]

2.6 ARTIFICIAL INTELLIGENCE AND EMERGENCY MANAGEMENT

As of 2023, the concept of applying Artificial Intelligence (AI) to emergency management is novel. Emergency managers can start planning documents, threat assessments, grant application language, and more – all using (mostly free) web-based artificial intelligence systems to generate the words, from the vast resources worldwide, on the internet. All of this, of course, is subject to verification and validation by Emergency Management Intelligence analysis and curation.

AI can also be used to provide quicker, sometimes more profound, information for Emergency Management to use in making response decisions, but that needs to be carefully vetted and validated beforehand. For example, deciding to dispatch generators to a hospital versus a shelter, when both have life safety adverse impacts of not having electricity, may benefit from a faster and more complete risk analysis and assessment performed by a computer, but the decision to take one action over the other should be made by a responsible human.[64]

2.7 EMERGENCY MANAGEMENT ACCREDITATION PROGRAM

In the United States, governmental organizations[65] and jurisdictions – especially states – have the voluntary opportunity to present their overall Emergency Management program for accreditation by an independent organization:

> The Emergency Management Accreditation Program (EMAP), as an independent non-profit organization, fosters excellence and accountability in Emergency Management and Homeland Security Programs by establishing credible standards applied in a peer-reviewed Assessment and Accreditation Process. With ongoing concerns about terrorism, pandemics, and catastrophic natural disasters, the nation's leaders and citizens acknowledge the need to strengthen emergency preparedness measures and response capabilities efficiently and effectively. The Emergency Management Standard by EMAP and the voluntary Assessment and Accreditation Process are intended to promote consistent quality in Emergency Management Programs, thus providing tangible benefits to the community and public infrastructure these Programs serve. Many Programs[66] utilize the Standards and the Assessment and Accreditation Process for strategic planning, improvement efforts, and resource allocations.
>
> **(EMAP, 2023, p. 4)**[67]

EMAP has several standards, all of which must be met for accreditation. **Standard 4.5.3** – which addresses the Emergency Operations Plan – has areas which are impacted by and from water-related threats and hazards:

- 2: agriculture and natural resources
- 5: critical infrastructure and key resource restoration
- 12: energy and utility services

- 15: firefighting/fire protection
- 16: food, water, and commodities distribution
- 20: mass care and sheltering
- 23: private sector coordination
- 27: transportation systems and resources (navigable waterways count as a transportation systems resource)

There is obvious – and purposeful – alignment between the EMAP standards and the ESFs, RSFs, and CLs. Meeting all of the standards of EMAP is a high bar – and one which professional emergency managers in government organizations should seek to achieve.

2.8 EMERGENCY MANAGEMENT ASSISTANCE COMPACT

One tool in a U.S. state or territory's Emergency Management toolbox for obtaining additional resources (people, systems, stuff, etc.) on any response is through mutual aid. Typically, one community has mutual aid agreements with its neighbors for Emergency Services (such as fire, police, emergency medical services, etc.). When states and territories need help themselves, before they ask FEMA (and sometimes at the same time), they can ask each other for mutual aid, through the Emergency Management Assistance Compact (EMAC):[68]

> EMAC has been ratified by U.S. Congress (PL 104–321) and is law in all 50 states, the District of Columbia, Puerto Rico, Guam, the U.S. Virgin Islands, and the Northern Mariana Islands. EMAC's Members can share resources from all disciplines, protect personnel who deploy, and be reimbursed for mission-related costs.
>
> **(EMAC, n.d., p. 1)**[69]

EMAC legislation provides liability protection, compensation/reimbursement, workman's compensation, and license reciprocity, among other features. A number of disciplines are covered by EMAC, including Animal Emergency, Emergency Medical Services, Fire and Hazardous Materials, Human Services, Incident Management, Law Enforcement, Mass Care, Public Health and Medical, National Guard, Public Works, Search and Rescue, and Telecommunicator Emergency Response.

2.9 DISASTERS AND CHILDREN

> Unless someone like you cares a whole awful lot, Nothing is going to get better. It's not.
>
> **Dr. Seuss (Theodor Geisel)**[70]

And finally, for this overview of basic Emergency Management, a section on disasters and their adverse impacts on children. The adverse impacts which any disaster has on children must be considered by emergency managers holistically – across all the other dimensions of ESFs, RSFs, CLs, POETE, etc. In the same way that Emergency Management needs to create physical and emotional safe spaces for Belonging, additional spaces need to be designated for children and families with children.

THE DISASTER OF CHILDREN DROWNING

In 2023, child drownings were still the leading cause of death in the United States for children ages 1–4.[71]

Please let that sink in for a moment.

It is certainly an open question as to whether emergency managers should get involved in the disaster cycle for drownings. There are pracademics[72] in Emergency Management, who make the case for direct engagement in the full spectrum of PPRRM[73] because this hazard fits much of the criteria of the scope of Emergency Management. Multiple agencies respond, including police, fire/EMS, hospitals, etc. Multiple agencies – including whole-community ones – support prevention and protection elements, such as teaching children how to swim and providing lifeguarding education for communities. And multiple organizations can collaborate, coordinate, cooperate, and communicate on mitigation strategies and courses of action to reduce this deadly threat.

If a life safety hazard exists in a community – especially one which adversely impacts socially vulnerable populations at a much higher level – emergency managers have a responsibility to be part of the solution.

In 2010, a National Commission on Children and Disasters was convened. Their report[74] to the President and Congress still has work to be done. FEMA and the Texas A&M Engineering Extension Service (TEEX)[75] both are working toward expanding an emergency manager's knowledge into children and disaster issues, threats, hazards, critical infrastructure (i.e., childcare and educational facilities), etc., as part of deliberative planning.[76]

> The points made about restoring those facilities alone, for first and emergency responders, should be a priority to sufficiently have the workforce needed to support the response and recovery from any disaster. The International Association of Emergency Managers – Children and Disasters Caucus is also working on a long-term project to amplify the disaster needs of infants, children, and young adults (from pre-K through college) to the same levels of Community Lifelines, Emergency Support Functions and/or Recovery Support Functions. Emergency managers need to consider their community's children as unique and vital in many ways.
>
> **(Prasad, 2021, p. 1)[77]**

And children with disabilities of any kind, connect these two global concerns together – the nearly exponential need to support children with disabilities[78] before, during, and after disasters. The U.S. Federal Laws which require an individualized educational program (IEP) for children with any type of disability – including cognitive, emotional, and behavioral ones – has had advocacy[79] made to extend these individualized services into emergency and disaster planning. The strong emergency manager will get ahead of this curve and collaborate with their local school district and others, to help ensure that all children are covered through the LIPER priorities.

All emergency managers should complete FEMA's "Community Preparedness: *Integrating the Needs of Children*" workshop,[80] the online self-study IS-366A: Planning for the Needs of Children in Disasters[81] course from FEMA, and the MGT-439 Pediatric Disaster Response and Emergency Preparedness[82] course from TEEX, to have a more comprehensive understanding of the complexities involved with disasters adversely affecting children.

FIGURE 2.16 Drawing by E.H. Shepard from *Winnie the Pooh*.

Source: Public domain.

CHILDREN AND DISASTER: SPECIFIC REFERENCES/ADDITIONAL READING

Arao, B., & Clemens, K. (2013). From Safe Spaces to Brave Spaces: A New Way to Frame Dialogue Around Diversity and Social Justice. In *The Art of Effective Facilitation* (1st ed., pp. 135–150). Routledge. https://doi.org/10.4324/9781003447580-11

Garrett, A. L. (2019). The Role of the Federal Government in Supporting Domestic Disaster Preparedness, Response, and Recovery. *Current Treatment Options in Pediatrics, 5*, 255–266. https://doi.org/10.1007/s40746-019-00162-7

Gerry, L. M. (2023). *National Geographic Readers: Water!* Penguin Random House.

National Commission on Children and Disasters. (2010). *2010 Report to the President and Congress.* Agency for Healthcare Research and Quality. See www.acf.hhs.gov/sites/default/files/documents/ohsepr/nccdreport.pdf

Prasad, M. (2021). Space Aliens – Emergency Management Roles & Responsibilities. *Domestic Preparedness Journal.* See https://domesticpreparedness.com/cbrne/space-aliens-emergency-management-roles-and-responsibilities

2.10 GENERAL EMERGENCY MANAGEMENT MAPPING AND TOOLS

In Table 2.5, please find *some* of the general emergency management mapping and geospatial intelligence system (GIS) tools. This is not a comprehensive list, but rather a starting point for a new emergency manager. All these websites are free to access from the internet and are current as of publication date of this book. Some are governmental and some not. No endorsement or financial incentives were involved in providing this list. There are other sites which have subscription costs, require additional GIS software[83] and/or specialized governmental authorization (for example, the U.S. Homeland Infrastructure Foundation-Level Datasets[84] or DamWatch®,[85] from the U.S. Engineering Services, contracted by the USACE). Most Emergency Management organizations already have access to these. There are also additional water-related websites listed in Chapter 3.

TABLE 2.5
General Emergency Management Mapping and Tools

Title/Description	Sponsor/Description	Website Address
(CAMEO) Software Suite Computer-Aided Management of Emergency Operations	U.S. Environmental Protection Agency	www.epa.gov/cameo/
Climate Risk and Resilience Portal (ClimRR)	Center for Climate Resilience and Decision Science – Argonne National Laboratory	https://climrr.anl.gov/
Climate.gov	NOAA	www.climate.gov/

(Continued)

TABLE 2.5 ***(Continued)***

General Emergency Management Mapping and Tools

Title/Description	Sponsor/Description	Website Address
Coastal Emergency Risks Assessment	University of Louisiana	https://coastalrisk.live/
Coastal Flood Risk – Flooding Risk Mapping and Flood Insurance Maps	FEMA	www.fema.gov/flood-maps/coastal
Current event maps of natural and man-made hazards	Radio Distress-Signaling and Infocommunications (RSOE)	https://rsoe-edis.org/eventMap
Dam Safety	Association of State Dam Safety Officials®	https://damsafety.org/
Dam Alliance	US Society on Dams	https://www.ussdams.org/
Fire and Smoke maps	The Gleaner	https://data.thegleaner.com/fires/
Fire/Smoke/Air Quality Map	Interagency Wildland Fire Air Quality Response Program	https://fire.airnow.gov/
Flood Risk Communication Toolkit	FEMA	www.fema.gov/es/floodplain-management/manage-risk/communication-toolkit-community-officials
Geospatial Resource Center	FEMA	https://gis-fema.hub.arcgis.com/
Hurricane Decision Support Tool – Hurrevac	National Hurricane Program	https://hvx.hurrevac.com/hvx/
Living Atlas	ESRI	https://livingatlas.arcgis.com/en/home/
Minimum Economic Recovery Standards (MERS) Interactive Handbook	SPHERE	https://handbook.hspstandards.org/mers
National Drought Mitigation Center	University of Nebraska	https://drought.unl.edu/
National Integrated Drought Information System	NOAA	www.drought.gov/
Natural Hazards Research Australia	Australian Government	www.naturalhazards.com.au/our-research-focus
OnTheMap for Emergency Management – combined natural hazard modeling with demographics impacted	U.S. Census Bureau	https://onthemap.ces.census.gov/em/
PrepToolkit and the National Preparedness System	FEMA	https://preptoolkit.fema.gov/
Rainfall Totals	CoCoRaHS Mapping System	https://maps.cocorahs.org/
Severe Weather	ESRI	www.esri.com/en-us/disaster-response/disasters/severe-weather
Social Vulnerability Index	U.S. Centers for Disease Control and Prevention	www.atsdr.cdc.gov/placeandhealth/svi

TABLE 2.5 *(Continued)*

General Emergency Management Mapping and Tools

Title/Description	Sponsor/Description	Website Address
Transit Security Report	Global Incidents Map	www.globalincidentmap.com/
U.S. Tornado Map – past 48 Hours	University of Michigan	https://tornadopaths.engin.umich.edu/
US Census Data	U.S. Census Bureau	https://data.census.gov/
US Current Flooding Map – GIS version of flood gauges and other data points	ESRI	https://bit.ly/ESRI_US_FLOOD
US Flight Tracker and Weather Radar overlay	Flightaware	https://flightaware.com/live/
US flight tracker, including cargo, small aircraft, military, and Helo operations	FlightTrader	www.flightradar24.com/37.93,-84.83/9
US Medicare Electricity-Dependent Map	U.S. Health and Human Services	https://empowerprogram.hhs.gov/empowermap
Warn, Alert, and Response Network	Public Broadcasting System	https://warn.pbs.org/
Water Forecast Services and Flood Inundation Maps – *experimental* collection	NWS	www.weather.gov/news/241101-national-water-prediction
Weather Radar	NWS	https://radar.weather.gov/
Web Tools	U.S. Geological Survey	www.usgs.gov/products/web-tools

REFERENCES/ADDITIONAL READING

Each chapter in the Hazards Parts 3, 4 and 5 will also have additional resources, references, and reading sections. Also, there are comprehensive footnotes throughout the book, which reference additional websites of interest for Emergency Managers.

Boyarksy, A. (2024). *Riding the Wave: Applying Project Management Science in the Field of Emergency Management*. Routledge. https://doi.org/10.1201/9781003201557

California Department of Water Resources. (n.d.). *California Water Watch*. State of California. Accessed January 13, 2024. See https://cww.water.ca.gov/

California Governor's Office of Emergency Services. (2023). *California State Emergency Plan Coordinating Draft*. The State of California. See www.caloes.ca.gov/wp-content/uploads/Preparedness/Documents/2023-SEP-Draft-Public-Review.pdf

Council on Environmental Quality. (2022). *Climate and Economic Justice Screening Tool*. Executive Office of the President of the United States. See https://screeningtool.geoplatform.gov/en/#3/33.47/-97.5

Fagel, M. J., Mathews, R. C., & Murphy, J. H. (Eds.). (2021). *Principles of Emergency Management and Emergency Operations Centers (EOC)* (2nd ed.). CRC Press. https://doi.org/10.4324/9781315118345

FEMA. (2015). *National Preparedness Goal Second Edition*. USDHS. See www.fema.gov/sites/default/files/2020-06/national_preparedness_goal_2nd_edition.pdf

FEMA. (2020). *Building Codes Save: A Nationwide Study. Losses Avoided as a Result of Adopting Hazard-Resistant Building Codes.* FEMA. See https://www.fema.gov/sites/default/files/2020-11/fema_building-codes-save_study.pdf

FEMA. (2023). *Achieving Equitable Recovery: A Post-Disaster Guide for Local Officials and Leaders.* FEMA. See www.fema.gov/sites/default/files/documents/fema_rr-508_EquityGuide_20231108_508Final.pdf

Humanitarian Coalition of Canada. (2023). *Crash Course: Tropical Storms 101.* Humanitarian Coalition. Accessed January 3, 2024. See www.humanitariancoalition.ca/crash-course-tropical-storms-101

IvyPanda. (2020). *Incident Command Models in the US, the UK, Qatar.* See https://ivypanda.com/essays/incident-command-models-in-the-us-the-uk-qatar/

Lasswell, P. D. (1936). *Politics: Who Gets What, When, How.* Whittlesey House.

McKinney, K. (2018). *Moment of Truth: The Nature of Catastrophes and How to Prepare for Them.* Savio Republic.

Molefi, N., O'Mara, J., & Richter, A. (2021). *Global Diversity, Equity & Inclusion Benchmarks: Standards for Organizations Around the World.* QED Consulting. See www.qedconsulting.com/component/content/article/104-services/products/161-global-diversity-equity-inclusion-benchmarks?Itemid=566.

National Governors Association. (2014). *Governor's Guide to Mass Evacuation.* NGA. See www.nga.org/wp-content/uploads/2018/08/GovGuideMassEvacuation.pdf

Phillips, B. D., Neal, D. M., & Webb, G. R. (2022). *Introduction to Emergency Management and Disaster Science* (3rd ed.). Routledge. See www.routledge.com/Introduction-to-Emergency-Management-and-Disaster-Science/Phillips-Neal-Webb/p/book/9780367898991

Sandman, P. M. (1988). Risk Communication: Facing Public Outrage. *Management Communication Quarterly*, 2(2), 235–238. https://doi.org/10.1177/0893318988002002006

The White House. (n.d.). *Justice40: A Whole-of-Government Initiative.* The White House. Accessed December 31, 2023. See www.whitehouse.gov/environmentaljustice/justice40/

* * * *

Emergency managers should take any of the free American Red Cross disaster courses available, but they should especially take the Red Cross' "Everyone is Welcome" course on inclusion, diversity, equity, and accessibility. It is available for both Red Cross volunteers and community partners. The web self-study version is 20 minutes long, and the in-person course is 45 minutes in duration. Contact your local American Red Cross chapter for details, at www.redcross.org.[86] There may be a similar course at other countries' Red Cross or Red Crescent Societies.[87]

NOTES

1 Auden, W. H. (1957). First Things First. *The New Yorker.* See www.newyorker.com/magazine/1957/03/09/first-things-first

2 FEMA. (2023). *Swift Current.* FEMA. Accessed December 8, 2023. See www.fema.gov/grants/mitigation/flood-mitigation-assistance/swift-current

3 See www.fema.gov/sites/default/files/2020-06/fema-local-mitigation-planning-handbook_03-2013.pdf
4 McKinney, K. (2018). *Moment of Truth: The Nature of Catastrophes and How to Prepare for Them.* Savio Republic, p. 220.
5 Sandman's work to add public outrage as a factor in risk analysis (Risk = Hazard + Outrage) has not been universally adopted in Emergency Management. In the highly volatile and expedited global communications world today, it should be factored in. See www.psandman.com/index.htm
6 See www.weforum.org/publications/global-risks-report-2024/
7 World Economic Forum. (2024). *Global Risks Report 2024: The Risks Are Growing – But So Is Our Capacity to Respond.* United Nations Office for Disaster Risk Reduction. Accessed January 20, 2024. See www.preventionweb.net/news/global-risks-report-2024-risks-are-growing-so-our-capacity-respond
8 Ma, N., & Liu, Y. (2020). Risk Factors and Risk Level Assessment: Forty Thousand Emergencies over the Past Decade in China. *Jamba (Potchefstroom, South Africa), 12*(1), 916. https://doi.org/10.4102/jamba.v12i1.916
9 See www.emacweb.org/
10 See https://domesticpreparedness.com/articles/key-bridge-collapse-unity-of-effort
11 Prasad, M. (2023). Global Pandemics Are Extinction-Level Events and Should Not Be Coordinated Solely through National or Jurisdictional Emergency Management. *Pracademic Affairs, 3.* Naval Postgraduate School/Center for Homeland Defense and Security. See www.hsaj.org/articles/22285
12 Dos Santos, V. M., & Son, C. (2023). Modern Firefighters' Three-Level Situation Awareness in Fire and Non-Fire Incidents. *Proceedings of the Human Factors and Ergonomics Society Annual Meeting.* https://doi.org/10.1177/21695067231192647
13 See https://oregonclho.org/
14 See www.cdc.gov/healthywater/emergency/pdf/19_302124-e_ewsp-grab-n-go-p.pdf
15 See www.jointcommission.org/standards/standard-faqs/ambulatory/emergency-management-em/000001216/
16 U.S. Department of Homeland Security. (2015). *National Preparedness Goal Second Edition.* FEMA. See www.fema.gov/sites/default/files/2020-06/national_preparedness_goal_2nd_edition.pdf
17 Jhung, M. A., Shehab, N., Rohr-Allegrini, C., Pollock, D. A., Sanchez, R., Guerra, F., & Jernigan, D. B. (2007). Chronic Disease and Disasters: Medication Demands of Hurricane Katrina Evacuees. *American Journal of Preventive Medicine, 33*(3), 207–210. https://doi.org/10.1016/j.amepre.2007.04.030
18 Prasad, M. (2023). Global Pandemics Are Extinction-Level Events and Should Not Be Coordinated Solely through National or Jurisdictional Emergency Management. *Pracademic Affairs, 3.* NPS Center for Homeland Defense and Security. See www.hsaj.org/articles/22285
19 See www.mtu.edu/geo/community/seismology/learn/earthquake-measure/magnitude/
20 See www.reuters.com/world/asia-pacific/magnitude-69-earthquake-strikes-banda-sea-region-indonesia-emsc-2023-11-08/
21 FEMA. (2018). *Threat and Hazard Identification and Risk Assessment (THIRA) and Stakeholder Preparedness Review (SPR) Guide Comprehensive Preparedness Guide (CPG) 201* (3rd ed.). FEMA. See www.fema.gov/sites/default/files/2020-07/threat-hazard-identification-risk-assessment-stakeholder-preparedness-review-guide.pdf
22 Boyarsky, A. (2024). *Riding the Wave: Applying Project Management Science in the Field of Emergency Management.* CRC Press.

23 Eisenhower, D. D. (1957). *Remarks at the National Defense Executive Reserve Conference*. Accessed December 8, 2023. See www.eisenhowerlibrary.gov/eisenhowers/quotes

24 See https://checkyourfact.com/2019/10/17/fact-check-benjamin-franklin-failing-prepare/

25 Gibson, V., & Johnson, D. (2016). CPTED, But Not as We Know It: Investigating the Conflict of Frameworks and Terminology in Crime Prevention through Environmental Design. *Security Journal*, *29*, 256–275. https://doi.org/10.1057/sj.2013.19

26 See www.fema.gov/sites/default/files/2020-08/fema430.pdf

27 Paraphrased from a quote attributed to George Patton, "You fight like you train".

28 See https://openwho.org/courses/wash-ntds; See https://openwho.org/courses/water-safety-planning

29 See https://teex.org/wp-content/uploads/MGT343%20Disaster%20Mgt%20for%20WWW%20Utilities.pdf

30 USDHS. (2023). *Homeland Security Exercise and Evaluation Program*. FEMA. Accessed December 9, 2023. See www.fema.gov/emergency-managers/national-preparedness/exercises/hseep

31 Prasad, M. (n.d.). *Emergency Action Plans – The Basics*. Barton Dunant. Accessed December 16, 2023. See https://blog.bartondunant.com/eap/

32 See www.fema.gov/emergency-managers/national-preparedness/frameworks/response

33 See www.fema.gov/sites/default/files/2020-06/national_disaster_recovery_framework_2nd.pdf

34 USDHS. (2016). *National Disaster Recovery Framework*. U.S. Department of Homeland Security. See https://www.fema.gov/sites/default/files/2020-06/national_disaster_recovery_framework_2nd.pdf

35 FEMA. (2022). *Recovery Support Function Leadership Group*. FEMA. See www.fema.gov/emergency-managers/national-preparedness/frameworks/national-disaster-recovery/rsflg

36 See www.fema.gov/emergency-managers/national-preparedness/mission-core-capabilities

37 See www.epa.gov/waterresilience/femas-community-lifelines-construct

38 See www.fema.gov/emergency-managers/practitioners/lifelines

39 See www.csis.org/programs/global-food-and-water-security-program

40 Hodges, L. R., & Larra, M. D. (2021). Emergency Management as a Complex Adaptive System. *Journal of Business Continuity & Emergency Planning*, *14*(4), 354–368. See https://pubmed.ncbi.nlm.nih.gov/33962703/

41 See www.whitehouse.gov/build/resources/bipartisan-infrastructure-law-tribal-playbook/

42 FEMA. (2016). *National Disaster Recovery Framework* (2nd ed.). U.S. Department of Homeland Security. See www.fema.gov/sites/default/files/2020-09/national_disaster_recovery_framework_2nd-edition.pdf

43 Phillips, B. D., Neal, D. M., & Webb, G. R. (2016). *Introduction to Emergency Management* (2nd ed.). CRC Press. https://doi.org/10.1201/9781315394701

44 See www.apu.apus.edu/online-bachelor-degrees/bachelor-of-arts-in-emergency-and-disaster-management/

45 See www.fema.gov/sites/default/files/2020-06/fema-mitigation-ideas_02-13-2013.pdf

46 FEMA. (2011). *A Whole Community Approach to Emergency Management: Principles, Themes, and Pathways for Action*. FEMA. See www.fema.gov/sites/default/files/2020-07/whole_community_dec2011__2.pdf

47 McKinney, K. (2018). *Moment of Truth: The Nature of Catastrophes and How to Prepare for Them*. Savio Republic.

48 Biden, J. (2021). *Executive Order on Diversity, Equity, Inclusion, and Accessibility in the Federal Workforce*. The White House. See www.whitehouse.gov/briefing-room/presidential-actions/2021/06/25/executive-order-on-diversity-equity-inclusion-and-accessibility-in-the-federal-workforce/

49 FEMA. (2023). *Inclusion, Diversity, Equity and Accessibility in Exercises Considerations and Best Practices Guide*. FEMA. See www.fema.gov/sites/default/files/documents/fema_inclusion-diversity-equity-accessibility-exercises.pdf

50 See www.digital-adoption.com/six-sigma/#:~:text=Agile,spread%20to%20other%20business%20discipline

51 Marcus, L. J., McNulty, E. J., Henderson, J. M., & Dorn, B. C. (2019). *You're It: Crisis, Change, and How to Lead When It Matters Most*. PublicAffairs.

52 de Waal, A., Weaver, M., Day, T., & van der Heijden, B. (2019). Silo-Busting: Overcoming the Greatest Threat to Organizational Performance. *Sustainability, 11*(23), 6860. https://doi.org/10.3390/su11236860

53 Fiest, K. M., & Krewulak, K. D. (2021). Space, Staff, Stuff, and System: Keys to ICU Care Organization During the COVID-19 Pandemic. *Chest, 160*(5), 1585–1586. https://doi.org/10.1016/j.chest.2021.07.001

54 See www.jesip.org.uk/wp-content/uploads/2022/03/JESIP-Joint-Doctrine-October-2021.pdf

55 See www.iloveqatar.net/guide/living/national-command-centre-ncc-qatar

56 See www.civildefence.govt.nz/assets/Uploads/CIMS-3rd-edition-FINAL-Aug-2019.pdf

57 See https://knowledge.aidr.org.au/resources/ajem-apr-2012-aiims-doctrine-have-we-got-the-fundamentals-right/

58 World Meteorological Organization. (2023). *Tropical Cyclone Naming*. World Meteorological Organization. Accessed January 29, 2024. See https://wmo.int/content/tropical-cyclone-naming

59 See www.epa.gov/hwp

60 See https://crsreports.congress.gov/product/pdf/R/R47032

61 See www.cdc.gov/surveillance/index.html

62 See www.nae.usace.army.mil/missions/public-services/flood-plain-management-services/

63 See https://cemir.org/faqs

64 Sahoh, B., & Choksuriwong, A. (2023). The Role of Explainable Artificial Intelligence in High-Stakes Decision-Making Systems: A Systematic Review. *Journal of Ambient Intelligence and Humanized Computing, 14*(6), 7827–7843. https://doi.org/10.1007/s12652-023-04594-w

65 See www.sam.usace.army.mil/Media/News-Stories/Article/1168126/districts-emergency-management-branch-earns-accreditation/

66 See www.dcwater.com/events/emergency-management-accreditation-program-emap-celebration

67 Emergency Management Accreditation Program. (2023). *Emergency Management Standard*. EMAP. See https://emap.org/wp-content/uploads/2023/04/EMAP-EMS-5-2022-Emergency-Management-Standard.pdf

68 See www.emacweb.org/

69 EMAC. (n.d.). *Emergency Management Assistance Compact*. NEMA. Accessed January 18, 2024. See www.emacweb.org/

70 Seuss, D. (1971). *The Lorax*. Random House Children's Books.

71 U.S. Consumer Product Safety Commission. (2023). *CPSC Issues New Drowning Report with Child Fatalities; Reminder for Extra Water Safety Vigilance*. U.S. Consumer Product Safety Commission. Accessed December 21, 2023. See www.cpsc.gov/Newsroom/News-Releases/2023/CPSC-Issues-New-Drowning-Report-with-Child-Fatalities-Reminder-for-Extra-Water-Safety-Vigilance

72 See https://bartondunant.com/smarter-water-watcher

73 Protection, Prevention (Preparedness), Response, Recovery, and Mitigation. Not an acronym used in Emergency Management but covered earlier in this chapter.

74 See www.acf.hhs.gov/sites/default/files/documents/ohsepr/nccdreport.pdf
75 See https://teex.org/
76 See www.dec.ny.gov/news/press-releases/2023/11/dec-announces-start-of-construction-of-project-to-protect-camp-hollis-from-future-flooding-and-high-water
77 Prasad, M. (2021). Space Aliens – Emergency Management Roles & Responsibilities. *Domestic Preparedness Journal.* See https://domesticpreparedness.com/cbrne/space-aliens-emergency-management-roles-and-responsibilities
78 See https://info.childcareaware.org/blog/explainer-equity-inclusion-for-children-with-disabilities-in-idea
79 See https://njcdd.org/school-safety-issues-affecting-students-with-disabilities/
80 See https://community.fema.gov/PreparednessCommunity/s/communitypreparedness?language=en_US
81 See https://training.fema.gov/is/courseoverview.aspx?code=IS-366.a&lang=en
82 See https://teex.org/wp-content/uploads/MGT-439-Pediatric-Disaster-Response.pdf
83 See www.arcgis.com/apps/MapSeries/index.html?appid=ea8b0eeb2e9c45b790329c0ed2fdc225
84 See https://hifld-geoplatform.opendata.arcgis.com/
85 See https://usengineeringsolutions.com/damwatch/
86 See www.redcross.org/
87 See www.ifrc.org/national-societies-directory

BIBLIOGRAPHY

Auden, W. H. (1957). First Things First. *The New Yorker.* See www.newyorker.com/magazine/1957/03/09/first-things-first

Biden, J. (2021). *Executive Order on Diversity, Equity, Inclusion, and Accessibility in the Federal Workforce.* The White House. See www.whitehouse.gov/briefing-room/presidential-actions/2021/06/25/executive-order-on-diversity-equity-inclusion-and-accessibility-in-the-federal-workforce/

Boyarsky, A. (2024). *Riding the Wave: Applying Project Management Science in the Field of Emergency Management.* CRC Press.

de Waal, A., Weaver, M., Day, T., & van der Heijden, B. (2019). Silo-Busting: Overcoming the Greatest Threat to Organizational Performance. *Sustainability, 11*(23), 6860. https://doi.org/10.3390/su11236860

Dos Santos, V. M., & Son, C. (2023). Modern Firefighters' Three-Level Situation Awareness in Fire and Non-Fire Incidents. *Proceedings of the Human Factors and Ergonomics Society Annual Meeting.* https://doi.org/10.1177/21695067231192647

Eisenhower, D. D. (1957). *Remarks at the National Defense Executive Reserve Conference.* Accessed December 8, 2023. See www.eisenhowerlibrary.gov/eisenhowers/quotes

EMAC. (n.d.). *Emergency Management Assistance Compact.* NEMA. Accessed January 18, 2024. See www.emacweb.org/

Emergency Management Accreditation Program. (2023). *Emergency Management Standard.* EMAP. See https://emap.org/wp-content/uploads/2023/04/EMAP-EMS-5-2022-Emergency-Management-Standard.pdf

Fakhruddin, B., Blanchard, K., & Ragupathy, D. (2020). Are We There Yet? The Transition from Response to Recovery for the Covid-19 Pandemic. *Progress in Disaster Science, 7,* 100102. https://doi.org/10.1016/j.pdisas.2020.100102

FEMA. (2011). *A Whole Community Approach to Emergency Management: Principles, Themes, and Pathways for Action.* FEMA. See www.fema.gov/sites/default/files/2020-07/whole_community_dec2011__2.pdf

FEMA. (2015). *National Preparedness Goal Second Edition.* U.S. Department of Homeland Security. See www.fema.gov/sites/default/files/2020-06/national_preparedness_goal_2nd_edition.pdf

FEMA. (2016a). *National Disaster Recovery Framework* (2nd ed.). U.S. Department of Homeland Security. See www.fema.gov/sites/default/files/2020-09/national_disaster_recovery_framework_2nd-edition.pdf

FEMA. (2016b). *National Disaster Recovery Framework.* FEMA. See www.fema.gov/emergency-managers/practitioners/national-disaster-recovery-framework

FEMA. (2018). *Threat and Hazard Identification and Risk Assessment (THIRA) and Stakeholder Preparedness Review (SPR) Guide Comprehensive Preparedness Guide (CPG) 201 3rd Edition.* FEMA. See www.fema.gov/sites/default/files/2020-07/threat-hazard-identification-risk-assessment-stakeholder-preparedness-review-guide.pdf

FEMA. (2019a). *IS-0200.c Basic Incident Command System for Initial Response.* FEMA. See https://emilms.fema.gov/is_0200c/groups/518.html

FEMA. (2019b). *National Response Framework.* U.S. Department of Homeland Security. See www.fema.gov/sites/default/files/2020-04/NRF_FINALApproved_2011028.pdf

FEMA. (2020). *FEMA Job Aid Increasing Resilience Using THIRA/SPR and Mitigation Planning.* FEMA. See www.fema.gov/sites/default/files/2020-09/fema_thira-hmp_jobaid.pdf

FEMA. (2022a). *E0237: Planning Process Theory and Application Student Manual.* FEMA.

FEMA. (2022b). *Recovery Support Function Leadership Group.* FEMA. See www.fema.gov/emergency-managers/national-preparedness/frameworks/national-disaster-recovery/rsflg

FEMA. (2023a). *Inclusion, Diversity, Equity and Accessibility in Exercises Considerations and Best Practices Guide.* FEMA. See www.fema.gov/sites/default/files/documents/fema_inclusion-diversity-equity-accessibility-exercises.pdf

FEMA. (2023b). *Swift Current.* FEMA. Accessed December 8, 2023. See www.fema.gov/grants/mitigation/flood-mitigation-assistance/swift-current

Fiest, K. M., & Krewulak, K. D. (2021). Space, Staff, Stuff, and System: Keys to ICU Care Organization During the COVID-19 Pandemic. *Chest, 160*(5), 1585–1586. https://doi.org/10.1016/j.chest.2021.07.001

Gibson, V., & Johnson, D. (2016). CPTED, But Not as We Know It: Investigating the Conflict of Frameworks and Terminology in Crime Prevention Through Environmental Design. *Security Journal, 29*, 256–275. https://doi.org/10.1057/sj.2013.19

Hodges, L. R., & Larra, M. D. (2021). Emergency Management as a Complex Adaptive System. *Journal of Business Continuity & Emergency Planning, 14*(4), 354–368. https://pubmed.ncbi.nlm.nih.gov/33962703/

Humanitarian Coalition of Canada. (2023). *Crash Course: Tropical Storms 101.* Humanitarian Coalition. Accessed January 3, 2024. See www.humanitariancoalition.ca/crash-course-tropical-storms-101

Jhung, M. A., Shehab, N., Rohr-Allegrini, C., Pollock, D. A., Sanchez, R., Guerra, F., & Jernigan, D. B. (2007). Chronic Disease and Disasters: Medication Demands of Hurricane Katrina Evacuees. *American Journal of Preventive Medicine, 33*(3), 207–210. https://doi.org/10.1016/j.amepre.2007.04.030

Ma, N., & Liu, Y. (2020). Risk Factors and Risk Level Assessment: Forty Thousand Emergencies Over the Past Decade in China. *Jamba (Potchefstroom, South Africa), 12*(1), 916. https://doi.org/10.4102/jamba.v12i1.916

Marcus, L. J., McNulty, E. J., Henderson, J. M., & Dorn, B. C. (2019). *You're It: Crisis, Change, and How to Lead When It Matters Most.* PublicAffairs.

McKinney, K. (2018). *Moment of Truth: The Nature of Catastrophes and How to Prepare for Them* (p. 220). Savio Republic.

Phillips, B. D., Neal, D. M., & Webb, G. R. (2016). *Introduction to Emergency Management* (2nd ed.). CRC Press. https://doi.org/10.1201/9781315394701

Prasad, M. (2021). Space Aliens – Emergency Management Roles & Responsibilities. *Domestic Preparedness Journal.* See https://domesticpreparedness.com/cbrne/space-aliens-emergency-management-roles-and-responsibilities

Prasad, M. (2023). Global Pandemics Are Extinction-level Events and Should Not Be Coordinated Solely through National or Jurisdictional Emergency Management. *Pracademic Affairs, 3.* Naval Postgraduate School/Center for Homeland Defense and Security. See www.hsaj.org/articles/22285

Prasad, M. (n.d.). *Emergency Action Plans – The Basics.* Barton Dunant. Accessed December 16, 2023. See https://blog.bartondunant.com/eap/

Sahoh, B., & Choksuriwong, A. (2023). The Role of Explainable Artificial Intelligence in High-Stakes Decision-Making Systems: A Systematic Review. *Journal of Ambient Intelligence and Humanized Computing, 14*(6), 7827–7843. https://doi.org/10.1007/s12652-023-04594-w

Seuss, D. (1971). *The Lorax.* Random House Children's Books.

U.S. Consumer Product Safety Commission. (2023). *CPSC Issues New Drowning Report with Child Fatalities; Reminder for Extra Water Safety Vigilance.* U.S. Consumer Product Safety Commission. Accessed December 21, 2023. See www.cpsc.gov/Newsroom/News-Releases/2023/CPSC-Issues-New-Drowning-Report-with-Child-Fatalities-Reminder-for-Extra-Water-Safety-Vigilance

US Department of Homeland Security. (2015). *National Preparedness Goal Second Edition.* FEMA. See www.fema.gov/sites/default/files/2020-06/national_preparedness_goal_2nd_edition.pdf

USDHS. (2023). *Homeland Security Exercise and Evaluation Program.* FEMA. Accessed December 9, 2023. See www.fema.gov/emergency-managers/national-preparedness/exercises/hseep

World Economic Forum. (2024). *Global Risks Report 2024: The Risks Are Growing – But So Is Our Capacity to Respond.* United Nations Office for Disaster Risk Reduction. Accessed January 20, 2024. See www.preventionweb.net/news/global-risks-report-2024-risks-are-growing-so-our-capacity-respond

World Meteorological Organization. (2023). *Tropical Cyclone Naming.* World Meteorological Organization. Accessed January 29, 2024. See https://wmo.int/content/tropical-cyclone-naming

Part 2

Threats around Water

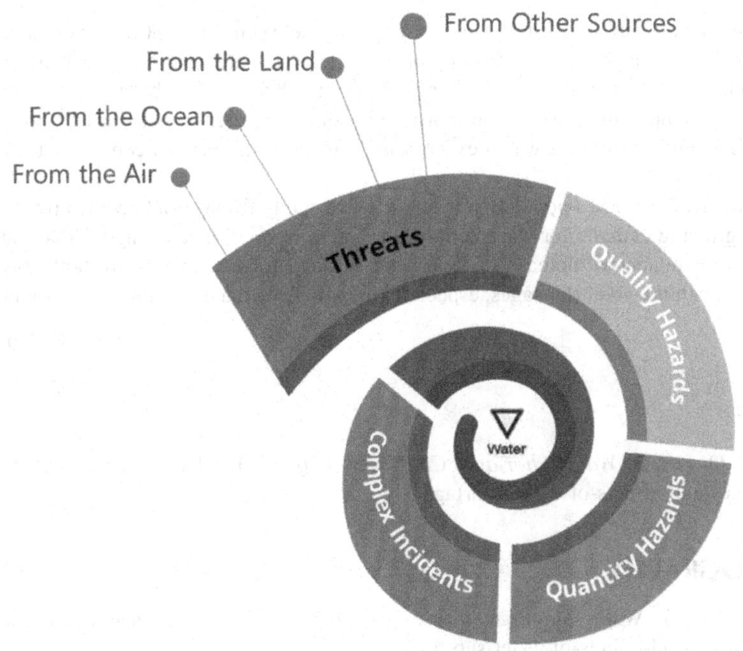

This part covers the chapters for many of the threats which can generate water-related hazards. The chapters in this part are organized by where the threat may come from: air, ocean, land, and other sources.

DOI: 10.1201/9781003474685-4

For example, dam breeches or river flooding are threats from land-based sources. A tropical storm is a threat from the ocean. Torrential rain and snowfall both are threats from the air. Threats do not always generate hazards. And sometimes there are threats which have built-in hazards, such as chemical spills or cyberattacks. This part overview will help align the Quality and Quantity hazards found in Parts 3 and 4 for adverse impacts to the Disaster Phase Cycles, Response's Incident Command System, Emergency Support Functions, Recovery Support Functions, and Community Lifelines. All these systems and constructs are quantified in each Hazard part.

Chapter 3 – Threats from the Air
Chapter 4 – Threats from the Ocean
Chapter 5 – Threats from the Land
Chapter 6 – Threats from Other Sources

Emergency managers should recognize that Climate Change is increasing the negative impacts from threats, many of which are natural. The intensity, duration, and frequency of water quantity hazards (both too much and too little) can also change the water cycle globally. This is water stress.

> While the amount of rainfall can naturally vary between different regions and times of year, climate change and rising global temperatures are altering rainfall patterns, which in turn, impact the quality and spatial distribution of global water resources. Warmer temperatures mean that moisture in soil evaporates at faster rates, and more frequent and severe heat waves exacerbate drought conditions and contribute towards water shortages.
>
> Today, the world's population is just short of eight billion people, which translates to a growing demand for water amid water stress from climate change. Urbanisation and an exponential increase in freshwater demand for households are both driving factors behind water shortages, especially in regions with a precarious water supply.
>
> **(Lai, 2022, p. 1)**[1]

NOTE

1 Lai, O. (2022). *Water Shortage: Causes and Effects.* Earth.org. See https://earth.org/causes-and-effects-of-water-shortage/

BIBLIOGRAPHY

Lai, O. (2022). Water Shortage: Causes and Effects. *Earth.org.* See https://earth.org/causes-and-effects-of-water-shortage/

3 Threats from the Air

A snow day literally and figuratively falls from the sky, unbidden, and seems like a thing of wonder. (Or cause for dismay: from here in Columbia County, the decision to keep New York City public schools open just seems mean.)

Susan Orlean,[1] writer

The chapters in this part are focused on threats. Threats are what can (but not always) generate hazards to people, places, and things. Threats are causal. And making a threat is very different from posing a threat. When it comes to water-related hazards, generated by water-related threats, this can be the difference between, for example, a blizzard watch and a blizzard warning.

Precipitation is the technical term for when moisture falls from the air. Threats from the air include excessive rain, hail, sleet, ice, snow, and everything wet in-between. This chapter will cover when water in any form comes down from the sky. This chapter will introduce some of the terminology and key phrases used when describing air-related threats.

3.1 AIR POLLUTION

One of the few human-generated threats from the air, air pollution[2] can turn ordinary rain into acid rain. There are other chemical pollutants[3] added to the air and then moved into the land, rivers, and streams through rain. Researchers in 2023, identified bacteria and other chemical compounds in sea spray – well beyond the short distance from the ocean, swimmers, and others:

> Rainfall in the US-Mexico border region causes complications for wastewater treatment and results in untreated sewage being diverted into the Tijuana River and flowing into the ocean in south Imperial Beach. This input of contaminated water has caused chronic coastal water pollution in Imperial Beach for decades. New research shows that sewage-polluted coastal waters transfer to the atmosphere in sea spray aerosol formed by breaking waves and bursting bubbles. Sea spray aerosol contains bacteria, viruses, and chemical compounds from the seawater.
>
> **(Monroe, 2023, p. 1)[4]**

3.2 AIR PRESSURE SYSTEMS

Air pressure systems – when either low or high barometric pressure helps or hinders wind flow across continents. In North America, one of these fast-moving systems

DOI. 10.1201/9781003474685-5

is called an "Alberta Clipper", named after a 19th-century merchant sailing vessel known for its speed:

> Alberta Clippers are caused by low-pressure systems that form when warm winds from the Pacific Ocean collide with the colder air over the Rocky Mountains. The system moves "leeward" (downwind, that is, to the East) from the mountains before getting swept up into the jet stream,[5] which carries it screaming down over the unsuspecting Midwestern United States before dumping it out into the Atlantic Ocean.
>
> Clipper systems usually bring bitter cold, with sharp temperature declines of as much as 30°F in a few hours' time, and powerful winds of up to 45 miles per hour. Alberta Clippers don't always bring much snow, though. The storms are low in moisture and usually only produce a few inches on their own, though this can increase drastically if the storm picks up moisture over the Great Lakes, resulting in "Lake Effect" snowfall of a foot or more.
>
> **(McLeod, 2023, p. 1)**[6]

Pressure systems, Fronts, and Jet Streams are meteorological science elements which emergency managers should be familiar[7] with, and Emergency Management teams should have curated Emergency Management Intelligence for these and other weather elements as part of their ongoing workforce's responsibilities.

3.3 ATMOSPHERIC RIVERS IN THE SKY

An atmospheric river (AR) is a flowing column of condensed water vapor in the atmosphere responsible for producing significant levels of rain and snow, especially in the western United States. When ARs move inland and sweep over the mountains, the water vapor rises and cools to create heavy precipitation. Though many ARs are weak systems that simply provide beneficial rain or snow, some of the larger, more powerful ARs can create extreme rainfall and floods capable of disrupting travel, inducing mudslides, and causing catastrophic damage to life and property.

A well-known example of a strong atmospheric river is called the "Pineapple Express" because moisture builds up in the tropical Pacific around Hawaii and can wallop the U.S. and Canada's West Coasts with heavy rainfall and snow.

Prevailing winds cross over warm bands of tropical water vapor to form this "river", which travels across the Pacific as part of the global conveyor belt. When it reaches the west coast, the Pineapple Express can dump as much as 5 inches of rain on California in one day.[8]

A strong AR transports an amount of water vapor roughly equivalent to 7.5–15 times the average flow of water at the mouth of the Mississippi River. ARs are a primary feature in the entire global water cycle and are tied closely to both water supply and flood risks, particularly in the western United States. Atmospheric rivers are long, narrow ribbons of moisture, typically found at about 10,000–15,000 ft (3,000–4,500 m) above the surface – they're like "rivers in the sky". They can travel thousands of miles and are responsible for 30–50% of the wet season precipitation along the U.S. West Coast. In California, atmospheric rivers have caused almost the entire state to receive 400–600% of its typical average rainfall

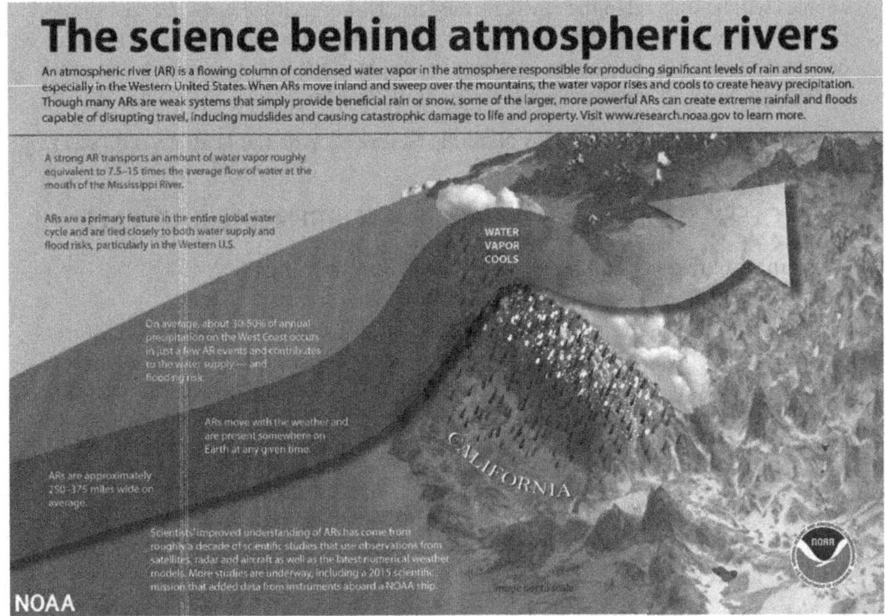

FIGURE 3.1 The science behind atmospheric rivers.

Source: NOAA Research. (2023). *Atmospheric Rivers: What Are They and How Does NOAA Study Them?* U.S. Department of Commerce. See https://research.noaa.gov/2023/01/11/atmospheric-rivers-what-are-they-and-how-does-noaa-study-them/

since Christmas 2022, and the southern Sierra Nevada appears to have recorded its largest ever snowpack – more than 250% above its seasonal average (that is over 600 inches – 15 m – of snow!).[9]

3.4 BOMBOGENESIS – OR "BOMB CYCLONES"

When rapid wind – a cyclone – occurs in the latitudes between the tropics and polar regions, bombogenesis can occur – when the cold air mass collides with a warm air mass, especially over warm ocean water. It is commonly called a bomb cyclone.[10] This is wind-only, so it is not a threat for this book.

3.5 EXCESSIVE HAIL

Hail is solid ice formed inside of a thunderstorm.

> Hail can damage aircraft, homes and cars,[11] and can be deadly to livestock and people. Hailstones are formed when raindrops are carried upward by thunderstorm updrafts into extremely cold areas of the atmosphere and freeze. Hailstones then grow by colliding with liquid water drops that freeze onto the hailstone's surface. If the water freezes instantaneously when colliding with the hailstone, cloudy ice will form as air

bubbles will be trapped in the newly formed ice. However, if the water freezes slowly, the air bubbles can escape and the new ice will be clear. The hail falls when the thunderstorm's updraft can no longer support the weight of the hailstone, which can occur if the stone becomes large enough or the updraft weakens.

(NOAA National Severe Storms Laboratory, n.d., p. 1)[12]

3.6 EXCESSIVE RAIN – PLUVIOSITY – TORRENTIAL RAINFALL

Heavy rain or pluviosity can be a life-safety threat, worldwide.[13]

> Rain is often considered an inconvenience, or at most a side-effect of a powerful storm, but heavy rainfall can cause serious damage and destruction. Common consequences of heavy (or "torrential") rain are flash floods and landslides, which can decimate houses and flood entire neighborhoods. Even more concerning is the increasing prevalence of these events, which many scientists link to climate change.
>
> **(Cohen and Shah, n.d., p. 1)**[14]

NEW YORK CITY'S "RAINPROOF NYC"

In 2024, New York City (NYC) announced a collaborative public–private partnership effort to combat the adverse effects of heavy rainfall, in many cases due to Climate Change. Aligning stormwater management, green infrastructure, cloudburst[15] projects, and Bluebelts ("ecologically rich and cost-effective drainage systems that naturally handle the runoff precipitation that falls on streets and sidewalks") (NYC, 2024, p. 1),[16] are all part of the plan:

> "Increased rainfall is increasingly impacting New Yorkers' way of life, and addressing the climate crisis must be a collaborative process that involves not only experts and elected officials, but also the community members who are directly impacted", said Chief Climate Officer and DEP Commissioner Rohit T. Aggarwala. "Creating community-specific solutions will allow us to more quickly fight the impacts of climate change. Working groups like this that involve community members who are directly involved in the work are a useful complement to the technical work".
>
> **(NYC, 2024, p. 1)**[17]

The possibility of adverse impacts – such as flooding from excessive rain on critical infrastructure such as dams and nuclear plants – is measured by Probable Maximum Precipitation (PMP).

3.7 EXCESSIVE SLEET/FREEZING RAIN/ICE STORMS

Excessive rain can immediately freeze – both in the air and on surfaces such as power lines, tree limbs, etc. This includes Black Ice which can occur on both roadways and frozen lakes, Ice Storms[18] which take down trees and powerlines, Ice Jams and Frost which has human and agricultural impacts.

3.8 EXCESSIVE SNOW (BLIZZARDS, LAKE EFFECT SNOW, WHITEOUTS, ETC.)

Many of these frozen water threats from the air have wind as an extenuating circumstance.

3.8.1 Blizzards

Blizzards are very dangerous winter snowstorms which have blowing snow from heavy winds.

> Officially, the National Weather Service defines a blizzard as a storm which contains large amounts of snow OR blowing snow, with winds in excess of 35 mph and visibilities of less than 1/4 mile for an extended period of time (at least 3 hours).
>
> Blizzard conditions often develop on the northwest side of an intense storm system. The difference between the lower pressure in the storm and the higher pressure to the west creates a tight pressure gradient, or difference in pressure between two locations, which in turn results in very strong winds. These strong winds pick up available snow from the ground, or blow any snow which is falling, creating very low visibilities and the potential for significant drifting of snow.
>
> **(NOAA, n.d.f, p. 1)**[19]

3.8.2 Lake Effect Snow

Lake Effect Snow[20] in the United States occurs near the Great Lakes states, as well as near the Great Salt Lake in Utah, when cold air moves across these large open-water

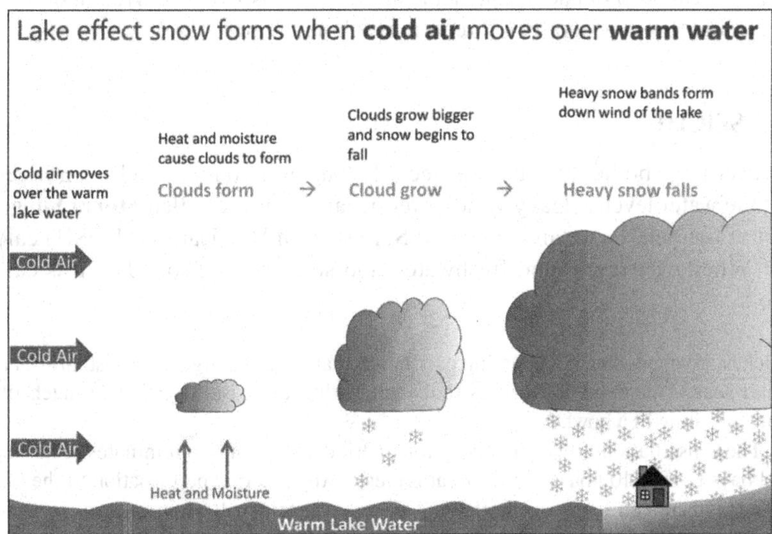

FIGURE 3.2 Lake effect snow forms when cold air moves over warm water.

Source: NOAA. (n.d.). *What Is a Lake Effect Snow*? U.S. Department of Commerce. Accessed March 1, 2024. See www.weather.gov/safety/winter-lake-effect-snow

bodies freshwater. Terms used to describe the threats from Lake Effect Snow include a Lake Effect Snow Squall. Other wind or squall types include Pre-frontal Squall Line, Pre-Hurricane Squall, Snow Squall, Squall, and Squall Line.[21]

This threat from the air can occur anywhere on Earth where there are large open bodies of water – including seawater areas, during warmer weather. This includes the Black Sea impacting Istanbul and northern Turkey,[22] in Northern Europe near the Baltic Sea, in Japan,[23] and even in Iran.[24]

3.8.3 Whiteouts

Whiteouts occur when drier snow which is more powder-like is blown around. This can cause very low visibility conditions for drivers. It does not have to be snowing for whiteout conditions to exist, and they can reduce visibility to near-zero distance.

3.9 NOR'EASTERS

> A Nor'easter is a storm along the East Coast of North America, so called because the winds over the coastal area are typically from the northeast. These storms may occur at any time of year but are most frequent and most violent between September and April. Some well-known Nor'easters include the notorious Blizzard of 1888, the "Ash Wednesday" storm of March 1962, the New England Blizzard of February 1978, the March 1993 "Superstorm" and the recent Boston snowstorms of January and February 2015.
>
> **(NOAA, n.d., p. 1)**[25]

Nor'easters can also impact Canada and similar storms have occurred in other parts of world, but this is generally a threat in the United States.

3.10 SEICHES

Wind over large bodies of water – especially long durations of wind – can adversely affect the water levels. Heavy winds over ocean water are called **Storm Surge** and is part of the Sea, Lake, and Overland Surges from Hurricanes (SLOSH) calculation.[26] When over raw water, freshwater, and some coastal sounds,[27] it is called a seiche.

> Seiches are phenomena of standing oscillation that occur in large lakes, estuaries, and small seas. This condition causes the water within the basin to oscillate much like water sloshing in a bowl.
>
> These oscillations are of relatively long period, extending from minutes in harbors and bays to over 10 hours in the Great Lakes.[28] Any external perturbation to the lake or embayment can force an oscillation. They result primarily from changes in atmospheric pressure and the resultant wind conditions and occur over the entire basin. In harbors, the forcing can be the result of short waves[29] and wave groups at the harbor entrance.
>
> **(USACE, n.d., p. 1)**[30]

Seiches can dry up a large lake on one end and overflow it on the other. Seiches produce water-level displacement (WLD). In the United States, the USACE performs research[31] on seiches and provides Protection and Prevention information to communities. The USACE produced a video[32] in 2024 explaining seiches, which can be found on YouTube.[33]

3.11 SNOW SQUALL

A snow squall is a very short duration (usually less than 1 hour) snowstorm, impacting a specific area:

> Snow squalls, often associated with strong cold fronts, are a key wintertime weather hazard. They move in and out quickly, and typically last less than an hour. The sudden white-out conditions combined with falling temperatures produce icy roads in just a few minutes. Squalls can occur where there is no large-scale winter storm in progress and might only produce minor accumulations. Snow squalls can cause localized extreme impacts to the traveling public and to commerce for brief periods of time. Unfortunately, there is a long history of deadly traffic accidents associated with snow squalls. Although snow accumulations are typically an inch or less, the added combination of gusty winds, falling temperatures and quick reductions in visibility can cause extremely dangerous conditions for motorists.
>
> If a snow squall warning is issued for your area, avoid or delay motor travel until the squall passes through your location. There truly is no safe place on the highway during a snow squall. However if you are already in transit and cannot exit the road in time, reduce your speed, turn on your headlights and hazard lights and allow plenty of distance between you and the car in front of you. It's also best not to slam on your brakes. With slick/icy roads, this could contribute to the loss of vehicle control and also increase the risk of a chain reaction crash.
>
> **(NOAA, n.d., p. 1)[34]**

3.12 SQUALLS

A Squall Line, a Bow Echo, or a Quasi-Linear Convective System (QLCS) are all forms of squalls:

> A "squall line" refers to a linearly-oriented zone of convection (i.e., thunderstorms). Squall lines are common across the United States east of the Rockies, especially during the spring when the atmosphere is most "dynamic". A "bow echo" or "bowing line segment" is an arched/bowed out line of thunderstorms, sometimes embedded within a squall line. All these terms fall under the more generic term Quasi-Linear Convective System (QLCS). Bow echoes, most common in the spring and summer, usually are associated with an axis of enhanced winds that create straight-line wind damage at the surface. In fact, bow echo-induced winds/downbursts account for a large majority of the structural damage resulting from convective non-tornadic winds. Transient tornadoes also can occur in squall lines, especially in association with bow echoes. These tornadoes, however, tend to be weaker and shorter-lived on average than those associated with supercell thunderstorms.
>
> **(NOAA, n.d., p. 1)[35]**

3.13 SUPERCELL THUNDERSTORMS

> Supercell thunderstorms are perhaps the most violent of all thunderstorm types, and are capable of producing damaging winds, large hail, and weak-to-violent tornadoes. They are most common during the spring across the central United States when moderate-to-strong atmospheric wind fields, vertical wind shear (change in wind direction and/or speed with height), and instability are present. The degree and vertical distribution of moisture, instability, lift, and wind fields have a profound influence on convective storm type, including supercells, multicells (including squall lines and bow echoes), ordinary/pulse storms, or a combination of storm types.
>
> **(NOAA, n.d., p. 1)**[36]

NOTES

1 Orlean, S. (2011). Snow Day. *The New Yorker.* See www.newyorker.com/culture/susan-orlean/snow-day-2
2 See www.history.com/topics/natural-disasters-and-environment/water-and-air-pollution
3 Kjellstrom, T., Lodh, M., McMichael, T., et al. (2006). Air and Water Pollution: Burden and Strategies for Control. In D. T. Jamison, J. G. Breman, A. R. Measham, et al. (Eds.), *Disease Control Priorities in Developing Countries* (2nd ed.). The International Bank for Reconstruction and Development/The World Bank. See www.ncbi.nlm.nih.gov/books/NBK11769/
4 Monroe, R. (2023). *Coastal Water Pollution Transfers to the Air in Sea Spray Aerosol and Reaches People on Land.* UCSanDiego. See https://scripps.ucsd.edu/news/coastal-water-pollution-transfers-air-sea-spray-aerosol-and-reaches-people-land
5 See www.farmersalmanac.com/what-exactly-is-the-jet-stream
6 McLeod, J. (2023). What's an Alberta Clipper? The Cold Facts. *Farmers' Almanac.* See www.farmersalmanac.com/what-alberta-clipper
7 See www.weather.gov/lmk/basic-fronts
8 See https://oceanservice.noaa.gov/facts/pineapple-express.html
9 See https://disasterphilanthropy.org/disasters/california-storms/
10 See https://oceanservice.noaa.gov/facts/bombogenesis.html
11 See https://denverite.com/2023/08/07/red-rocks-ampitheatre-hail-storm-louis-tomlinson/
12 NOAA National Severe Storms Laboratory. (n.d.). *Severe Weather 101: Hail Basics.* U.S. Department of Commerce. Accessed March 1, 2024. See www.nssl.noaa.gov/education/svrwx101/hail/
13 See https://en.mehrnews.com/news/208564/Heavy-rains-in-Brazil-leave-at-least-7-dead
14 Cohen, E., & Shah, A. (n.d.). *Torrential Rain.* SkyDayProject/Only One Sky NFP. Accessed March 1, 2024. See https://skydayproject.com/torrential-rain
15 See www.nyc.gov/office-of-the-mayor/news/023-23/mayor-adams-construction-new-cloudburst-resiliency-projects-better-manage-intense
16 New York City (NYC). (2023). *Marking 11 Years Since Superstorm Sandy, Mayor Adams Celebrates Completion of $110 Million New Creek Bluebelt Expansion to Prevent Flooding on Staten Island.* NYC. See www.nyc.gov/office-of-the-mayor/news/832-23/marking-11-years-since-superstorm-sandy-mayor-adams-celebrates-completion-110-million-new
17 Mayor's Office of Climate & Environmental Justice. (2024). *Adams Administration Launches "Rainproof NYC," Public–Private Partnership to Develop Innovative Solutions for Increased Heavy Rainfall.* NYC. See https://climate.cityofnewyork.us/press-release-rainproof-nyc/

18 See www.fema.gov/disaster/898
19 NOAA. (n.d.). *Winter Storms and Blizzards.* U.S. Department of Commerce. Accessed March 1, 2024. See www.weather.gov/fgz/WinterStorms
20 See www.weather.gov/safety/winter-lake-effect-snow
21 See https://forecast.weather.gov/glossary.php?word=SQUALL
22 See http://meetings.copernicus.org/www.cosis.net/abstracts/EGU06/05142/EGU06-J-05142.pdf
23 See www.researchgate.net/publication/325825413
24 See https://web.archive.org/web/20080724020750/http://news.nationalgeographic.com/news/2008/01/photogalleries/snow-pictures/
25 NOAA. (n.d.). *What Is a Nor'Easter?* U.S. Department of Commerce. Accessed March 1, 2024. See www.weather.gov/safety/winter-noreaster
26 See www.nhc.noaa.gov/surge/slosh.php
27 See www.nps.gov/articles/coastal-geohazards-seiches.htm
28 See www.iwr.usace.army.mil/Missions/Coasts/Tales-of-the-Coast/Americas-Coasts/Great-Lakes-Coast/
29 See www.iwr.usace.army.mil/Missions/Coasts/Tales-of-the-Coast/Coastal-Dynamics/Waves/
30 USACE. (n.d.) *Seiches.* USACE. Accessed February 4, 2024. See www.iwr.usace.army.mil/Missions/Coasts/Tales-of-the-Coast/Coastal-Dynamics/Wind/Seiches/
31 See www.13abc.com/2024/01/27/new-research-done-lake-erie-reveals-frequency-seiche/
32 See https://youtu.be/qFJ6WkQ-VK0?si=_5cOUvt5zUyCiHXd
33 USACE/Buffalo District. (2024). *Lake Erie Seiche Study Presentation.* USACE. Accessed February 4, 2024. See https://youtu.be/qFJ6WkQ-VK0?si=_5cOUvt5zUyCiHXd
34 NOAA. (n.d.). *Snow Squall.* U.S. Department of Commerce. Accessed March 1, 2024. See www.weather.gov/safety/winter-snow-squall
35 NOAA. (n.d.). *Squall Line/Bow Echo/QLCS.* U.S. Department of Commerce. Accessed March 1, 2024. See www.weather.gov/lmk/squallbow
36 NOAA. (n.d.). *Supercell Structure and Dynamics.* U.S. Department of Commerce. Accessed March 1, 2024. See www.weather.gov/lmk/supercell/dynamics

BIBLIOGRAPHY

Cohen, E., & Shah, A. (n.d.). *Torrential Rain.* SkyDayProject/Only One Sky NFP. Accessed March 1, 2024. See https://skydayproject.com/torrential-rain

Kjellstrom, T., Lodh, M., McMichael, T., et al. (2006). Air and Water Pollution: Burden and Strategies for Control. In D. T. Jamison, J. G. Breman, A. R. Measham, et al. (Eds.), *Disease Control Priorities in Developing Countries* (2nd ed.). The International Bank for Reconstruction and Development/The World Bank. See www.ncbi.nlm.nih.gov/books/NBK11769/

Mayor's Office of Climate & Environmental Justice. (2024). *Adams Administration Launches "Rainproof NYC," Public-Private Partnership to Develop Innovative Solutions for Increased Heavy Rainfall.* NYC. See https://climate.cityofnewyork.us/press-release-rainproof-nyc/

McLeod, J. (2023). *What's an Alberta Clipper? The Cold Facts.* Farmers' Almanac. See www.farmersalmanac.com/what-alberta-clipper

Monroe, R. (2023). *Coastal Water Pollution Transfers to the Air in Sea Spray Aerosol and Reaches People on Land.* UCSanDiego. See https://scripps.ucsd.edu/news/coastal-water-pollution-transfers-air-sea-spray-aerosol-and-reaches-people-land

New York City (NYC). (2023). *Marking 11 Years Since Superstorm Sandy, Mayor Adams Celebrates Completion of $110 Million New Creek Bluebelt Expansion to Prevent Flooding on Staten Island*. NYC. See www.nyc.gov/office-of-the-mayor/news/832-23/marking-11-years-since-superstorm-sandy-mayor-adams-celebrates-completion-110-million-new

NOAA. (n.d.a). *Snow Squall*. U.S. Department of Commerce. Accessed March 1, 2024. See www.weather.gov/safety/winter-snow-squall

NOAA. (n.d.b). *Squall Line/Bow Echo/QLCS*. U.S. Department of Commerce. Accessed March 1, 2024. See www.weather.gov/lmk/squallbow

NOAA. (n.d.c). *Supercell Structure and Dynamics*. U.S. Department of Commerce. Accessed March 1, 2024. See www.weather.gov/lmk/supercell/dynamics

NOAA. (n.d.d). *What Is a Lake Effect Snow?* U.S. Department of Commerce. Accessed March 1, 2024. See www.weather.gov/safety/winter-lake-effect-snow

NOAA. (n.d.e). *What Is a Nor'Easter*? U.S. Department of Commerce. Accessed March 1, 2024. See www.weather.gov/safety/winter-noreaster

NOAA. (n.d.f). *Winter Storms and Blizzards*. U.S. Department of Commerce. Accessed March 1, 2024. See www.weather.gov/fgz/WinterStorms

NOAA National Severe Storms Laboratory. (n.d.). *Severe Weather 101: Hail Basics*. U.S. Department of Commerce. Accessed March 1, 2024. See www.nssl.noaa.gov/education/svrwx101/hail/

NOAA Research. (2023). *Atmospheric Rivers: What Are They and How Does NOAA Study Them?* U.S. Department of Commerce. See https://research.noaa.gov/2023/01/11/atmospheric-rivers-what-are-they-and-how-does-noaa-study-them/

Orlean, S. (2011). Snow Day. *The New Yorker*. See www.newyorker.com/culture/susan-orlean/snow-day-2

USACE. (n.d.). *Seiches*. USACE. Accessed February 4, 2024. See www.iwr.usace.army.mil/Missions/Coasts/Tales-of-the-Coast/Coastal-Dynamics/Wind/Seiches/

USACE/Buffalo District. (2024). *Lake Erie Seiche Study Presentation*. USACE. Accessed February 4, 2024. See https://youtu.be/qFJ6WkQ-VK0?si=_5cOUvt5zUyCiHXd

4 Threats from the Ocean

Water, water, every where,
And all the boards did shrink;
Water, water, every where,
Nor any drop to drink.

– Samuel Taylor Coleridge, *The Rime of the Ancient Mariner*[1]

The chapters in this part are focused on threats. Threats are what can (but not always) generate hazards to people, places, and things. Threats are causal. And making a threat is very different from posing a threat. When it comes to water-related hazards, generated by water-related threats, this can be the difference between, for example, a tropical storm watch and a hurricane warning. This chapter will introduce some of the terminology and key phrases used when describing ocean-related threats – those from the ocean which can generate *water-related* hazards for which Emergency Management may be involved in any part of the disaster phase cycle.

Climate Change has a significant role in Ocean threats. The polar icebergs/glaciers are melting at an alarming rate – they are melting too fast – and this directly impacts Sea Level Rise, an Ocean-based threat. The water **quantity** hazards, such as flooding,[2] from this and other ocean-based threats will be covered in Part 4.

Ocean Acidification from increased carbon dioxide in the atmosphere is a chronic/systemic disaster. Yes, humans caused it – and humans can fix it – but not through Emergency Management.

Pollution – especially oil and chemical spills – can occur in the ocean – historic examples[3] include the Exxon *Valdez*,[4] BP's *Deepwater Horizon*,[5] and the MT *Hebei Spirit*[6] – and while there may appear to be no reasonable ways to fully mitigate against these threats and the environmental and economic hazards they generate, there are Response and Recovery aspects for the environmental impacts to the ecology, land, and water systems. Climate Change and the cascading threat of Sea Level Rise can also globally worsen ocean-based hazardous waste incidents.[7] The water **quality** hazards from these and other ocean-based threats will be covered in Part 4.

Overfishing, which can also be a source of pollution,[8] maritime transport, offshore drilling, deep sea mining, and even ocean noise – can all be harmful threats to the environment, the planet, etc. Similar to ocean acidification, none of these should have Emergency Management applied as the sole methodology of Protection, Prevention, Response, Recovery, or Mitigation.

DOI: 10.1201/9781003474685-6

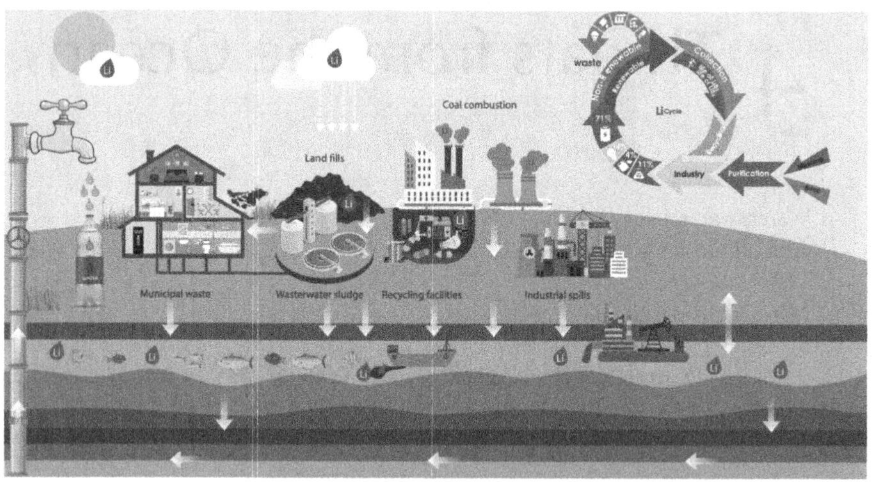

FIGURE 4.1 Illustration of the Li cycle (Li contents come from the Li-ion batteries wastes, coal combustion, municipal waste etc.) with various transport routes along groundwater pathways from possible domestic and industrial sources to human consumption through drinking water.

Source: Adeel, M., Zain, M., Shakoor, N., et al. (2023). Global Navigation of Lithium in Water Bodies and Emerging Human Health Crisis. *NPJ Clean Water*, *6*, 33. https://doi.org/10.1038/s41545-023-00238-w, p. 4.

4.1 HUMAN-MADE INCIDENTS

As noted above, there are numerous short-term and long-term human-made incidents which can cause a threat from the ocean. Included in this are threats from the ocean from human-made activities which are not considered incidents – such as ocean-water desalination processing into potable water source and underwater drilling. Lithium mining, used in battery production, as well as the proper disposal of old batteries, etc. has become an emerging environmental concern to our water systems.

> The Lithium (Li) industry has been expanding worldwide, over the last decades, and projections expect an increasing demand for its production in the coming years. It has been identified as an emerging pollutant and it occurs widely in aquatic environments, raising concern about its effects on ecosystems. Besides the increasing research on this topic, there is still limited understanding and discussion on the marine and coastal implications of Li occurrence.
>
> **(Barbosa et al., 2023, p. 1)**[9]

4.2 SEA LEVEL RISE

> Sea-level rise is one of the most significant effects of climate change. High projected rates of future sea-level rise have captured the attention of the world. Particularly, countries which are located in low-lying areas as well as small islands are concerned

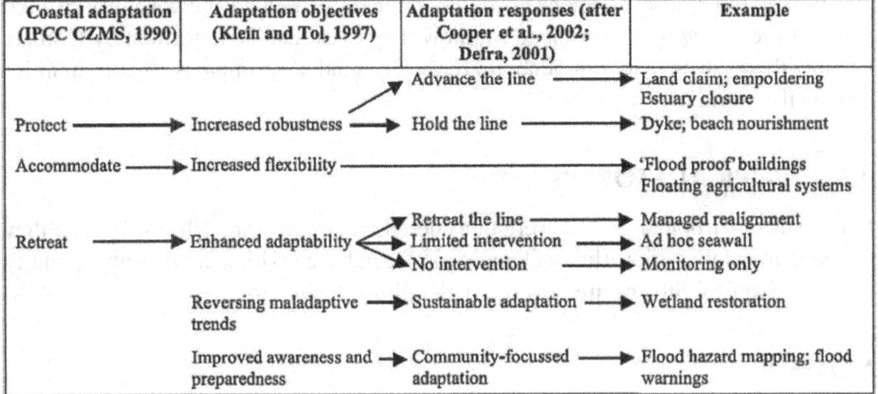

FIGURE 4.2 Evolution of planned coastal adaptation practice.

Source: Mimura, N. (2013). Sea-Level Rise Caused by Climate Change and Its Implications for Society. *Proceedings of the Japan Academy. Series B, Physical and Biological Sciences, 89*(7), 281–301. https://doi.org/10.2183/pjab.89.281, p. 296.

> that their land areas would be decreased due to inundation and coastal erosion and, at worst, a large proportion of their population may be forced to migrate to other countries. Therefore, this issue has resulted in heightened attention internationally, as the effects of climate change become apparent.
>
> **(Mimura, 2013, p. 281)[10]**

4.3 SLOSH

One storm surge model used in the United States by the National Weather service in advance of tropical storms is the Sea, Lake, and Overland Surges from Hurricanes – or SLOSH model.

> The model is applied to 38 specific coastal areas, called basins, along the Atlantic and Gulf of Mexico coasts of the U.S.; Oahu, Hawaii; Puerto Rico; and the Virgin Islands. SLOSH is also used to create simulation studies to assist in the "hazards analysis" portion of hurricane evacuation planning by the Federal Emergency Management Administration (FEMA), the U.S. Army Corps of Engineers, and state and local emergency managers.
>
> **(Glahn et al., 2009, p. 1)[11]**

4.4 TIDAL WAVES/ROGUE WAVES

Tidal and Rogue waves are large waves, usually originating in the ocean. There has been some tidal wave action on large lakes, such as the U.S. Great Lakes – but not to the extremes of these waves in the oceans. Both Tidal and Rogue waves are covered in detail, in Chapter 13.

There is quite a bit of science[12] and mathematics[13] behind the causality of tidal waves. Emergency managers need to know they exist, can be amplified by Climate Change themselves, and can generate devastating adverse impacts to communities around the world.

4.5 TROPICAL STORMS

For this book, tropical storms were considered a complex threat/hazard, since they are generated by multiple threat elements, and can have multiple quality and quantity hazards. Tropical Storms are covered in detail in Chapter 14.

NOTES

1 Coleridge, S. T., & Noyes, C. E. (Eds.). (1900). *The Rime of the Ancient Mariner.* Globe School Book Company. See www.loc.gov/item/00003326/
2 See www.channelnewsasia.com/asia/philippines-riverside-community-schools-students-flooding-climate-change-sea-levels-3979616
3 Barron, M. G., Vivian, D. N., Heintz, R. A., & Yim, U. H. (2020). Long-Term Ecological Impacts from Oil Spills: Comparison of *Exxon Valdez, Hebei Spirit*, and Deepwater Horizon. *Environmental Science & Technology*, *54*(11), 6456–6467. https://doi.org/10.1021/acs.est.9b05020
4 Piatt, J. F., Lensink, C. J., Butler, W., Kendziorek, M., & Nysewander, D. R. (1990). Immediate Impact of the "Exxon Valdez" Oil Spill on Marine Birds. *The Auk*, *107*(2), 387–397. https://doi.org/10.2307/4087623
5 Lichtveld, M., Sherchan, S., Gam, K. B., Kwok, R. K., Mundorf, C., Shankar, A., & Soares, L. (2016). The Deepwater Horizon Oil Spill through the Lens of Human Health and the Ecosystem. *Current Environmental Health Reports*, *3*(4), 370–378. https://doi.org/10.1007/s40572-016-0119-7
6 Yim, U. H., Kim, M., Ha, S. Y., Kim, S., & Shim, W. J. (2012). Oil Spill Environmental Forensics: The *Hebei Spirit* Oil Spill Case. *Environmental Science & Technology*, *46*(12), 6431–6437. https://doi.org/10.1021/es3004156
7 See www.axios.com/2023/12/10/sea-levels-hazardous-waste-climate-change
8 See https://oceanliteracy.unesco.org/threats-to-the-ocean/
9 Barbosa, H., Soares, A. M. V. M., Pereira, E., & Freitas, R. (2023). Lithium: A Review on Concentrations and Impacts in Marine and Coastal Systems. *Science of the Total Environment*, *857*, 159374. https://doi.org/10.1016/j.scitotenv.2022.159374
10 Mimura, N. (2013). Sea-Level Rise Caused by Climate Change and Its Implications for Society. *Proceedings of the Japan Academy. Series B, Physical and Biological Sciences*, *89*(7), 281–301. https://doi.org/10.2183/pjab.89.281
11 Glahn, B., Taylor, A., Kurkowski, N., & Shaffer, W. A. (2009). *The Role of the SLOSH Model in National Weather Service Storm Surge Forecasting.* U.S. Department of Commerce. See www.weather.gov/media/mdl/Vol-33-Nu1-Glahn.pdf
12 Marghany, M. (2021). Chapter 6 – Novel Relativistic Theories of Ocean Wave Nonlinearity Imagine Mechanism in Synthetic Aperture Radar. In M. Marghany (Ed.), *Nonlinear Ocean Dynamics* (pp. 163–190). Elsevier. https://doi.org/10.1016/B978-0-12-820785-7.00009-5
13 Halsne, T., Benetazzo, A., Barbariol, F., Christensen, K. H., Carrasco, A., & Breivik, Ø. (2024). Wave Modulation in a Strong Tidal Current and Its Impact on Extreme Waves. *Journal of Physical Oceanography*, *54*(1), 131–151. https://doi.org/10.1175/JPO-D-23-0051.1

5 Threats from the Land

We think of our land and water and human resources not as static and sterile possessions but as life-giving assets to be administered by wise provision for future days.

– Franklin D. Roosevelt[1]

The chapters in this part are focused on threats. Threats are what can (but not always) generate hazards to people, places, and things. Threats are causal. And making a threat is very different from posing a threat. When it comes to water-related hazards generated by water-related threats, this can be the difference between a flood watch and a flood warning. Threats from the land can be natural threats or human-made ones. Or both. When a lake with a dam on one side starts to accumulate too much water – for whatever reason – overflow is one possible hazard, dam releases are others, and a dam breech can also be another hazard.

Threats from rivers, tributaries, streams, etc. are effectively threats from the land. Rivers will bend and move over time. This chapter will introduce some of the terminology and key phrases used when describing land-related threats.

5.1 AGRICULTURE

Farming is performed on approximately half of the land in the United States. Working farms have a significant impact on water quality and water quantity in many parts of the country.

> Agricultural operations can have significant effects on water quality, due to the extent of farm activities on the landscape, the soil-disturbing nature of those activities, and associated impacts from sediment, nutrients, pesticides, and herbicides. The National Water Quality Assessment[2] shows that agricultural runoff is the leading cause of water quality impacts to rivers and streams, the third leading source for lakes, and the second largest source of impairments to wetlands. About a half million tons of pesticides, 12 million tons of nitrogen, and 4 million tons of phosphorus fertilizer are applied annually to crops in the continental United States. Soil erosion, nutrient loss, bacteria from livestock manure, and pesticides constitute the primary stressors to water quality.
>
> **(USEPA, n.d., p. 1)[3]**

DOI: 10.1201/9781003474685-7

5.2 AVALANCHES

Avalanches of any kind are certainly a threat, and those made of snow and ice will be covered in this book. The U.S. National Weather Service describes avalanches as follows:

> A rapid flow of snow down a hill or mountainside. Although avalanches can occur on any steep slope given the right conditions, certain times of the year and types of locations are naturally more dangerous. While avalanches are sudden, there are typically a number of warning signs you can look for or feel before one occurs. In 90 percent of avalanche incidents, the snow slides are triggered by the victim or someone in the victim's party. Avalanches kill more than 150 people worldwide each year.
>
> **(NWS, n.d., p. 1)**[4]

While there is avalanche weather information available in selected locations around the United States, from the National Weather Service, in the United States, there is no National Center for avalanches, as there is for Storm Prediction[5] or Hurricanes.[6] Other countries, such as Switzerland,[7] have centralized avalanche monitoring and alerts. There is also a European Avalanche Monitoring Service at www.avalanches.org.[8]

COLORADO'S AVALANCHE INFORMATION CENTER

Colorado has an Avalanche Information Center, which is a private–public partnership, housed in their Department of Natural Resources, between that

North American Public Avalanche Danger Scale

Avalanche danger is determined by the likelihood, size, and distribution of avalanches.
Safe backcountry travel requires training and experience. You control your risk by choosing when, where, and how you travel.

Danger Level	Travel Advice	Likelihood	Size and Distribution
5 - Extreme	**Extraordinarily dangerous avalanche conditions.** Avoid all avalanche terrain.	Natural and human-triggered avalanches certain.	Very large avalanches in many areas.
4 - High	**Very dangerous avalanche conditions.** Travel in avalanche terrain not recommended.	Natural avalanches likely; human-triggered avalanches very likely.	Large avalanches in many areas; or very large avalanches in specific areas.
3 - Considerable	**Dangerous avalanche conditions.** Careful snowpack evaluation, cautious route-finding, and conservative decision-making essential.	Natural avalanches possible; human-triggered avalanches likely.	Small avalanches in many areas; or large avalanches in specific areas; or very large avalanches in isolated areas.
2 - Moderate	**Heightened avalanche conditions on specific terrain features.** Evaluate snow and terrain carefully; identify features of concern.	Natural avalanches unlikely; human-triggered avalanches possible.	Small avalanches in specific areas; or large avalanches in isolated areas.
1 - Low	**Generally safe avalanche conditions.** Watch for unstable snow on isolated terrain features.	Natural and human-triggered avalanches unlikely.	Small avalanches in isolated areas or extreme terrain.

FIGURE 5.1 North American Public Avalanche Danger Scale.

Source: Colorado Avalanche Information Center. (n.d.). *Avalanche Danger Scale*. Accessed March 4, 2024. See https://avalanche.state.co.us/forecasts/tutorial/danger-scale. Used with permission.

department, Colorado's Department of Transportation, and a nonprofit group called the Friends of the CAIC.[9] The group provides forecasts, collects observations, tallies avalanche-related accidents, and more. They also created a five-level avalanche danger scale:

> Avalanche control and mitigation in Colorado is critical work. Whether you're backcountry skiing, splitboarding, or just passing through, CDOT's highly trained winter operations team is on the job. Our crews drop charges from helicopters, ski into the backcountry, and launch shells from air cannons and Howitzers to trigger slides in avalanche paths that threaten our state highways. If you're recreating in the mountains, avalanche safety is serious business. When it comes to traffic safety in the mountains, that's our business.
>
> **(CDOT Colorado Avalanche Operations, 2024, p. 1)[10]**

5.3 CROSS-BORDER WATER SUPPLY ISSUES

There can be numerous opportunities for positive collaboration and coordination between nations, when a water source (lake, river, etc.) is contained or flows between two or more countries. The opposite[11] can be true as well, too. Potable water supply – whether used partially for electricity production by one nation, and at the same time for human consumption and agricultural use by another country – must be a shared resource and responsibility.[12]

There are case examples of both some historical cross-border water supply issues between countries and internal water supply concerns within a country (water rights of sovereign tribal nations, noted in Chapter 7, for example) in this book.

5.4 DAMS

Dams store water, wastewater, or other liquid materials. They can be used for flood control, energy production, potable water supply, recreation, pollution control, and other purposes. There are more than 100,000 dams in the United States, including in Puerto Rico – and over 60% of them are privately owned. The Dam sector includes dams, navigation locks, and levees. As of November 2021 – for those dams in the National Inventory of Dams – more than 8,000 were classified as high hazard[13] – meaning any failure or misoperation will likely result in loss of human life. This is not a grading of the condition of the dam, but rather the threat.

5.5 DEFORESTATION

Deforestation can have an adverse effect on potable water supply. There is no consensus on this premise between researchers. Some believe that reducing vegetation (i.e., trees) can increase streamflow.[14] Others[15] have found the opposite. There may be critical distinctions between arid land on some continents and that found on others, when it comes to how deforestation impacts water supply.

High hazard dams are those where failure or mis-operation will likely cause loss of human life. Hazard classification refers to the potential consequences of a dam's failure, **not the condition of the dam**.

Top 10 states with the most high hazard dams

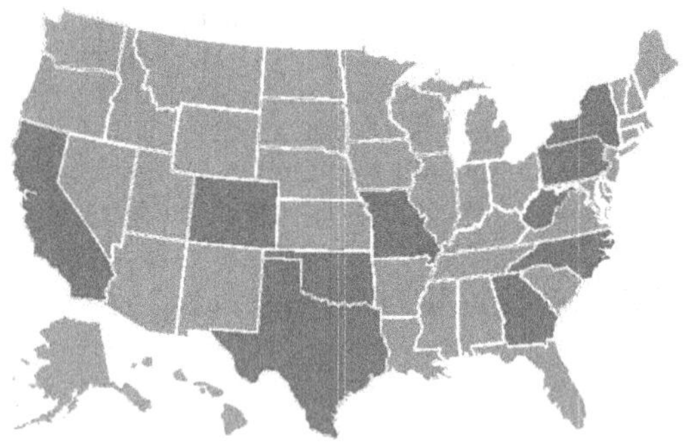

	State	# High Hazard Dams
1	MO	1,463
2	TX	1,411
3	NC	1,307
4	CA	805
5	PA	797
6	GA	630
7	CO	453
8	OK	449
9	WV	432
10	NY	424

FIGURE 5.2 High hazard dams.

Source: CISA. (2021). *Dam Sector Profile*. USDHS. See www.cisa.gov/sites/default/files/publications/dams-sector-profile-112221-508.pdf, p. 3. Public domain.

5.6 DROUGHT AND ARIDIFICATION

The concerns that droughts are occurring more intensely and more frequently is a threat for emergency managers to recognize as hazard generating to a community's water supply and to farming,[16] for example. Droughts can come and go, the permanent drying out of land – making vegetation growth nearly impossible – is aridification.[17] Chapter 10 covers droughts.

5.7 FLOODING

Too much rainfall can cause flooding of bodies of water. So can snowmelt. When the land cannot hold or absorb this excess water, flooding occurs:

> Flooding is when the water level or discharge in a body of water reaches or exceeds a certain threshold. Mountain torrents, rivers and lakes are no longer able to absorb the volume of water and burst their banks. When critical levels are reached, factors such as terrain and the permanent and temporary protective measures in place determine the severity of a flood.[18]

Terms such as "100-year or 500-year floods" can be misleading[19] to the public, as they are now occurring year-over-year. A new term in use is a "mega-flood",[20] and they can be predicted to some degree, when viewing on a larger scale (i.e., continents instead of individual countries). The year 2021 saw both Germany and Belgium experience mass casualty flooding incidents in the Rhine basin. A post-incident review of the data indicated this level of flooding could have been predicted.

5.8 GROUND BLIZZARDS

The heavy accumulation of snow on the ground already coupled with strong winds can generate a ground blizzard,[21] with whiteout conditions and quickly lowering temperatures from cold fronts, which come into an area many times unexpectedly.

5.9 GROUND CONTAMINATION

Also known as soil contamination. Contamination of the ground or soil can impact groundwater.

> In many parts of the world groundwater is the main source of water for day to day use. Wells dug by hand, or boreholes drilled by machine, down into the saturated layer under the ground yield water for domestic use, irrigation and industry, whilst groundwater is also used as a source of energy to heat our homes. Over a third of the world's population is supplied with drinking water from groundwater and, of the 700 million people worldwide who don't have an adequate water supply at present, most will have to be supplied from groundwater in the future. Groundwater also meets over 40% of irrigation water demand and provides about a quarter of all industrial supplies.
>
> **(International Association of Hydrogeologists, n.d., p. 1)**[22]

5.10 GROUNDWATER MISMANAGEMENT

Advances in the technology used for drilling worldwide – and the required groundwater treatment – have had both negative and positive impacts.[23] For example:

> The permeable reactive barrier (PRB) is proven as a promising technology for groundwater treatment by an interaction between the reactive material and the contaminant when the dissolved compounds migrate. In the permeable reactive barrier (PRB), water moves in a natural gradient, and no further energy is used to achieve the treatment.[24] The PRB is classified as in situ treatment, and the contaminant is transformed in the contaminated site into less toxic or immovable forms. The key benefits of the PRB innovation are minimal maintenance costs and long durability.
>
> **(Al-Hashimi et al., 2021, p. 1)**[25]

Urbanization,[26] industrialization,[27] and other human factors[28] also are sources of groundwater mismanagement threats.

5.11 LAND DEGRADATION

Deforestation, drought, groundwater mismanagement, and other threats can generate a threat to the land itself, in terms of its ability to support agriculture. This can also include "salinization, waterlogging, alkalization, and soil compaction" (Kawy and Darwish, 2019, p. 1).[29]

5.12 LAND USE

Land use can generate threats to waterways. This can include farm effluent, erosion, sealed surfaces, water usage, and land drainage/flood protection.[30]

> Urban sprawl (particularly the paving of large segments of the landscape) can have significant and usually negative impacts on water resources. Although growth and land use change may be inevitable in many communities, the way in which growth takes place affects its impact on water quality. With careful planning and a commitment to protect streams, rivers, and ground water, land use practices can be implemented that balance the need for jobs and economic development with protection of the natural environment. Development that takes place without such considerations, however, can lead to significant degradation of streams and ground water, and loss of aquatic life.
>
> **(Frankenberger, n.d., p. 1)**[31]

Land and water "grabbing" – the large-scale acquisition of both land and water rights by large multinational corporations – can also add to threats against sustainable long-term development and a community's water supply system.[32]

"MALL SNOW" COULD LAST FOR MONTHS

The combination of heavy snow, below freezing temperatures, ice, road salt – which can reduce the freezing point of ice[33] – and debris can generate snow piles which do not melt for months: defying rising temperatures and snowmelts elsewhere. The phenomenon, usually found in mall parking lots, has been called "mall snow"[34] – and can have built-in hazards, since it is relatively unknown what is in the large pile of snow, including chemicals and sharp objects. Usually, mall snow is created by large snowplows, bulldozers, and back hoes, which move the snow from the mall parking lot into large piles in corners. These large piles do not melt as fast as homeowner-based snow-shoveled piles. Stormwater management, workforce safety, and other elements need to be considered collaboratively between the community and the mall property owners.

5.13 PERMAFROST

While the duration of snow cover has an impact, this book will not cover permafrost, as it is predominantly frozen ground material (rocks, debris, moraine, etc.). Hazards from permafrost on land include landslides and debris flows.[35] Whether "Mall Snow" (see above) could be considered permafrost is certainly a humorous question.

5.14 POLLUTION

In some arid areas, wastewater – often a valuable resource when recycled – is used to grow crops. Pathogens in that water can cause cholera or diarrhoea, though farmers are often not aware of those potential consequences.

(Washing or boiling vegetables greatly reduces the risk of illness.)

The problem can be exacerbated by flooding, which can inundate sewage systems or stores of fertilizer, polluting both surface water and groundwater. Fertilizer run-off, can cause algal blooms in lakes, killing fish. Storm run-off and forest fires are further risks to farming and food security.

In some places around the world, pollution is also seeping into groundwater, with potential long-term impacts on crops, though more research is needed to establish the precise effects on plants and human health.

(UN Environment Programme, 2022, p. 1)[36]

5.15 SALTWATER INTRUSION

Saltwater Intrusion is a complex threat/hazard. It can be generated from multiple threats occurring simultaneously or cascading, and has multiple hazards related to water sources. Saltwater Intrusion is covered in this book in Chapter 17.

NOTES

1 Roosevelt, F. D. (1935). *Message to Congress on the Use of Our National Resources*. The American Presidency Project. See www.presidency.ucsb.edu/documents/message-congress-the-use-our-national-resources
2 See www.epa.gov/waterdata/attains
3 USEPA. (n.d.). *Nonpoint Source: Agriculture*. USEPA. Accessed March 4, 2024. See www.epa.gov/nps/nonpoint-source-agriculture
4 NWS. (n.d.). *Avalanche Safety*. NOAA/NWS. Accessed January 20, 2024. See www.weather.gov/safety/winter-avalanche#:~:text=Know%20the%20three%20factors%20required,and%20wind%20are%20common%20triggers
5 See www.spc.noaa.gov/
6 See www.nhc.noaa.gov/
7 See www.natural-hazards.ch/home/current-natural-hazards/avalanches.html
8 See www.avalanches.org/
9 See https://support.friendsofcaic.org/
10 *Colorado Department of Transportation (CDOT) Colorado Avalanche Operations*. State of Colorado. Accessed January 20, 2024. See https://youtu.be/tM7ih401y14?si=jFy4o5RdIWSRCbab
11 See https://interestingengineering.com/video/the-engineering-behind-cloud-seeding-the-art-of-creating-rain
12 Zakeri, B., Hunt, J. D., Laldjebaev, M., Krey, V., Vinca, A., Parkinson, S., & Riahi, K. (2022). Role of Energy Storage in Energy and Water Security in Central Asia. *Journal of Energy Storage*, *50*, 104587. https://doi.org/10.1016/j.est.2022.104587
13 See www.cisa.gov/sites/default/files/publications/dams-sector-profile-112221-508.pdf

14 Andréassian, V. (2004). Waters and Forests: From Historical Controversy to Scientific Debate. *Journal of Hydrology*, *291*(1–2), 1–27.

15 Mapulanga, A. M., & Naito, H. (2019). Effect of Deforestation on Access to Clean Drinking Water. *Proceedings of the National Academy of Sciences*, *116*(17), 8249–8254. https://doi.org/10.1073/pnas.1814970116

16 See www.unep.org/news-and-stories/story/five-threats-water-sustains-our-farms

17 See www.vaildaily.com/news/aridification-in-the-west-is-here-to-stay-so-where-do-we-go-from-here/

18 See www.slf.ch/en/natural-hazards/floods/

19 See https://theconversation.com/whats-a-100-year-flood-a-hydrologist-explains-162827

20 See www.homelandsecuritynewswire.com/dr20231124-how-megafloods-can-be-predicted

21 See www.weather.gov/safety/winter-ground-blizzard

22 International Association of Hydrogeologists. (n.d.). *Groundwater: More about the Hidden Treasure*. International Association of Hydrogeologists. Accessed March 4, 2024. See https://iah.org/education/general-public/groundwater-hidden-resource

23 Madhnure, P. (2014). Groundwater Exploration and Drilling Problems Encountered in Basaltic and Granitic Terrain of Nanded District, Maharashtra. *Journal of the Geological Society of India*, *84*(3), 341–351. https://doi.org/10.1007/s12594-014-0138-7

24 Faisal, A. A., Jasim, H. K., Naji, L. A., Naushad, M., & Ahamad, T. (2021). Cement Kiln Dust-Sand Permeable Reactive Barrier for Remediation of Groundwater Contaminated with Dissolved Benzene. *Separation Science and Technology*, *56*(5), 870–883.

25 Al-Hashimi, O., Hashim, K., Loffill, E., Marolt Čebašek, T., Nakouti, I., Faisal, A. A. H., & Al-Ansari, N. (2021). A Comprehensive Review for Groundwater Contamination and Remediation: Occurrence, Migration and Adsorption Modelling. *Molecules (Basel, Switzerland)*, *26*(19), 5913. https://doi.org/10.3390/molecules26195913

26 Liu, Y.-R., van der Heijden, M. G. A., Riedo, J., Sanz-Lazaro, C., Eldridge, D. J., Bastida, F., . . . & Delgado-Baquerizo, M. (2023). Soil Contamination in Nearby Natural Areas Mirrors that in Urban Greenspaces Worldwide. *Nature Communications*, *14*(1), 1706. https://doi.org/10.1038/s41467-023-37428-6

27 Maddela, N. R., Ramakrishnan, B., Kakarla, D., Venkateswarlu, K., & Megharaj, M. (2022). Major Contaminants of Emerging Concern in Soils: A Perspective on Potential Health Risks. *RSC Advances*, *12*(20), 12396–12415. https://doi.org/10.1039/d1ra09072k

28 Hilpert, M., & Breysse, P. N. (2014). Infiltration and Evaporation of Small Hydrocarbon Spills at Gas Stations. *Journal of Contaminant Hydrology*, *170*, 39–52. https://doi.org/10.1016/j.jconhyd.2014.08.004

29 Kawy, W. A. M. A., & Darwish, K. M. (2019). Assessment of Land Degradation and Implications on Agricultural Land in Qalyubia Governorate, Egypt. *Bulletin of the National Research Centre*, *43*(1), 70. https://doi.org/10.1186/s42269-019-0102-1

30 See www.sciencelearn.org.nz/image_maps/91-land-use-impacts-on-waterways

31 Frankenberger, J. (n.d.). *Land Use & Water Quality*. Purdue University. Accessed March 4, 2024. See https://engineering.purdue.edu/SafeWater/watershed/landuse.html

32 Rulli, M. C., Saviori, A., & D'Odorico, P. (2013). Global Land and Water Grabbing. *Proceedings of the National Academy of Sciences*, *110*(3), 892–897. https://doi.org/10.1073/pnas.1213163110

33 See https://sciencenotes.org/why-salt-makes-ice-colder-how-cold-ice-gets/

34 See https://news.yahoo.com/dirty-parking-lot-snow-pile-180602756.html

35 See www.slf.ch/en/permafrost/

36 UN Environment Programme. (2022). *Five Threats to the Water that Sustains Our Farms*. United Nations. See www.unep.org/news-and-stories/story/five-threats-water-sustains-our-farms

BIBLIOGRAPHY

Al-Hashimi, O., Hashim, K., Loffill, E., Marolt Čebašek, T., Nakouti, I., Faisal, A. A. H., & Al-Ansari, N. (2021). A Comprehensive Review for Groundwater Contamination and Remediation: Occurrence, Migration and Adsorption Modelling. *Molecules (Basel, Switzerland)*, *26*(19), 5913. https://doi.org/10.3390/molecules26195913

Andréassian, V. (2004). Waters and Forests: From Historical Controversy to Scientific Debate. *Journal of Hydrology*, *291*(1–2), 1–27.

CISA. (2021). *Dam Sector Profile*. USDHS. See www.cisa.gov/sites/default/files/publications/dams-sector-profile-112221-508.pdf

Colorado Avalanche Information Center. (n.d.). *Avalanche Danger Scale*. Accessed March 4, 2024. See https://avalanche.state.co.us/forecasts/tutorial/danger-scale

Colorado Department of Transportation (CDOT) Colorado Avalanche Operations. State of Colorado. Accessed January 20, 2024. See https://youtu.be/tM7ih401y14?si=jFy4o5RdIWSRCbab

Faisal, A. A., Jasim, H. K., Naji, L. A., Naushad, M., & Ahamad, T. (2021). Cement Kiln Dust-Sand Permeable Reactive Barrier for Remediation of Groundwater Contaminated with Dissolved Benzene. *Separation Science and Technology*, *56*(5), 870–883.

Frankenberger, J. (n.d.). *Land Use & Water Quality*. Purdue University. Accessed March 4, 2024. See https://engineering.purdue.edu/SafeWater/watershed/landuse.html

Hilpert, M., & Breysse, P. N. (2014). Infiltration and Evaporation of Small Hydrocarbon Spills at Gas Stations. *Journal of Contaminant Hydrology*, *170*, 39–52. https://doi.org/10.1016/j.jconhyd.2014.08.004

International Association of Hydrogeologists. (n.d.). *Groundwater – More About the Hidden Treasure*. International Association of Hydrogeologists. Accessed March 4, 2024. See https://iah.org/education/general-public/groundwater-hidden-resource

Kawy, W. A. M. A., & Darwish, K. M. (2019). Assessment of Land Degradation and Implications on Agricultural Land in Qalyubia Governorate, Egypt. *Bulletin of the National Research Centre*, *43*(1), 70. https://doi.org/10.1186/s42269-019-0102-1

Liu, Y.-R., van der Heijden, M. G. A., Riedo, J., Sanz-Lazaro, C., Eldridge, D. J., Bastida, F., . . . & Delgado-Baquerizo, M. (2023). Soil Contamination in Nearby Natural Areas Mirrors That in Urban Greenspaces Worldwide. *Nature Communications*, *14*(1), 1706. https://doi.org/10.1038/s41467-023-37428-6

Maddela, N. R., Ramakrishnan, B., Kakarla, D., Venkateswarlu, K., & Megharaj, M. (2022). Major Contaminants of Emerging Concern in Soils: A Perspective on Potential Health Risks. *RSC Advances*, *12*(20), 12396–12415. https://doi.org/10.1039/d1ra09072k

Madhnure, P. (2014). Groundwater Exploration and Drilling Problems Encountered in Basaltic and Granitic Terrain of Nanded District, Maharashtra. *Journal of the Geological Society of India*, *84*(3), 341–351. https://doi.org/10.1007/s12594-014-0138-7

Mapulanga, A. M., & Naito, H. (2019). Effect of Deforestation on Access to Clean Drinking Water. *Proceedings of the National Academy of Sciences*, *116*(17), 8249–8254. https://doi.org/10.1073/pnas.1814970116

NWS. (n.d.). *Avalanche Safety*. NOAA/NWS. Accessed January 20, 2024. See www.weather.gov/safety/winter-avalanche#:~:text=Know%20the%20three%20factors%20required,and%20wind%20are%20common%20triggers

Roosevelt, F. D. (1935). *Message to Congress on the Use of Our National Resources*. The American Presidency Project. See www.presidency.ucsb.edu/documents/message-congress-the-use-our-national-resources

Rulli, M. C., Saviori, A., & D'Odorico, P. (2013). Global Land and Water Grabbing. *Proceedings of the National Academy of Sciences*, *110*(3), 892–897. https://doi.org/10.1073/pnas.1213163110

UN Environment Programme. (2022). *Five Threats to the Water That Sustains Our Farms.* United Nations. See www.unep.org/news-and-stories/story/five-threats-water-sustains-our-farms

USEPA. (n.d). *Nonpoint Source: Agriculture.* USEPA. Accessed March 4, 2024. See www.epa.gov/nps/nonpoint-source-agriculture

Zakeri, B., Hunt, J. D., Laldjebaev, M., Krey, V., Vinca, A., Parkinson, S., & Riahi, K. (2022). Role of Energy Storage in Energy and Water Security in Central Asia. *Journal of Energy Storage*, *50*, 104587. https://doi.org/10.1016/j.est.2022.104587

6 Threats from Other Sources

> Maybe this is what the future will look like: fresh, clean water will be so rare it will be guarded by armies. Water as the next oil – the next resource worth going to war over.
>
> **– Body Shop founder Anita Roddick[1]**

The chapters in this part are focused on threats. Threats are what can (but not always) generate hazards to people, places, and things. Threats are causal. And making a threat is very different from posing a threat. When it comes to water-related hazards which are generated by human-made threats, this can be the difference between an alert that a system is vulnerable to an attack and the attack itself. Other source threats (those not from the air, land, or ocean) can be a natural threat or a human-made one. Or both. When a cyberattack disables the controls on a dam, and the flooding impacts the town below, that is an example of a complex set of hazards generated from a human-made threat.

Threats from Other Sources can also be made in the air, on the land, and on or from the oceans. An airplane can crash into a water treatment plant, construction workers can crack a water main, or a whale could wash up on the beach. Inaction by humans, failing to maintain levees, dams, water treatment plants, etc. can be the cause for water-related hazards. The same is true for failing to clear ice dams on rivers. Regardless of whether the hazard started naturally or not, if humans have an opportunity to prevent additional adverse impacts and/or protect populations from them, they should.

While this chapter could attempt to cover every possible threat not already noted, it will not. However, this chapter will introduce some of the terminology and key phrases used when describing *some* of the Other Sources for threats to water, which align to the three hazard Parts of the book.

6.1 HUMAN DECISIONS

6.1.1 CYBERTHREATS

Water systems are a combination of physical devices and computer systems which may control them. The networked electronic connections from system to system, and site to site involve the internet and the Internet-of-Things (IoT).

> In the past, water system security was achieved largely through isolation, limiting access to control components. However, with the emergence of IoT, water systems, as

DOI: 10.1201/9781003474685-8

> with other critical infrastructure services, are increasingly using a smart systems philosophy. This promotes the incorporation of IoT and analytics into industrial control systems (ICS) to improve the sensing and control capacity and ensure better integration with business processes. Collectively, this is known as the Industrial Internet of Things (IIoT), often labelled Industry 4.0, in which IoT is applied to industrial applications. It relies on connecting multiple layers of cyber – physical systems to facilitate autonomous decentralised decision-making and to improve the use of real-time data and predictive analytics to promote reliability, efficiency and productivity. With these technological advances, water systems that collect, treat, transport and distribute water to customers are undergoing a similar transformation, becoming highly connected and facing new technological challenges in the drive to provide safe water reliably.
>
> **(Tuptuk et al., 2021, p. 1)**[2]

Cyberthreats do not have to occur at a water or wastewater treatment facility in order to adversely affect a community's water supply. There are many points along the supply chain, which can generate a threat and hazards to potable water production.

6.1.2 Direct (Purposeful or Accidental)

Deciding to change a water supply system from one source to another, such as what was done in Flint, Michigan, in 2014, created long-term water-related health hazards for that community.

On April 25, 2014, the City of Flint, Michigan, changed their municipal water supply source from the Detroit-supplied Lake Huron water to the Flint River. The switch caused water distribution pipes to corrode and leach lead and other contaminants into municipal drinking water. In October 2016, Flint residents were advised not to drink the municipal tap water unless it had been filtered through a NSF International approved filter certified to remove lead. Although the city reconnected to the original Detroit water system that same month, the potential damage was already done, and a state of emergency was declared on January 16, 2016.[3]

6.1.3 Supply-Chain Management/Logistics Threats

There are several economic and business threats which can generate water-related hazards. For example, if there are supply chain management issues obtaining filters for water treatment plants, those shortages can disrupt the water treatment systems and threaten potable water generation. Work stoppages, strikes, labor disputes which create sabotage incidents – all of these also have the possibility of threatening a community's water supply.

6.2 OBSOLESCENCE OF WATER SUPPLY SYSTEMS

A big factor in water main breaks – and internal water pipe bursts – is older, unmaintained/uninspected water supply systems. Many times water pipes are buried under roadways decades ago, when the vehicular traffic was less frequent – and maybe not as heavy as today. Asphalt cracks, then salt and water seepage, freezing and

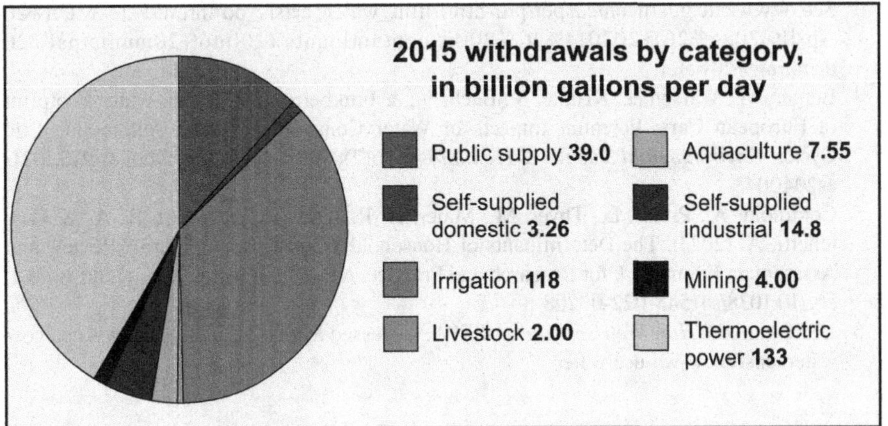

FIGURE 6.1 U.S. 2015 water-use withdrawals by category.

Source: USGS. Public domain.

thawing over years; all of this can create sinkholes and disturb the ground around a water pipe.

Pipe bursts inside of buildings can threaten critical infrastructure, as well as cause building evacuations. If the building is a school or a hospital, these can have cascading hazards, as well.

6.3 WATER USAGE

How much water is consumed by humans, for what purpose,[4] and where[5] – are all factors which can add to water scarcity. Water demand can be a threat itself.

> In the US, we are lucky to have easy access to some of the safest treated water in the world – just by turning on the tap. We wake up in the morning, take a shower, brush our teeth, grab a cup of coffee, and head out for the day. Water is an important part of our daily lives and we use it for a wide variety of purposes, but do we really understand how much we use?
>
> The average American family uses more than 300 gallons of water per day at home. Roughly 70 percent of this use occurs indoors.
>
> Nationally, outdoor water use accounts for 30 percent of household use yet can be much higher in drier parts of the country and in more water-intensive landscapes. For example, the arid West has some of the highest per capita residential water use because of landscape irrigation.
>
> **(USEPA, n.d., p. 1)[6]**

NOTES

1 Roddick, A. (2001). *Business as Unusual.* Thorsons.
2 Tuptuk, N., Hazell, P., Watson, J., & Hailes, S. (2021). A Systematic Review of the State of Cyber-Security in Water Systems. *Water, 13*(1), 81. https://doi.org/10.3390/w13010081

3 See www.cdc.gov/nceh/casper/pdf-html/flint_water_crisis_pdf.html#:~:text=On%20April%2025%2C%202014%2C%20the,contaminants%20into%20municipal%20drinking%20water
4 Berger, M., Warsen, J., Krinke, S., Bach, V., & Finkbeiner, M. (2012). Water Footprint of European Cars: Potential Impacts of Water Consumption along Automobile Life Cycles. *Environmental Science & Technology*, *46*(7), 4091–4099. https://doi.org/10.1021/es2040043
5 Cominola, A., Preiss, L., Thyer, M., Maier, H. R., Prevos, P., Stewart, R. A., & Castelletti, A. (2023). The Determinants of Household Water Consumption: A Review and Assessment Framework for Research and Practice. *NPJ Clean Water*, *6*(1), 11. https://doi.org/10.1038/s41545-022-00208-8
6 USEPA. (n.d.). *How We Use Water*. USEPA. Accessed March 5, 2024. See www.epa.gov/watersense/how-we-use-water

BIBLIOGRAPHY

Berger, M., Warsen, J., Krinke, S., Bach, V., & Finkbeiner, M. (2012). Water Footprint of European Cars: Potential Impacts of Water Consumption Along Automobile Life Cycles. *Environmental Science & Technology*, *46*(7), 4091–4099. https://doi.org/10.1021/es2040043

Cominola, A., Preiss, L., Thyer, M., Maier, H. R., Prevos, P., Stewart, R. A., & Castelletti, A. (2023). The Determinants of Household Water Consumption: A Review and Assessment Framework for Research and Practice. *NPJ Clean Water*, *6*(1), 11. https://doi.org/10.1038/s41545-022-00208-8

Roddick, A. (2001). *Business as Unusual*. Thorsons.

Tuptuk, N., Hazell, P., Watson, J., & Hailes, S. (2021). A Systematic Review of the State of Cyber-Security in Water Systems. *Water*, *13*(1), 81. https://doi.org/10.3390/w13010081

USEPA. (n.d.). *How We Use Water*. USEPA. Accessed March 5, 2024. See www.epa.gov/watersense/how-we-use-water

Part 3

Quality Hazards with Water

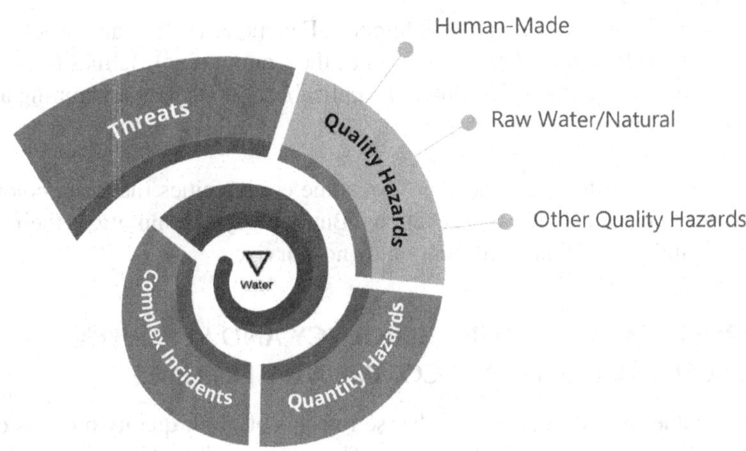

This part has three chapters and covers quality hazards with water-related threats:

- Human-Made Hazards to Water Sources (Chapter 7)
- Natural Water Source Contamination (Chapter 8)
- Water Source Quality Hazards from Other Incidents and Threats (Chapter 9)

Chapter 7 has material on contaminating raw water or natural water sources, including lakes, aquifers, groundwater, etc. Issues for human-made water supply *systems* contamination can be found in Part 5. Regardless of the water quality **threat**, the actions for preparedness/prevention/protection, response, recovery, and mitigation will most likely be similar, when it comes to specific quality **hazards** with water. And while there are specific sections in Part 5 for water treatment plants, wastewater treatment

DOI: 10.1201/9781003474685-9

plants, and other urban/suburban infrastructure systems, the threats which can impact those human-made facilities and systems can also generate hazards for groundwater and raw water sources, including wells, cisterns, aquifers, and watersheds used directly by the public for both drinking water, recreational, and agricultural use.

Each chapter in this Quality Hazards part – and all the other chapters in the Quantity Hazards and Complex Incidents parts – will contain the following:

- An overview of the specific hazard and its adverse impacts to communities
- Case examples for historic reference (many of these will hopefully guide officials as to what ***not*** to do, as lessons learned from other's missteps and misfortunes)
- Additional References and Reading, more specific to the hazard

The overview section in Part 3 will also have a common set of the following:

- Adverse impacts of these types of **Quality Hazards** to the following:
  - The Disaster Phase Cycles (the before, during, and after incidents occur)
  - The Response's Incident Command System itself (each jurisdiction around the world may be different, this book uses the U.S. NIMS/ICS as the baseline)
  - The U.S.-based Emergency Support Functions (ESFs) and Recovery Support Functions (RSFs), as well as the Community Lifelines (CLs)
- A summary of the POETE process, applicable to deliberative planning and problem-solving

Water Quality is a life safety element – in some communities the taste, color, and odor are not as critical as in other locations. But in every community, if their drinking water is unhealthy, that community may not survive.

ADVERSE IMPACTS TO THE EMERGENCY AND RECOVERY SUPPORT FUNCTIONS AND COMMUNITY LIFELINES

Following table shows some of the adverse impacts of water quality hazards on the U.S.-based Emergency Support Functions. There are similar tables in Parts 4 and 5 for the other hazard sections of this book.

Water-Related Threats and Hazards to the ESFs

Emergency Support Function (ESF)	Example Supporting Actions or Capabilities Related to Water-Related Threats or Hazards
#1 – Transportation	While an adverse impact to water quality system generally will not generate targeted hazards directly to the Transportation ESF, this ESF and the agencies, departments, staff, etc. who support it will have additional responsibilities and courses of action related to the emergency distribution of potable water to the public. Transportation of bulk water (via tankers, on pallets carried on trucks, etc.) will impact this ESF. Also, public transportation routes (bus, paratransit, etc.) will need to be examined, to pass to the Points of Distribution or PODs.

Water-Related Threats and Hazards to the ESFs *(Continued)*

Emergency Support Function (ESF)	Example Supporting Actions or Capabilities Related to Water-Related Threats or Hazards
#2 – Communications	There are generally no direct adverse impacts to this ESF (in terms of Communications Systems) from a water quality hazard, but the additional crisis communications to the public about alternative sources for potable water, water quality alerts and warnings, etc. will need to be made. Agencies, departments, staff, etc. should be on alert for false information about the water quality and points-of-distribution, and counter against them, if found. A higher volume of calls into Public Safety Access Points or PSAPs will occur and need to be managed and triaged. As this may be longer-term, rerouting some call volume to 2–1–1 should be considered. Details about internal (for non-public use) points-of-distribution of water will need to be communicated internally within the operation, possibly through the Safety Officer.
#3 – Public Works and Engineering	**This ESF is significantly impacted by this hazard. Subject-matter experts from the water system authority, treatment plant management, etc. may be in unified command and/or acting as liaison officers to support the restoration of full capability of the water supply system**.
#4 – Firefighting	If the incident shuts down a water treatment plant, reducing or eliminating water pressure to the community's fire hydrant system, significant adverse impacts to that community's firefighting capability will occur. The reverse is also true, if firefighting efforts overtax potable water systems, they may generate public health issues for the surrounding community.[1]
#5 – Information and Planning	Essential Elements of Intelligence/Information about alternative water supply sources (both fixed and mobile), water quality testing results, distribution status, restoration timelines, and more – are all needed by ESF#5 to work through the operational planning cycle (see Chapter 2 for details).
#6 – Mass Care, Emergency Assistance, Temporary Housing, and Human Services	While sheltering, feeding, and family reunification will most likely not be adversely impacted by a water quality hazard impacting a community's water supply system, the fourth element of ESF#6: distribution of emergency supplies is definitely a course of action/response mission, which will most likely be implemented if the incident lasts more than a preplanned time, MOUs/MOAs have triggers, or leadership orders emergency distribution of water to the public.
#7 – Logistics	Temporary supply of bottled water, potable water, etc. are all needed. This may involve distribution at PODs, door-to-door, or both. Logistical supply chain integrity and security may be needed, cross-walked with ESF#13.
#8 – Public Health and Medical Services	As previously noted, every incident, emergency, or disaster has a public health and medical services impact. Water quality adversely impacts hospitals, clinics, assisted living facilities, nursing homes, dialysis centers,[2] blood/plasma collection sites, and more. While many of these licensed facilities have mandates for contingency planning – covering water quality concerns, as well as others – those plans are generally short-term (incidents lasting hours or days) and not long-term (weeks into months).
#9 – Search and Rescue	This ESF will most likely not be adversely impacted by these water quality hazards.
#10 – Oil and Hazardous Materials Response	As noted in the case studies in the chapters in this part, there can be incidents where oil and hazardous materials make their way into the streams and rivers, and therefore adversely impact downstream raw water and water sources. The adverse health impacts from those human-made hazards, as well as heavy metals, elements occurring naturally, and microorganisms/parasites may require missions for cleanup from this ESF. The high priority of this cleanup and remediation of the oil and/or hazardous materials is both a life safety and an incident stabilization course of action.

(Continued)

Water-Related Threats and Hazards to the ESFs *(Continued)*

Emergency Support Function (ESF)	Example Supporting Actions or Capabilities Related to Water-Related Threats or Hazards
#11 – Agriculture and Natural Resources	Crop irrigation, livestock watering, parks and recreation areas can all be adversely impacted by this hazard. For example, a harmful algal bloom on a lake should generate a no swimming or boating prohibition.
#12 – Energy	While quality may not be a concern, if the water supply is reduced or cut off to a nuclear power plant, that can have significant impact on the Energy ESF. The same is true for hydroelectric power generation.
#13 – Public Safety and Security	May need to support ESF#6 and ESF#7 in the distribution of emergency supplies (potable water). Also, first responders and other operational groups will need their own point-of-distribution for water, separate from the public ones.
#14 – Cross-Sector Business and Infrastructure	Water Management Districts may be a separate legal entity from the local government, and have their own emergency management staff, protocols, legal requirements, etc. How water quality problems with water sources are caused – and solved – may depend on how staff at water treatment plants, wastewater treatment plants, water supply systems, etc. collaborate, coordinate, cooperate, and communicate day-to-day with local emergency management officials in the jurisdiction. The same is true for exercising the emergency management scenarios, including human-made hazards to water sources. Recognizing in advance that there are differences in the actions of for-profit organizations from nonprofit organizations and governmental organizations is key. For example, in many jurisdictions if after some period of time (usually days) the water supply remains unhealthy, alternative processes for the rerouting of water systems and/or purchase, delivery, and distribution of potable water will most likely be necessary. Who is responsible for these missions needs to be established in advance and memorialized in formal plan documents.
#15 – External Affairs	Coordinate with ESF#2 on both internal and external messaging regarding water quality status, restoration of systems status, alternative sources of potable water, etc.

Source: Barton Dunant.

Recovery Support Functions

Similarly, following table shows some of the adverse impacts of water quality hazards on the U.S.-based Recovery Support Functions (RSF). There are similar tables in Parts 3 and 4 for the other hazard sections of this book.

Recovery Support Functions

Federal Recovery Support Functions (RSFs)

Economic

Potable Water is a commodity, sometimes at the cost of what is ethical and humane.[3] This can lead governments to make decisions about who gets access to potable water and for what purposes, as different priorities than the LIPER order. The sooner that restoration of the water supply system of a community is completed, the sooner the adverse impacts to the economy of that community will be alleviated. If cascading incidents occur (a warehouse fire is unable to be extinguished quickly, because

Recovery Support Functions *(Continued)*

Federal Recovery Support Functions (RSFs)

Economic (*continued*)

of low/no water pressure in nearby fire hydrants), the economic loss of that business – and jobs – could be massive to a community. There is also reputational risk to communities with natural water supply contamination incidents, as the public may view these incidents as chronic or systemic, and therefore unable to be mitigated against.

Health and Social Services

Hospitals and Healthcare facilities – including walk-in facilities such as clinics and dialysis centers – both are adversely impacted by a lack of potable water. Recovery for this support function is critical. The general public needs quality water for activities of daily living. Recovery work needs to include temporary solutions for potable water needs, as well as restoration of water supply systems for long-term recovery.

Community Assistance

This RSF includes strengthening, repairing, and even building relationships between local, state, tribal, and territorial stakeholders and partners. Watershed recovery master plans[4] would be an example of where the Community Assistance federal resources – on a Presidentially Declared Disaster – could utilize this Community Assistance RSF, for long-term water quality Recovery and Mitigation. Other elements of a community (social, educational, etc.) will need Recovery support for water quality hazard incidents as well.

Infrastructure Systems

As noted in ESF#3 Public Works and Engineering, recovery of this infrastructure element *if* adversely impacted by any water quality hazard, to support the restoration of full capability of the water supply system is also critical. Most of the recovery focus on emergency management – and the impacted community's leadership – will be on this RSF.

Housing

Long-term housing is not possible without potable water systems. Generally, the housing RSF is activated when housing is damaged or destroyed by an incident (such as a hurricane or typhoon), but it may be necessary to call upon additional state/territorial/tribal nation and federal resources to support alternative housing – especially for socially vulnerable populations – due to a long-term lack of potable water.

Natural and Cultural Resources

- While watersheds, rivers, estuaries, etc. can be equally adversely impacted by natural or human-made causes for water quality and quantity threats, the reputational risk to a community must also be taken into account by emergency management, in coordination (and cooperation) with the entity who may have caused the disaster. For example, tourism, fisheries, and other economic activity may not return to an area which was previously designated (or perceived as) contaminated. Emergency managers must be concerned and mission against these adverse impacts, in addition to those which clean up the site and restore the water quality. An example of this type of immediate response and long-term emergency management recovery work can be found in the 2010 Deepwater Horizon disaster.
- If infrastructure – such as older water pipes – to cultural resources and sites needs to be replaced because of the water quality hazards from an incident, this may take extended time and financial resources to remediate and restore capability to the cultural resource. Sites may have to be closed or temporarily restricted in access (i.e., potable water brought in, as well as port-a-potties, if freshwater systems are compromised). For natural resources adversely impacted by water quality hazards, this same type of closure to access may be necessary. This can involve long-term cleanup and removal of material (soil, contaminants, etc.) in and around parks, recreational areas, watershed areas, reservoirs, lakes, rivers, streams, etc. – even complete reworking of water systems for those sites.

Source: FEMA.

Generally, adverse impacts from water-related threats and hazards target the following RSFs, but as potable water (systems and for hydration) is part of multiple Community Lifelines (see Chapter 2), the lack of potable water can impact all of the RSFs.

COMMUNITY LIFELINES

As noted in Chapter 2, the Community Lifelines (CLs) are a highly useful list of EEIs, which can be visualized as a dashboard. Water Quality hazards will have implications for **Food**, **Hydration**, **Shelter**, **Health and Medical**, and of course **Water Systems CLs**, but there can be impacts to the other CLs, from water-quality hazard:

Safety and Security (Law Enforcement/Security, Fire Service, Search and Rescue, Government Service, Community Safety): Human-made incidents will always have an intelligence/investigation aspect, including those which adversely impact raw water sources, water sources, dams, rivers, watersheds, etc. There should always be a concern for a nexus to terrorism[5] (i.e., the deliberate attempt to destroy a water system or poison a water supply) and this can have multinational impacts as well. Law Enforcement and Security will also be tasked with protective missions during the Response phase (guarding water PODs, repair work, etc.). This book also previously noted the impact to the Fire Service with low/no water pressure, due to quality problems with the water system. There may need to be missions for protection of water sources/supplies for use in fire service response.

Food, Hydration, Shelter (Food, Hydration, Shelter, Agriculture): Obvious impacts to this Community Lifeline for Water as a component of Hydration, and there may also be impacts to sheltering, if homes must be evacuated. Also, people will evacuate their homes days after the incident, if the quality of the water supply at their homes is adversely impacted. For example, after a hurricane or winter storm passes, if that community's water treatment plant was damaged, residents may not return to their homes – and choose to stay at shelters, for feeding and hydration assistance.

Health and Medical (Medical Care, Public Health, Patient Movement, Medical Supply Chain, Fatality Management): Aligns to both Emergency Support Function ESF#8) Public Health missions and the Health and Social Services RSF missions. As noted in the ESFs and RSFs, this Community Lifeline is significantly adversely impacted by any disruption to the potable water supply systems, which may happen with water quality hazards.

Energy (Power Grid, Fuel): Water and Energy have many nexus points, all of which need emergency management intelligence. Hydroelectric power generation[6] by its very name is dependent on a continuous and sufficient flow of water and must have its own Emergency Management protocols and procedures.[7] The process of generating electricity at traditional coal or other fuel electricity-generating plants creates large needs for cooling by water. When the water supplies to these plants are reduced or shut down, that can adversely impact that plant's operations.[8] The same is true for nuclear power plants, as well,[9] plus they have the added potential for catastrophic failures, as part of complex/cascading incidents.[10]

Communications (Infrastructure, Responder Communications, Alerts Warnings and Messages, Finance, 911 and Dispatch): One aspect of crisis communications for Public Information Officers, related to water supply systems, is to preplan (and exercise) a potential switch from "virgin" water sources to reclaimed wastewater, posttreatment, and processing. For some communities, this will need to be communicated well in advance of any possible switch to alternative systems and process flows, if an incident generates a need for a switch in water sources.

Transportation (Highway/Roadway/Motor Vehicle, Mass Transit, Railway, Aviation, Maritime): Water quality issues can have an adverse impact on this CL, and the negative status of this CL can impact the others as it relates to providing potable water to a community. If water quality becomes a hazard in a community, the full capabilities of the Transportation CL will be needed for transporting potable water to Points of Distribution, and in bulk transport to critical infrastructure/key resource sites, such as hospitals and dialysis centers. If navigable waterways are blocked or hampered by a water quality hazard, this CL will be impacted as well. Critical repair parts, filters, staff, etc. need to be transported to water and wastewater treatment sites, to make repairs and work toward restoration of the **Water Systems** CL.

Hazardous Material (Facilities, HAZMAT, Pollutants, Contaminants): Obvious impacts to this Community Lifeline for Pollutants and/or Contaminants, and there will of course be a HAZMAT response to the incident. There may be longer-term impacts to facilities, if a contaminated waterway supports those Critical Infrastructure/Key Resources (CIKR), from a needed clean water supply basis. Considerations on how natural freshwater supplies are utilized in dialysis centers and even nuclear power plants, for example.

Water Systems (Potable Water Infrastructure, Wastewater Management):[11] Understanding how wastewater runoff (i.e., treated water) is returned back into the environment (and to locations such as raw water drinking sources) is important for emergency managers to communicate to their local governmental officials. Additionally, the wastewater to drinking water movement (in California for example, as of 2023) will be a big mental public communications moment for drinking potable water which originated from reclaimed wastewater.

There may also be longer-term impacts to the food supply if the waterway also supports farming and/or agricultural production. Considerations on how natural freshwater supplies are utilized in rice farming, for example. Also, harmful algal blooms[12] can produce toxins which can make people and animals sick. These cyanobacteria can occur in either freshwater or saltwater sources.

NOTES

1 See https://pidwater.com/wqadvisory/100-boil-water-notice
2 Centers for Medicare and Medicaid Services. (n.d.). *Emergency Preparedness for Dialysis Facilities: A Guide for Chronic Dialysis Facilities*. U.S. Centers for Medicare & Medicaid Services. Accessed November 20, 2023. See www.cms.gov/medicare/end-stage-renal-disease/esrdnetworkorganizations/downloads/emergencypreparednessforfacilities2.pdf
3 Ibrahim, I. A. (2022). Water as a Human Right, Water as a Commodity: Can SDG6 Be a Compromise? *The International Journal of Human Rights*, *26*(3), 469–493. https://doi.org/10.1080/13642987.2021.1945582
4 See www.fema.gov/emergency-managers/practitioners/recovery-resilience-resource-library/boulder-county-creek-recovery
5 See www.csis.org/programs/global-food-and-water-security-program
6 See https://api.pageplace.de/preview/DT0400.9781003834731_A47367807/preview-9781003834731_A47367807.pdf
7 Jing, W., & Xiazhong, Z. (2011). The Study on Emergency Management System of Super-giant Water Resources and Hydropower Projects. *Procedia Environmental Sciences*, *11*, 1142–1146.
8 DeNooyer, T. A., Peschel, J. M., Zhang, Z., & Stillwell, A. S. (2016). Integrating Water Resources and Power Generation: The Energy – Water Nexus in Illinois. *Applied Energy*, *162*, 363–371.
9 Mccall, J., Macknick, J., & Macknick, J. (2016). *Water-Related Power Plant Curtailments: An Overview of Incidents and Contributing Factors*. https://doi.org/10.2172/1338176.
10 Aldrich, D. (2012). Networks of Power. *Natural Disaster and Nuclear Crisis in Japan*, 127–139.
11 See www.fema.gov/sites/default/files/documents/fema_p-2181-fact-sheet-4-1-drinking-water-systems.pdf
12 See www.cdc.gov/habs/index.html

BIBLIOGRAPHY

Aldrich, D. (2012). Networks of Power. *Natural Disaster and Nuclear Crisis in Japan*, 127–139.

Centers for Medicare and Medicaid Services. (n.d.). *Emergency Preparedness for Dialysis Facilities: A Guide for Chronic Dialysis Facilities*. U.S. Centers for Medicare & Medicaid Services. Accessed November 20, 2023. See www.cms.gov/medicare/end-stage-renal-disease/esrdnetworkorganizations/downloads/emergencypreparednessforfacilities2.pdf

DeNooyer, T. A., Peschel, J. M., Zhang, Z., & Stillwell, A. S. (2016). Integrating Water Resources and Power Generation: The Energy – Water Nexus in Illinois. *Applied Energy*, *162*, 363–371.

Ibrahim, I. A. (2022). Water as a Human Right, Water as a Commodity: Can SDG6 Be a Compromise? *The International Journal of Human Rights*, *26*(3), 469–493. https://doi.org/10.1080/13642987.2021.1945582

Jing, W., & Xiazhong, Z. (2011). The Study on Emergency Management System of Supergiant Water Resources and Hydropower Projects. *Procedia Environmental Sciences*, *11*, 1142–1146.

Mccall, J., Macknick, J., & Macknick, J. (2016). *Water-related Power Plant Curtailments: An Overview of Incidents and Contributing Factors*. https://doi.org/10.2172/1338176

7 Human-Made Hazards to Water Sources

> With every baby bottle, every sippy cup, every glass of water, my kids were being poisoned by a neurotoxin in the drinking water.
>
> Mona Hanna-Attisha, MD, author of *What the Eyes Don't See: A Story of Crisis, Resistance, and Hope in an American City*, on the Flint, Michigan water crisis.[1]

This chapter covers how human-made hazards can adversely impact potable/drinking water sources. This can be anything from accidental discharges of pollutants into a storm drain, to domestic violent extremists sabotaging a watershed.

This chapter – and all the other chapters in the three Hazards sections – will contain the following:

- An overview of the specific hazard and its adverse impacts to communities
- Case examples for historic reference (many of these will hopefully guide officials as to what ***not*** to do, as lessons learned from other's missteps and misfortunes)
- Adverse Impacts of this specific hazard series to the following:
  - The Disaster Phase Cycles (the before, during, and after incidents occur)
  - The Response's Incident Command System itself (each jurisdiction around the world may be different, this book uses the U.S. NIMS/ICS as the baseline)
- A summary of the POETE process, applicable to deliberative planning and problem-solving
- Additional references/reading material

Besides local jurisdictional emergency management groups, this hazard will need to be co-managed by water management districts, hospitals and other healthcare facilities, colleges/universities, etc. in a coordinated and collaborative way. All will have unmet needs, regulatory requirements, etc. which they will consider as a priority – and some of those priorities may conflict and compete for Response and Recovery resource allocation. Unified Command should always prioritize the LIPER (see Chapter 2), when making decisions.

7.1 OVERVIEW

If you live "off the grid"[2] in terms of *all* your infrastructure needs – including potable water and wastewater processing infrastructure – meaning you collect and filter

DOI: 10.1201/9781003474685-10

rainwater,[3] have a solar-powered or manual pump well, produce no gray or blackwater, or combinations of all – you probably will not need help from an emergency manager, even if your neighbors lose access to their raw water sources, or their own water supply and wastewater processing. There are obviously global communities where any clean, safe, potable (i.e., drinkable) water supply system is not the norm[4] and this "Utopian"[5] version of water management is in fact mismanaged or not managed at all. These communities suffer chronic and systemic disasters associated with their water quality (see Chapter 2 for more on chronic or systemic disasters), and their water quality is *not* a threat or hazard for Emergency Management to solve **by itself**. Anywhere there is inaccessibility to clean potable water on a regular, normalized basis **is a crisis unto itself**.

As compared to a solo survivalist mentality,[6] most people in developed countries count on clean water coming from their faucets and taps, every hour of every day. Many consider access to potable water as a global human right.[7] This book cannot cover the depth of challenging issues related to creating clean water infrastructure around the globe. It is worth restating again: The lack of accessible clean drinking water to more than 750 million people around the globe is an *extinction level event* – one which can literally wipe out whole populations, unless a whole-of-government/whole-community approach to solutions is urgently undertaken. This may include roles and missions for Emergency Management in the short term, but like other Type 0 Extinction Level Events such as a global pandemic,[8] it cannot fall solely to emergency managers to resolve. In fact, fixing the adverse impacts to any nation's existing water sources also requires whole-of-government/whole-community work and courses of action. Emergency managers can respond to and begin to start the recovery processes; but it will take many more hands and massive funding to get through mitigation work to prepare for future adverse impacts to those same systems.

Human-made freshwater source contamination can be accidental or intentional. The question of causality or even criminality associated with freshwater source contamination is immaterial to Emergency Management's Response and Recovery operations, except that there may be law enforcement intelligence and investigation elements imbedded within the Incident Command System (ICS) established, specifically, for responding to and recovery from the adverse impacts of the incident. If this is the case, then actions of the other Command and General Staff branches/sections may need to be adjusted and/or expanded to support local, state, and/or federal law enforcement intelligence and investigation elements. There may be a parallel incident command system established by law enforcement officials or regulatory entities at the water supply system (as shown in Figure 7.1) – and there may or may not be a liaison officer (LNO) linking the two – or more – ICSs.

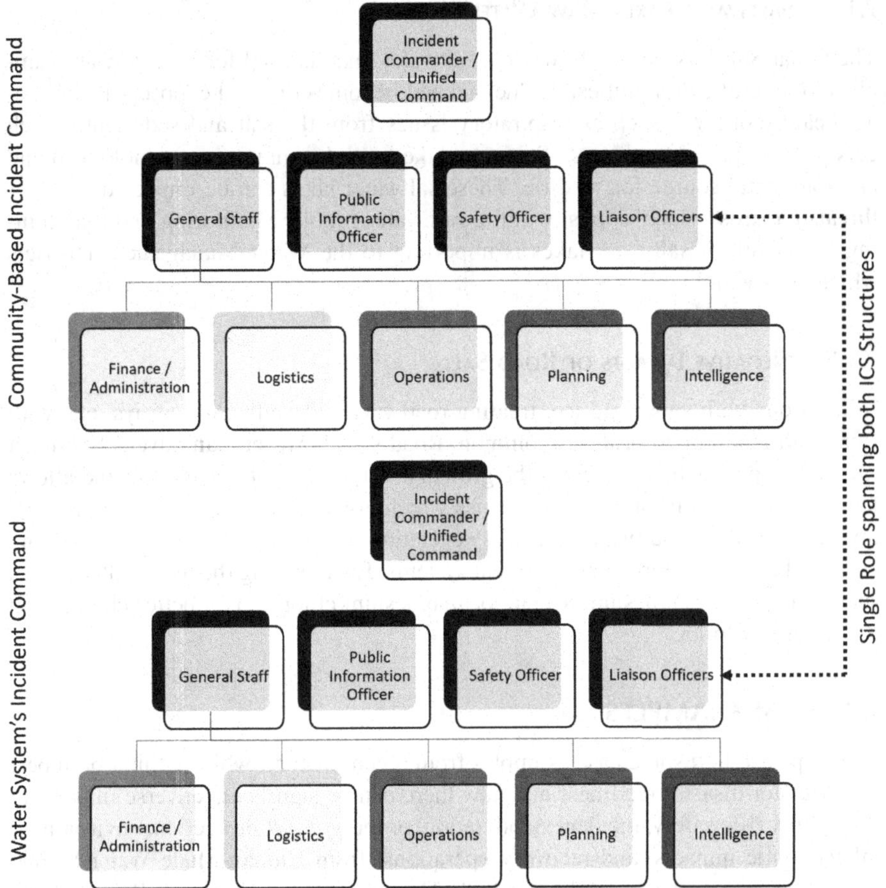

FIGURE 7.1 Liaison Officers role in spanning multiple Incident Command Systems.

Source: Barton Dunant. Used with permission.

Each incident, emergency, disaster, etc. is different. Emergency managers should be involved in the full disaster cycle planning, organization, equipping, training, and exercising for raw water, potable water, and/or drinking water treatment distribution and storage systems in their respective communities (including those which provide support or services to their jurisdiction's water system from another jurisdiction – such as a the **commodity impacts** to and from neighboring community's shared water treatment plant, watershed areas, dams, etc.).

In addition to the obvious pollutants to raw water and other water sources from chemicals and other human-made sources, there are natural elements which can be adverse to potable water. Most of those are covered in the next chapter – "Natural Water Source Contamination". Two hazards related specifically to salt are noted below in this chapter.

7.1.1 Saltwater Lakes: Raw Water Sources

The Great Salt Lake of Utah, while a saltwater lake, is used for potable water supplies to several communities. As the lakebed becomes dry and exposed, it can create health concerns such as respiratory issues from the salt and sediment, which gets picked up in dust storms.[9] Saltwater lakes are also an important biological and environmental source for wildlife. These saltwater lakes can be impacted by other threats/hazards – such as post wildfire rainwater runoff. Monitoring and maintaining the health of saltwater lakes is important to the Water Management Districts which use them.

7.1.2 Negative Impacts of Road Salt

In locations with snow and ice, treating roadways, sidewalks, driveways, etc. with Road Salt (sodium chloride) is common. Road salt, however, can adversely impact raw water sources by hampering the growth and breeding of aquatic life and allows for increased growth of parasitic species which are saltwater tolerant, such as mosquitoes and algae. Saltwater can kill vegetation and is highly corrosive – adding costs and processes for water treatment systems. Encouraging the use of alternatives such as calcium magnesium acetate or magnesium chloride is a better choice than the sodium chloride.[10]

7.2 CASE EXAMPLES

Below, please find some case examples from recent history, which exhibit both best practices for disaster readiness and how there can be significant adverse impacts to the LIPER (life safety, incident stabilization, property/asset protection, environmental/economic impacts, and recovery operations) from Human-Made Water Source Contamination, a water quality hazard. This is an area where the Preparedness Phase work can be improved through collaboration between the local community's Emergency Management entity and the Water Utility's own team. Many water utility systems need better standardized damage assessment criteria – and a clearer understanding of their role in the Community Lifelines – to align with the overall needs of the community.

7.2.1 Australia: Getting to Potable Water First

Equity concerns for rural and impoverished areas of Australia, are highlighted in 2022 research on access to quality drinking water in the remote and regional areas of that country. Australia has drinking water guidelines[11] which must be met throughout the nation, yet researchers have found that at least 25,000 people across 99 different locations (with populations less than 1,000 people) had accessed water services that did not comply with the regulations for baseline standards.

Additionally, another 408 regional and remote areas – serving a population of over 627,000 people – did not meet the higher "aesthetic determinants of good water quality across safety, taste and physical characteristics".[12] Australia lacks a national

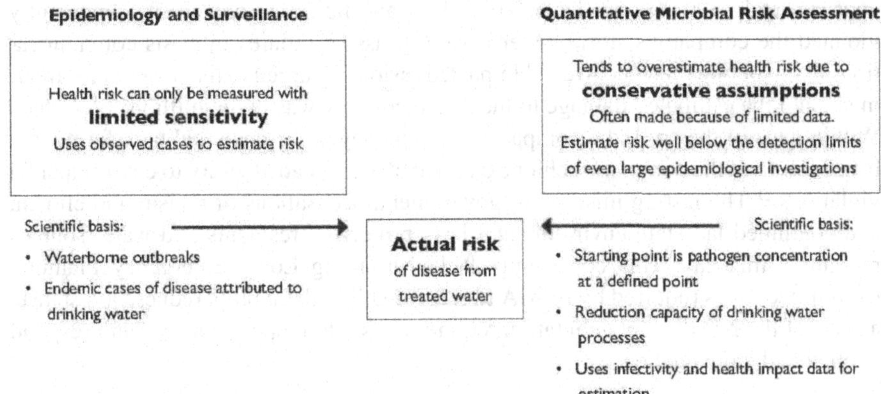

FIGURE 7.2 Quantitative Microbial Risk Assessment (QMRA) and epidemiological approaches to characterizing risks from drinking water sources.

Source: NHMRC, NRMMC. (2022). *Australian Drinking Water Guidelines Paper 6 National Water Quality Management Strategy.* National Health and Medical Research Council, National Resource Management Ministerial Council, Commonwealth of Australia, Canberra. See www.nhmrc.gov.au/file/18462/download?token=nthI3esn, p. 73. Public domain.

database, national standards, etc. to manage clean water quality and enforce compliance with standards – especially in rural and remote areas of the country, which also accounts for a vast majority of indigenous people and people with lower incomes (and apparently quality of life) than in urban population centers of Australia.

7.2.2 Bridge Collapses over Freshwater Sources

Many emergency managers are unaware that the United States Coast Guard has incident command jurisdiction[13] on any navigable waterways – including those *within* a state's borders. This is in addition to their responsibility to support notifications on railway incidents and responses to oil and other pollutant spills.

In 2012, there was a train derailment in New Jersey, which occurred over a navigable waterway, and the U.S. Coast Guard took command[14] (much to the surprise of local and state first responders,[15] and to some extent because of failures to protect workers on-scene).[16] These chemical spills into water can elicit massive response efforts from multiple federal agencies and have the potential for long-term health hazards to the local population.[17] In 2024, the Francis Scott Key Bridge in Baltimore, Maryland, was struck by a container ship,[18] causing the entire bridge to collapse into the river, killing six people and disrupting port operations and roadway transportation.

7.2.3 Elk River, West Virginia

In 2014, West Virginia state inspectors discovered a significant leak of unknown chemicals from a tank farm into the Elk River, which provides drinking water to approximately 300,000 residents. The spill – and failure to contain it – apparently violated the company's storm water permit. Massive failures in crisis communications to the public then occurred.[19] Impacted residents surged to the hospital for treatment, and the estimated damage to the local economy was in the millions of dollars. Within days of the spill, the company which managed the chemical tank farm went bankrupt[20] and the company's chief executive officer pleaded guilty to environmental violations.[21] The lasting mistrust of government, accusations of racism and elitism, and continued lax state environmental laws protecting residents and water sources remain to this day. This community lacked a strong Local Emergency Planning Committee[22] – as required by FEMA to receive U.S Federal preparedness assistance, as part of the NIMS implementation requirements – to help prepare for and respond to chemical emergencies.

7.2.4 New Freeport, Pennsylvania: "Orphan" Wells – Watershed Risk

Orphan wells are abandoned oil and gas wells, which were left – many times in a state of disrepair or not cleaned up – by companies which no longer exist. These neglected wells leech chemicals and oil into watersheds, existing nearby water wells/cisterns and can also emit methane in high concentrations, causing explosions. The U.S. State of Pennsylvania has identified more than 27,000 of these orphan wells, but "researchers estimate there could be as many as 760,000 in the state alone".[23]

In the United States, the U.S. Coast Guard operates the National Pollution Funds Center,[24] which enforces the Oil Pollution Act[25] and oversees oil transportation in navigable waterways. They also maintain the National Response Center,[26] which is a telephone support hotline for pollution and railroad incidents, The Coast Guard is also part of the U.S. National Response Team,[27] which plans for and responds to oil and hazardous substances incidents.

7.2.5 Hoosick Falls, New York: Long-Term Recovery

The water source in Hoosick Falls, New York, was contaminated with chemicals (perfluorooctanoic acid), as noted in a court settlement document,[28] by three major manufacturers. There were questions about whether the local government officials failed to inform and warn the public about drinking the contaminated water and the settlement documents include provisions for property tax abatements, as well as financial settlements and ongoing medical monitoring.[29]

7.2.6 U.S. Nationwide Recognition of Per- and Polyfluoroalkyl Substances (PFAS)

The United States Environmental Protection Agency (EPA) is taking steps to address the concern of per- and polyfluoroalkyl substances, also known as PFAS in the nation's watersheds, cisterns, and wells. PFAS are based on organic chemicals but are modified by humans and found in products such as fire retardants, and even nonstick coatings on pots and pans. They are in a class of contaminants into the potable water sources, described as "forever chemicals", because they do not quickly breakdown naturally into nonharmful elements.[30] The EPA has learned so far the following:

> PFAS are widely used, long-lasting chemicals, components of which break down very slowly over time.
>
> Because of their widespread use and their persistence in the environment, many PFAS are found in the blood of people and animals all over the world and are present at low levels in a variety of food products and in the environment.
>
> PFAS are found in water, air, fish, and soil at locations across the nation and the globe.
>
> Scientific studies have shown that exposure to some PFAS in the environment may be linked to harmful health effects in humans and animals.[31]
>
> There are thousands of PFAS chemicals, and they are found in many different consumer, commercial, and industrial products. This makes it challenging to study and assess the potential human health and environmental risks.

PFAS contamination of, exposure to, harmful impacts from, etc. can be an excellent analogy for many other types of human-contaminated water sources. As shown in Table 7.1, the exposure to PFAS can occur in other ways, besides via the drinking water systems.

Some of the probable and possible health impacts from PFAS exposure are shown in Table 7.2.

7.2.7 Water Tower/Cooling Tower Transmissions of *Legionella*

First discovered in the United States in 1976, the *Legionella* bacteria species produces both a serious type of pneumonia called Legionnaires' Disease and a less lethal infection known as Pontiac Fever. It is named after an outbreak at a Philadelphia, PA, convention of the American Legion. Pontiac Fever occurred as early as 1968 (from people working at and visiting a Pontiac, Michigan health department office), but was not connected to *Legionella*, until research was done after the 1976 cases.[32]

> *Legionella* bacteria are found naturally in freshwater environments, like lakes and streams. The bacteria can become a health concern when they grow and spread in human-made building water systems:
>
> - Showerheads and sink faucets
> - Cooling towers (structures that contain water and a fan as part of centralized air-cooling systems for buildings or industrial processes)
> - Hot tubs
> - Decorative fountains and water features
> - Hot water tanks and heaters
> - Large, complex plumbing systems
>
> Home and car air-conditioning units do not use water to cool the air, so they are not a risk for *Legionella* growth.

TABLE 7.1
Places and Objects Where PFAS Can Be Found

PFAS can be present in our water, soil, air, and food as well as in materials found in our homes or workplaces:

- *Drinking water* – in public drinking water systems and private drinking water wells.
- *Soil and water at or near waste sites* – at landfills, disposal sites, and hazardous waste sites such as those that fall under the federal Superfund and Resource Conservation and Recovery Act programs.
- *Fire extinguishing foam* – in aqueous film-forming foams (or AFFFs) used to extinguish flammable liquid-based fires. Such foams are used in training and emergency response events at airports, shipyards, military bases, firefighting training facilities, chemical plants, and refineries.
- *Manufacturing or chemical production facilities that produce or use PFAS* – for example, at chrome plating, electronics, and certain textile and paper manufacturers.
- *Food* – for example, in fish caught from water contaminated by PFAS and dairy products from livestock exposed to PFAS.
- *Food packaging* – for example, in grease-resistant paper, fast food containers/wrappers, microwave popcorn bags, pizza boxes, and candy wrappers.
- *Household products and dust* – for example, in stain and water-repellent used on carpets, upholstery, clothing, and other fabrics; cleaning products; nonstick cookware; paints, varnishes, and sealants.
- *Personal care products* – for example, in certain shampoo, dental floss, and cosmetics.
- *Biosolids* – for example, fertilizer from wastewater treatment plants that is used on agricultural lands can affect groundwater and surface water and animals that graze on the land.

Source: United States Environmental Protection Agency. (2023). *Our Current Understanding of the Human Health and Environmental Risks of PFAS.* United States Environmental Protection Agency. Accessed November 20, 2023. See www.epa.gov/pfas/ our-current-understanding-human-health-and-environmental-risks-pfas

TABLE 7.2
Health Effects of PFAS

What We Know about Health Effects

Current peer-reviewed scientific studies have shown that exposure to certain levels of PFAS may lead to the following:

- Reproductive effects such as decreased fertility or increased high blood pressure in pregnant women.
- Developmental effects or delays in children, including low birth weight, accelerated puberty, bone variations, or behavioral changes.
- Increased risk of some cancers, including prostate, kidney, and testicular cancers.
- Reduced ability of the body's immune system to fight infections, including reduced vaccine response.
- Interference with the body's natural hormones.
- Increased cholesterol levels and/or risk of obesity.

TABLE 7.2 ***(Continued)***
Health Effects of PFAS

Additional Health Effects are Difficult to Determine

Scientists at EPA, in other federal agencies, and in academia and industry are continuing to conduct and review the growing body of research about PFAS. However, health effects associated with exposure to PFAS are difficult to specify for many reasons:

- There are thousands of PFAS with potentially varying effects and toxicity levels, yet most studies focus on a limited number of better known PFAS compounds.
- People can be exposed to PFAS in different ways and at different stages of their life.
- The types and uses of PFAS change over time, which makes it challenging to track and assess how exposure to these chemicals occurs and how they will affect human health.

Source: United States Environmental Protection Agency. (2023). *Our Current Understanding of the Human Health and Environmental Risks of PFAS.* United States Environmental Protection Agency. Accessed November 20, 2023. See www.epa.gov/pfas/ our-current-understanding-human-health-and-environmental-risks-pfas

> However, *Legionella* can grow in the windshield wiper fluid tank of a vehicle (such as a car, truck, van, school bus, or taxi), particularly if the tank is filled with water and not genuine windshield cleaner fluid.
>
> After *Legionella* grows and multiplies in a building water system, water containing *Legionella* can spread in droplets small enough for people to breathe in. People can get Legionnaires' disease or Pontiac fever when they breathe in small droplets of water in the air that contain the bacteria.
>
> Less commonly, people can get sick by aspiration of drinking water containing *Legionella*. This happens when water accidentally goes into the lungs while drinking. People at increased risk of aspiration include those with swallowing difficulties.
>
> In general, people do not spread Legionnaires' disease and Pontiac fever to other people. However, this may be possible under rare circumstances.
>
> **(CDC, 2021a, p. 1)**[33]

The U.S. CDC has many resources for emergency managers to utilize to communicate threats and hazards to the public, including infographics such as the one in Figure 7.2, on a full-cycle basis.

7.3 IMPACTS TO THE DISASTER PHASE CYCLES

As shown in Table 7.3, there are specific impacts to all four of the disaster cycle phases from human-made hazards to water sources.

7.4 ADVERSE IMPACTS TO THE INCIDENT COMMAND SYSTEM

For human-made water source contamination incidents, here are some example impacts to any jurisdiction's incident management system or systems. Not only are there different missions for different hazards – and levels of adverse impacts from said hazards – but there can also be direct life safety and other impacts to the

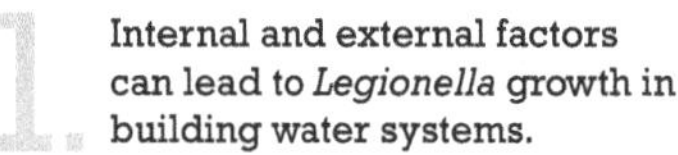

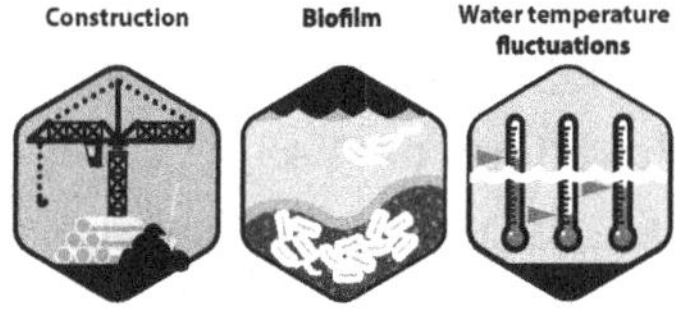

2. *Legionella* grows best in large, complex water systems that are not adequately maintained.

3. Water containing *Legionella* is aerosolized through devices.

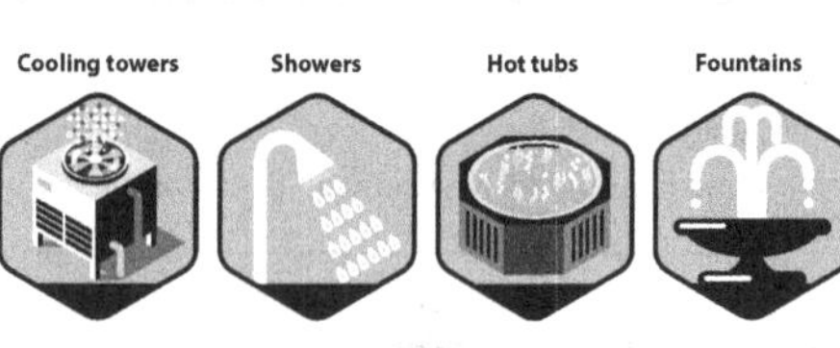

4. People can get sick when they breathe in small droplets of water or accidently swallow water containing *Legionella* into the lungs. Those at increased risk are adults 50 years or older, current or former smokers, and people with a weakened immune system or chronic disease.

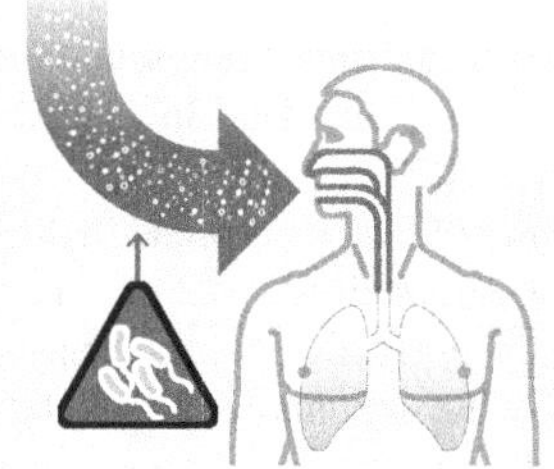

www.cdc.gov/legionella

03/30/21

FIGURE 7.3 How *Legionella* affects building water systems and people.

Source: Centers for Disease Control and Prevention. (2021). *How Legionella Affects Building Water Systems and People*. Accessed December 1, 2023. See www.cdc.gov/legionella/infographics/legionella-affects-water-systems.pdf

responders themselves. For example, in this chapter, the fact that the drinking water in the impacted community is unsafe to drink means that the Response team itself needs to utilize an alternative water source themselves – including for personal hygiene.

Please note, as described in the introduction, this book is not designed to provide a roadmap to each organizational construct of an incident management system as to "who" needs to perform "what", but rather highlight the adverse impacts which can occur; and disaster phase cycle missions (before, during and after) which can reduce or eliminate the adverse impacts from this threat/hazard. Target audiences on the practitioner side can include local emergency management officials, water district managers, dam operators, and others outside of governmental organizations. For Human-Made Hazards to Water Sources, Table 7.4 details what may be occurring:

TABLE 7.3
Impacts to the Disaster Phase Cycles from Human-Made Hazards to Water Sources

Preparedness/Protection/Prevention

There are steps and courses of action that emergency managers can take to protect water sources, including public messaging about watershed protection, alerting officials when threats are noticed, etc. Communities – especially geographic areas which have high socially vulnerable individuals (SVIs) populations – need preparedness messaging and training to respond themselves to human-made hazards to water sources. Understanding a community's Essential Elements of Intelligence (or EEIs) – as organized through its Community Lifelines, is one way to quantify and categorize the areas which need preparedness – through a process of becoming more ready to Respond, as well as protecting against and preventing adverse impacts from quality hazards. This includes having potable water on-hand as part of their sheltering-in-place kits and pre-identifying alternative locations to obtain potable water. Diverting or evacuating to alternative dialysis centers – and not the local hospital, which may be adversely impacted as well – for treatment, if necessary, and other personal health issues should be reminders to the public that they have the ultimate responsibility for their own life safety first.

Response

Emergency Management's response courses of action for a human-made hazard to the local water sources and water supply system, includes providing alternative sources of potable water, crisis communications about the status of the water, and public health syndromic surveillance. This is not a comprehensive list, but a reminder that there are significant response actions needed for this hazard, to be coordinated through the local emergency management agency, as much as possible (your jurisdiction may be different). Do not rely on the water systems authority to provide all the resources and response capability – in fact, their representative should be a liaison officer (or possibly part of unified command) to your jurisdiction's incident command system, not the other way around.

Recovery

There will be cleanup and remediation work, possible infrastructure[34] repair work, supply-chain restoration, financial reimbursement to government by culpable companies and other entities, continued health monitoring for the public, continued crisis communications, cooperation with investigators, including state/tribal/territorial and even federal level officials, and more. The more the standardized – for example, use of the Community Lifelines for metrics – the better to transition to and from outside support, such as Incident Management Teams (IMTs). Most likely, in this longer-term time frame for recovery, another incident of scale will occur and consume the time and staffing of emergency managers. Self-care and wellness are critical for not only first responders, but everyone involved in the disaster phase cycle.

Mitigation

As noted previously, there is a high probability that this type of incident will not receive any U.S. Federal Governmental subsidy (i.e., cost-share match) through the *Stafford Act.* There may be other funding available from other sources, but – generally speaking – there will not be federal mitigation funding, as there is for *Stafford Act.* U.S. Presidentially declared disasters or emergencies. State/Tribal/Territorial legislation may mandate mitigation actions by the organization operating the water supply system, or there may not be such legal mandates. emergency managers may have to utilize other funding sources to provide much needed mitigation work to reduce the possibility of a similar adverse set of hazards from another human-made hazard to water sources. If there was a Declared Disaster in the same jurisdiction which impacted the same water supply system from a natural hazard, it is possible that mitigation work to help prevent and protect against the adverse impacts of that natural hazard *and also* prevent and protect against the adverse impacts of a human-made hazard. In the same way that tank field containment alarms help notify officials more quickly when a storm ruptures a chemical holding tank, they will work if a terrorist attacks the same tank and creates a breech.

Source: Author.

TABLE 7.4
Adverse Impacts to the Incident Command System

Command (including SO, PIO, and LNOs)

Safety Officers (SOs) need emergency management intelligence about this threat, as soon as possible. As the water sources may be contaminated, potable water is needed for responders, emergency operations center staff, etc. This may extend beyond just drinking water into handwashing, bathing, etc. These workforce safety elements are paramount. While there are always pathogens and waterborne diseases which do not get filtered out of a water treatment process – extra attention – and advanced notice to deploying response teams – of this **hardship** should be made.

Public Information Officers (PIOs) need to urgently activate accurate templated messages to the public which provide critical crisis communications in the appropriate languages and modalities for oversaturation. If the public has a "boil water" warning needed, for example, it needs to be communicated far and wide, and quickly. See the Shelter-in-Place section in Chapter 2 and also the Additional References/Reading section links for more specifics. See also John Manuel's article on the Elk River and emergency risk communications noted in the References/Additional Reading section at the end of this chapter.

Liaison Officers (LNOs) from both the water supply companies/entities and potentially the organizations deemed responsible for the contamination will need to be prepared to financially commit to the Incident/Unified Command their support quickly, in order to effect solutions on a timely basis. In the United States, there may also be LNOs from federal agencies (such as the U.S. Environmental Protection Agency), who need to represent the full capabilities and capacities of their organizations, absent a Presidentially Declared Disaster or Emergency, where a unified Federal Coordinating Officer is then assigned to support the State (or Tribal) Coordinating Officer.

Intelligence

Sources of Emergency Management Intelligence for incidents where hazards are from human-made water sources will need to include more than just law-enforcement related ones. Additional subject-matter expertise of water systems officials from this jurisdiction, higher levels of government and possibly derived from mutual-aid compacts (i.e., EMAC), may be needed.

Finance/Administration

- Budgets for alternative water storage, delivery, and distribution may not be adequate. Emergency declarations at the local jurisdiction may not be sufficient to support emergency contracts, etc. Overtime and other staffing costs may exceed what is covered in normal fiscal budgets and/or emergency assistance reimbursements from higher governmental entities.
- If it is determined that a private entity is responsible for the human-made contamination of the water sources, the costs for remediation and cleanup may be borne by that entity. Governmental Finance/Administration officials will need to manage costs and expenses in such a way as to obtain reimbursement on a timely basis. (*Note the case study example above, where the company filed for bankruptcy days after the incident was discovered by inspectors.*) Additional elements of penalties and fines are not normally the responsibility of emergency management, but the organizations who are support partners to a jurisdiction's emergency management organization (such as the governmental environmental protection agency) may have dual responsibilities of both supporting the response/recovery work and issuing penalties and/or fines to the entity responsible for the adverse impacts caused by them. Emergency Management staff at those agencies must clearly divest themselves from any direct missions involved in the investigation or prosecution of malfeasance or criminal activity.

TABLE 7.4 *(Continued)*

Adverse Impacts to the Incident Command System

Command (including SO, PIO, and LNOs)

Logistics

With adverse impacts from water sources to water supply systems, command may designate points of distribution (PODs) for potable water to be donated to the public. This can take many forms (water from a truck where the public must bring their own containers, individual size bottled water cases which have weight and are usually distributed to residents via vehicle-based, gallon bottles of water, etc.). How this water is ordered/requisitioned is critical for logistics to align to the needs of the operation. Many times, there are also donations of water – from both corporations and the general public – which have to be managed as in-kind donations, probably through a non-governmental organization. In the event the incident is Stafford Act qualified for Public Assistance reimbursement of costs – these material donations, as well as hours of donated labor from volunteers, can be applied to public assistance projects.[35]

Operations

- Firefighting may be impacted – for other threats/hazards, for example, responding to a fire in the same jurisdiction as the impacted water supply system. Low or no water pressure adversely affects the water available through fire hydrants.

Planning

There are no specific adverse impacts to the Planning section, from this hazard. There are, however, additional planning activities, aligned planners from other incident commands, etc. associated with this hazard. Full-cycle planning through Mitigation work, will most likely involve large infrastructure projects which may require Planning elements maintain the operational tempo/battle rhythm beyond shorter-term incidents, such as winter storms, minor river flooding, etc.

Source: Author.

7.4.1 POETE Process Elements for This Hazard

Table 7.5 shows the POETE elements for **Human-Made Hazards to Water Sources**.

7.5 CHAPTER SUMMARY/KEY TAKEAWAYS

Water Quality issues have significant impacts on hospitals, dialysis centers, nursing homes/assisted living centers, daycare sites, etc. Boil alerts mean massive protocol changes at hospitals – longer than the alert itself. Community preparedness messaging can be very effective in minimizing adverse impacts from incidents involving this hazard. This should be accomplished through a whole-of-government/whole-community effort. A community cannot overcommunicate hazards associated with threats to their potable water sources and supply systems, so messaging via public media, community/civic/religious groups, libraries, social media, etc. – all should be maximized using consistent messaging before, during, and after any incidents. If the water supply system was damaged itself, longer-term remediation, repair, restoration, *and additional* elements which protect it from harm[36]/prevent adverse impacts in the future should be incorporated into Mitigation work. On U.S. Presidentially declared emergencies and disasters, the partial (usually 75%) Federal funding for this work, may be available.

TABLE 7.5
POETE Process Elements for This Hazard

Planning	While there will most likely not be a separate annex or appendix to any jurisdiction's emergency operations plan for this specific hazard, there should be elements of other main plan portions and annexes/appendices which cover the planning needed for a human-made hazard to water sources. Each Water Supplier may have a separate emergency response plan, with separate mandates from the Emergency Plans in the communities they serve. From this chapter, there are a number of adverse impacts, historical references of past incidents to learn from, and elements of exercises which can be performed, in order to facilitate positive changes and modifications to preparedness, response, recovery, mitigation, and continuity of operations plans to benefit everyone.
Organization	The *trained* staff needed to effect plans, using proper equipment should be exercised and evaluated on a regular basis each year. Actual incident response and recovery can help determine elements of POETE, including how these types of incidents are organized for staffing. Also, as part of both deployments and exercises, staff involved should be evaluated for both further training needed and the possibility of advancement in their disaster state roles to leadership positions.
Equipment	Equipment needed for this hazard specifically includes anything needed to remediate/repair the water systems plant. Also needed is any special equipment (traffic controls, pallet jacks, water tanker trucks, etc.) needed at – and in support of – emergency potable water distribution points. This can also include at warehouses and supply depots.
Training	In addition to standardized Incident Command System training, emergency managers should consider water quality-specific courses along the full disaster phase cycle, and not just for Response.
Exercises and evaluation	A full cycle of discussion and performance exercises should be designed for human-made hazards to Water Sources – both on their own and as part of complex coordinated attack scenarios. These exercises should involve the whole community in exercise planner's invitations for both additional exercise planners/evaluators and actual players, especially since many water supply system operators are separate legal entities from the local governments they serve. In the same way there are adversely impacted ESFs, RSFs, and CLs as noted in this chapter, those elements should be exercised and evaluated so that the elements of the POETE cycle are reviewed, and revised as needed (i.e., agreed upon for change, as part of the Improvement Plan components).

Source: Author.

7.5.1 What to Read Next

The next chapter will continue the theme of Water Source Contamination, but with the threat occurring naturally, instead of being human made. Cross-check both chapters, since some hazard impacts may be the same, and some may differ. Also, be prepared for the possibility of a complex or cascading incident: It may start out as a human-made hazard to a water source, but then may produce different hazards, including natural ones, somewhere else. As noted in the introduction to this chapter, water sources can extend into other geographies, other jurisdictions, and even other nations. Historically, this aspect has generated complex or cascading incidents.

NOTES

1 See https://lowninstitute.org/as-the-guardians-of-health-we-cant-look-away-mona-hanna-attisha-accepts-the-bernard-lown-award/
2 See https://www-businessinsider-com.cdn.ampproject.org/c/s/www.businessinsider.com/live-off-the-grid-tiny-house-no-electricity-2023-12?amp
3 See https://news.mit.edu/2015/mexican-village-solar-power-purify-water-1008
4 Neno, S. (2012). World Water Day: A Global Awareness Campaign to Tackle the Water Crisis. *Biotechnology Journal*, *7*(4), 473–474. https://doi.org/10.1002/biot.201200100
5 O'Kane, J. P. (2022). Eutopian and Dystopian Water Resource Systems Design and Operation – Three Irish Case Studies. *Hydrology*, *9*(9). https://doi.org/10.3390/hydrology9090159
6 See https://engineering.princeton.edu/news/2023/02/08/solar-powered-gel-filters-enough-clean-water-meet-daily-needs
7 United Nations. (2010). *The Human Right to Water*. United Nations. Accessed November 10, 2023. See www.un.org/waterforlifedecade/pdf/facts_and_figures_human_right_to_water_eng.pdf
8 Prasad, M. (2023). Global Pandemics are Extinction-Level Events and Should Not Be Coordinated Solely through National or Jurisdictional Emergency Management. *Pracademic Affairs*, *3*. See www.hsaj.org/articles/22285
9 See www.science.org/content/article/utah-s-great-salt-lake-has-lost-half-its-water-thanks-thirsty-humans
10 See www.epa.gov/saferchoice/products
11 Australian Government National Health and Medical Research Council. (n.d.). *Australian Drinking Water Guidelines*. Accessed November 10, 2023. See www.nhmrc.gov.au/about-us/publications/australian-drinking-water-guidelines
12 Australian National University. (2022). *Remote Australians Lack Access to Quality Drinking Water*. The Australia National University. Accessed November 10, 2023. See www.anu.edu.au/news/all-news/remote-australians-lack-access-to-quality-drinking-water
13 See https://www.keybridgeresponse2024.com/
14 See www.usatoday.com/story/news/nation/2012/12/16/derailment-chemical-tank-nj-creek/1774053/
15 See www.wkbn.com/news/local-news/east-palestine-train-derailment/east-palestine-train-derailment-compares-to-2012-incident-in-new-jersey/
16 See www.ntsb.gov/investigations/AccidentReports/Reports/RAR1401.pdf
17 See https://whyy.org/articles/train-derailment-rail-safety-paulsboro-nj-norfolk-southern-east-palestine/
18 See https://domesticpreparedness.com/articles/key-bridge-collapse-transportation-infrastructure-and-global-supply-chain
19 Manuel, J. (2014). Crisis and Emergency Risk Communication: Lessons from the Elk River Spill. *Environmental Health Perspectives*, *122*(8), A214–A219. https://doi.org/10.1289/ehp.122-A214
20 Visser, N. (2014). Freedom Industries, Company Behind West Virginia Chemical Spill, Files For Bankruptcy. *HuffPost*. Accessed November 10, 2023. See www.huffpost.com/entry/freedom-industries-bankruptcy-west-virginia-chemical-spill_n_4619385
21 Sheppard, K. (2016). You Can Pollute Drinking Water for 300,000 People and Get Just One Month in Prison. *HuffPost*. Accessed November 10, 2023. See www.huffpost.com/entry/west-virginia-chemical-spill-sentencing_n_56cf3ecbe4b0bf0dab311a4b
22 See www.fema.gov/pdf/emergency/nims/lepc_comp_fs.pdf

23 Cavazuti, L. (2023). "Orphan Well" Mishap Stirs Suspicion and Frustration in Tiny Pennsylvania Town. *NBC News.* Accessed November 20, 2023. See www.aol.com/orphan-well-mishap-stirs-suspicion-110000229.html
24 See www.uscg.mil/Mariners/National-Pollution-Funds-Center/About-NPFC/
25 See www.uscg.mil/Mariners/National-Pollution-Funds-Center/NPFC-Laws-and-Regulations/
26 See https://nrc.uscg.mil/
27 See www.nrt.org/nrt/Contact.aspx
28 KCC. (2023). *Hoosick Falls PFOA Settlement Website.* KCC Class Action Services LLC. Accessed November 10, 2023. See www.hoosickfallspfoasettlement.com/
29 Lyons, B. J. (2021). Deadlines Set for Hoosick Falls Class-Action Settlement. *Times Union/Hearst.* Accessed November 20, 2023. See www.timesunion.com/state/article/Deadlines-set-for-Hoosick-Falls-class-action-16424725.php
30 Goodwille, K. (2023). *EPA Discovers "Forever Chemicals" in Pacific Northwest Water Systems.* KING-TV. Accessed November 23, 2023. See www.king5.com/article/tech/science/environment/epa-discovers-forever-chemicals-local-water-systems/281-025a5f49-4591-42e4-a3b6-678364e5897c
31 United States Environmental Protection Agency. (2023). *PFAS Explained.* United States Environmental Protection Agency. Accessed November 20, 2023. See www.epa.gov/pfas/pfas-explained
32 Winn, W. C., Jr (1988). Legionnaires Disease: Historical Perspective. *Clinical Microbiology Reviews, 1*(1), 60–81. https://doi.org/10.1128/CMR.1.1.60
33 Centers for Disease Control and Prevention. (2021). *Legionella (Legionnaires' Disease and Pontiac Fever).* Centers for Disaster Control and Prevention. Accessed December 1, 2023. See www.cdc.gov/legionella/about/causes-transmission.html
34 See https://domesticpreparedness.com/articles/key-bridge-collapse-transportation-infrastructure-and-global-supply-chain
35 See https://www.govinfo.gov/content/pkg/FR-2013-11-01/pdf/2013-26018.pdf
36 See https://www.epa.gov/ground-water-and-drinking-water/national-primary-drinking-water-regulations

REFERENCES/ADDITIONAL READING

Beck, E. C. (1979). *The Love Canal Tragedy.* EPA Journal. See https://www.epa.gov/archive/epa/aboutepa/love-canal-tragedy.html

Brockovich, E. (2021). *Superman's Not Coming: Our National Water Crisis and What We the People Can Do About It.* Knopf Doubleday Publishing Group.

Centers for Disease Control. (n.d.). *Public Health 101 – Public Health Surveillance Resources and Additional Reading.* See www.cdc.gov/training/publichealth101/documents/surveillance-references.pdf

Duan, W., He, B., Takara, K., Luo, P., Nover, D., Sahu, N., & Yamashiki, Y. (2013). Spatiotemporal Evaluation of Water Quality Incidents in Japan Between 1996 and 2007. *Chemosphere, 93.* https://doi.org/10.1016/j.chemosphere.2013.05.060

Manuel, J. (2014). Crisis and Emergency Risk Communication: Lessons from the Elk River Spill. *Environmental Health Perspectives, 122*(8), A214–A219. https://doi.org/10.1289/ehp.122-A214

Martinet, M. (2025). *Fighting With FEMA: A Practical Regulations Handbook.* Routledge. https://doi.org/10.4324/9781003487869

Prasad, M. (2022). Templated Crisis Communication for People with Disabilities, Access and Functional Needs. *Journal of International Crisis and Risk Communication Research, 5*(2), 233–254. https://doi.org/10.30658/jicrcr.5.2.6

Ruckart, P. Z., Ettinger, A. S., Hanna-Attisha, M., Jones, N., Davis, S. I., & Breysse, P. N. (2019). The Flint Water Crisis: A Coordinated Public Health Emergency Response and Recovery Initiative. *PubMed Central*. https://doi.org/10.1097/phh.0000000000000871

South Carolina Emergency Management Division. (n.d.). *Dam Failure*. South Carolina Emergency Management Division. Accessed February 2, 2024. See www.scemd.org/prepare/types-of-disasters/dam-failure/

United Nations Department of Economic and Social Affairs (UNDESA). (2015). *International Decade for Action 'Water for Life' 2005–2015*. See www.un.org/waterforlifedecade/index.shtml

US Army Corp of Engineers. (2019). *A Guide to Public Alerts and Warnings for Dam and Levee Emergencies*. USACE. See www.publications.usace.army.mil/Portals/76/Users/182/86/2486/EP%201110-2-17.pdf?ver=2019-06-20-152050-550

White, G., & Damon, M. (2022). *The Worth of Water: Our Story of Chasing Solutions to the World's Greatest Challenge*. Penguin Publishing Group.

BIBLIOGRAPHY

Australian Government National Health and Medical Research Council. (n.d.). *Australian Drinking Water Guidelines*. Accessed November 10, 2023. See www.nhmrc.gov.au/about-us/publications/australian-drinking-water-guidelines

Australian National University. (2022). *Remote Australians Lack Access to Quality Drinking Water*. The Australia National University. Accessed November 10, 2023. See www.anu.edu.au/news/all-news/remote-australians-lack-access-to-quality-drinking-water

Cavazuti, L. (2023). 'Orphan Well' Mishap Stirs Suspicion and Frustration in Tiny Pennsylvania Town. *NBC News*. Accessed November 20, 2023. See www.aol.com/orphan-well-mishap-stirs-suspicion-110000229.html

Centers for Disease Control and Prevention. (2021a). *How Legionella Affects Building Water Systems and People*. Accessed December 1, 2023. See www.cdc.gov/legionella/infographics/legionella-affects-water-systems.pdf

Centers for Disease Control and Prevention. (2021b). *Legionella (Legionnaires' Disease and Pontiac Fever)*. Centers for Disaster Control and Prevention. Accessed December 1, 2023. See www.cdc.gov/legionella/about/causes-transmission.html

Goodwille, K. (2023). *EPA Discovers 'Forever Chemicals' in Pacific Northwest Water Systems*. KING-TV. Accessed November 23, 2023. See www.king5.com/article/tech/science/environment/epa-discovers-forever-chemicals-local-water-systems/281-025a5f49-4591-42e4-a3b6-678364e5897c

KCC. (2023). *Hoosick Falls PFOA Settlement Website*. KCC Class Action Services LLC. Accessed November 10, 2023. See www.hoosickfallspfoasettlement.com/

Lyons, B. J. (2021). Deadlines Set for Hoosick Falls Class-Action Settlement. *Times Union/Hearst*. Accessed November 20, 2023. See www.timesunion.com/state/article/Deadlines-set-for-Hoosick-Falls-class-action-16424725.php

Manuel, J. (2014). Crisis and Emergency Risk Communication: Lessons from the Elk River Spill. *Environmental Health Perspectives*, *122*(8), A214–A219. https://doi.org/10.1289/ehp.122-A214

Neno, S. (2012). World Water Day: A Global Awareness Campaign to Tackle the Water Crisis. *Biotechnology Journal*, *7*(4), 473–474. https://doi.org/10.1002/biot.201200100

NHMRC, NRMMC. (2022). *Australian Drinking Water Guidelines Paper 6 National Water Quality Management Strategy*. National Health and Medical Research Council, National Resource Management Ministerial Council, Commonwealth of Australia, Canberra. See www.nhmrc.gov.au/file/18462/download?token=nthI3esn

O'Kane, J. P. (2022). Eutopian and Dystopian Water Resource Systems Design and Operation – Three Irish Case Studies. *Hydrology*, *9*(9). https://doi.org/10.3390/hydrology9090159

Prasad, M. (2023). Global Pandemics Are Extinction-level Events and Should Not Be Coordinated Solely Through National or Jurisdictional Emergency Management. *Pracademic Affairs*, *3*. See www.hsaj.org/articles/22285

Sheppard, K. (2016). You Can Pollute Drinking Water for 300,000 People and Get Just One Month in Prison. *HuffPost*. Accessed November 10, 2023. See www.huffpost.com/entry/west-virginia-chemical-spill-sentencing_n_56cf3ecbe4b0bf0dab311a4b

United Nations. (2010). *The Human Right to Water*. United Nations. Accessed November 10, 2023. See www.un.org/waterforlifedecade/pdf/facts_and_figures_human_right_to_water_eng.pdf

United States Environmental Protection Agency. (2023a). *PFAS Explained*. United States Environmental Protection Agency. Accessed November 20, 2023. See www.epa.gov/pfas/pfas-explained

United States Environmental Protection Agency. (2023b). *Our Current Understanding of the Human Health and Environmental Risks of PFAS*. United States Environmental Protection Agency. Accessed November 20, 2023. See www.epa.gov/pfas/our-current-understanding-human-health-and-environmental-risks-pfas

Visser, N. (2014). Freedom Industries, Company Behind West Virginia Chemical Spill, Files for Bankruptcy. *HuffPost*. Accessed November 10, 2023. See www.huffpost.com/entry/freedom-industries-bankruptcy-west-virginia-chemical-spill_n_4619385

Winn, W. C., Jr. (1988). Legionnaires Disease: Historical Perspective. *Clinical Microbiology Reviews*, *1*(1), 60–81. https://doi.org/10.1128/CMR.1.1.60

8 Natural Water Source Contamination

> The Mississippi River will always have its own way; no engineering skill can persuade it to do otherwise.
>
> **– (Attributed to) Mark Twain**

This chapter covers how natural hazards can adversely impact raw water/natural water sites, which are the sources for potable/drinking water supply systems and may also be navigable waterways. These hazards can be anything from saltwater inundation (covered in more detail in Chapter 17) to naturally occurring elements and microorganisms contaminating a potable water source. Most nations utilize navigable waterways within their borders for commerce, potable water supply, and tourism.

This chapter – and all the other chapters in the three Hazards sections – will contain the following:

- An overview of the specific hazard and its adverse impacts to communities
- Case examples for historic reference (many of these will hopefully guide officials as to what ***not*** to do, as lessons learned from other's missteps and misfortunes)
- Adverse Impacts of this specific hazard series to the following:
  - The Disaster Phase Cycles (the before, during, and after incidents occur)
  - The Response's Incident Command System itself (each jurisdiction around the world may be different, this book uses the U.S. NIMS/ICS as the baseline)
- A summary of the POETE process, applicable to deliberative planning and problem-solving
- Additional references/reading material

Besides local jurisdictional emergency management groups, this hazard will need to be managed by water management districts, hospitals and other healthcare facilities, colleges/universities, etc. in a coordinated and collaborative way. All will have unmet needs which they will consider as a priority – and some of those priorities may conflict and *compete* for prioritization with each other for Response and Recovery resource allocation. Unified Command should always prioritize the LIPER (see Chapter 2), when making decisions.

DOI: 10.1201/9781003474685-11

8.1 OVERVIEW

The National Academy of Medicine recommends[1] the daily consumption of 108–156 ounces (3.19–4.61 liters) of water per person per day. This varies with gender, age, climate, and physical activity. Also, overall average daily use of potable water, including for bathing and toileting, can bump this amount up to 13.2–26.4 ***gallons of clean water*** (50–100 liters) per person per day, according to the World Health Organization.[2]

Freshwater or natural water sources – including groundwater – contamination as a hazard will most likely be generated from a natural threat (i.e., drought, harmful algal bloom). There will probably not be a question of causality or even criminality associated with freshwater source contamination from natural threats. However, human-made actions such as strip-mining fracking near waterways may expose naturally occurring elements which are harmful. Dredging, installing dams, canals, etc. may stir up sediment with biological hazards or enhance its growth beyond what normal filtration processes can handle. As with other hazards, the question of causality is immaterial to Emergency Management's Response and Recovery operations, except that there may be law enforcement intelligence and investigation elements imbedded within the Incident Command System (ICS) established, specifically, for responding to and recovery from the adverse impacts of the incident. If this is the case, then actions of the other Command and General Staff branches/sections may need to be adjusted and/or expanded to support law enforcement intelligence and investigation elements. This could be investigators for environmental protection laws, for example. There may be a parallel incident command system established by law enforcement officials and there may or may not be a liaison officer (LNO) linking the two – or more – ICSs. Each incident, emergency, disaster, etc. is different. Emergency managers should be involved in the full disaster cycle planning, organization, equipping, training, and exercising for freshwater, potable water, drinking water supply systems, and/or navigable waterways in their respective communities (including those which provide support or services to their jurisdiction's water system from another jurisdiction – such as a neighboring community's shared water treatment plant, watershed areas, dams, etc.).

8.1.1 Saltwater Intrusion/Inundation into Freshwater Sources/Supplies

Drought conditions can lower river levels, and where those rivers are flowing into the oceans (as estuaries), if the levels are not high enough, effectively the seawater (which is saltwater) can flow upstream into areas where water treatment plants and other river pumping stations are located. If those systems do not have enough (or any) desalination processes, the potable water supply will become contaminated, and unsafe to consume. This will generate a public health hazard, can also affect agriculture in their use of the water supply, and of course the ecology of the river itself. *As this is a complex or cascading set of hazards from a threat (drought causes lower river levels, causes saltwater inundation, causes water treatment plant and other potable water system contamination), it will be covered in more detail in Chapter 16.*

8.1.2 Groundwater Contamination from Naturally Occurring Heavy Metals and Other Elements

Groundwater contamination is a much more significant hazard than surface water contamination. The hazard is invisible and can generate chronic/systemic health issues for people and animals and may be difficult to detect (Chakraborti et al., 2015).[3] *Please note that human-made chemical contaminations are covered in Chapter 7.*

> The natural sources of groundwater contamination include seawater, brackish water, surface waters with poor quality, and mineral deposits. These natural sources may become serious sources of contamination if human activities upset the natural environmental balance, such as depletion of aquifers leading to saltwater intrusion, acid mine drainage as a result of exploitation of mineral resources and leaching of hazardous chemicals as a result of excessive irrigation.
>
> **(Li et al., 2021, p. 3)**[4]

8.1.2.1 Microorganisms and Parasites

The consistent and successful removal of harmful microorganisms – including bacteria – and parasites in water treatment systems – including wastewater treatment – relies mostly on expansive filtration processes. There are a wide range of systems used around the world – and this book does not have any recommendations for one system over another. Water system operators should fully understand what is needed for their communities, and more importantly, the consequence management aspects of supply chain management/security, in the event they cannot get replacement equipment or supplies, a cyberattack impacts their water treatment plant, etc.

8.2 CASE EXAMPLES

Below, please find some case examples from recent history, which exhibit both best practices for disaster readiness and how there can be significant adverse impacts to the LIPER (life safety, incident stabilization, property/asset protection, environmental/economic impacts, and recovery operations) from Natural Water Source Contamination, a water quality hazard.

Water Systems operators – including their Emergency Management teams – will need a strong understanding of the monitoring, filtration methods, process flow, contingency measures, countermeasures, etc. needed to help keep their systems safe and secure. An example of the history of elemental monitoring and filtration flow can be found in Figure 8.1.

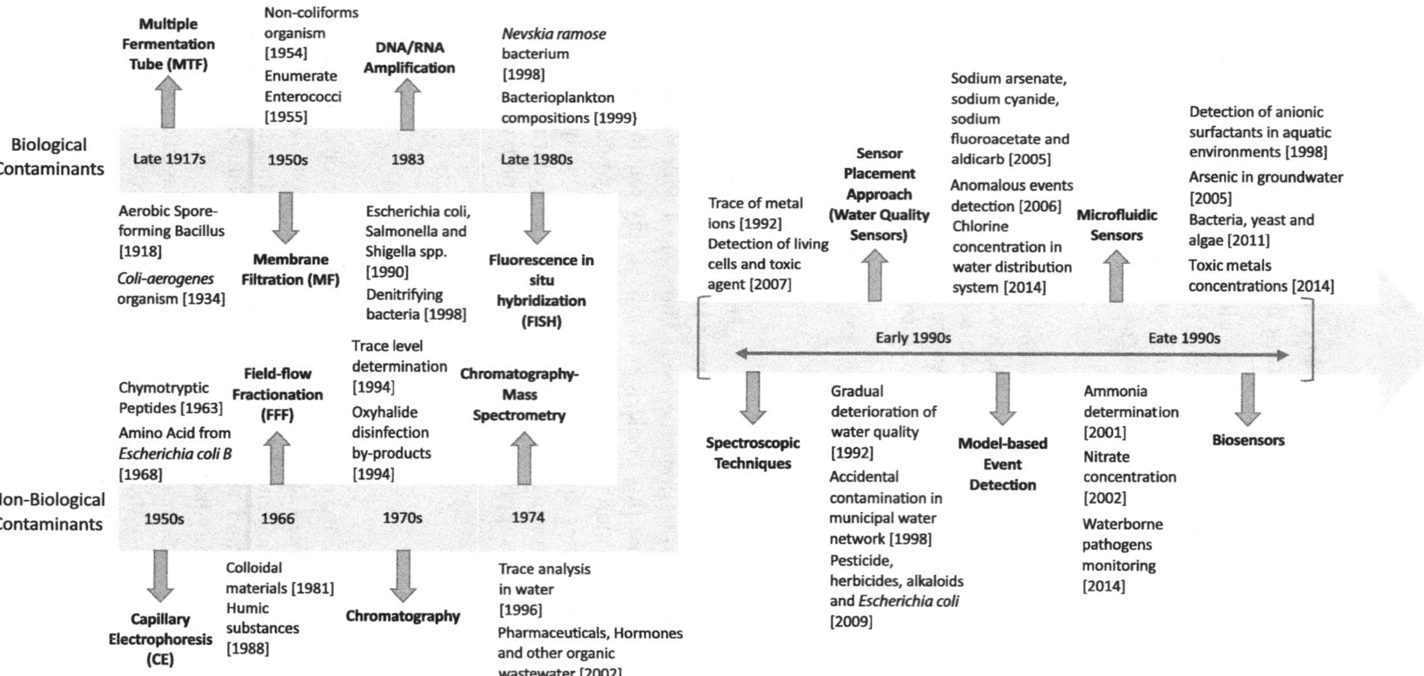

FIGURE 8.1 Evolution of contaminant detection techniques in water analysis application.

Source: Reprinted from Zulkifli, S. N., Rahim, H. A., & Lau, W. J. (2018). Detection of Contaminants in Water Supply: A Review on State-of-the-Art Monitoring Technologies and Their Applications. *Sensors and actuators B: Chemical*, *255*, 2657–2689. https://doi.org/10.1016/j.snb.2017.09.078. With permission from Elsevier.

8.2.1 Ambient Background and Anthropogenic Influxes of Arsenic in Florida Groundwater

While in the state of Florida in the United States arsenic has naturally occurred in the bulk rock concentrations, the addition of more arsenic via human-made activities can reach the level of a public health risk.

> The anthropogenic sources of arsenic in soils include: pesticides (used in Florida beginning in the 1890s), fertilizers, chromated copper arsenate (CCA)-treated wood, soil amendments, cattle-dipping vats, chicken litter, sludges from water treatment plants, and others.
>
> **(Missimer et al., 2018, p. 1)[5]**

8.2.2 Arsenic Contamination of Drinking Water Sources in Bangladesh

Bangladesh has used tube-wells to draw water from underground aquifers, for potable water sourcing since the 1940s. In the 1970s, the gastrointestinal parasites and other hazardous microorganisms in the surface water (i.e., rivers and streams) were a significant source of disease and death. That country increased its use of tube wells to reduce the use of surface water for human consumption but did not test for arsenic in the water. In 1983, arsenic-induced skin cancer was discovered and associated with those tube wells.

> The discovery of arsenic contamination of groundwater in many nations, including Argentina, Chile, China, India, Mexico, Taiwan, Thailand, the United States and, now, Bangladesh shows that this is a global problem. All groundwater sources used for drinking-water should be tested for arsenic. A retrospective look at the situation in Bangladesh is instructive in that a declaration of a public health emergency might have expedited a more rapid response to the problem.
>
> **(Smith et al., 2000, p. 1099)[6]**

8.2.3 Surveillance for Public Health Hazards through Wastewater Monitoring Systems

The COVID-19 global pandemic helped accelerate the use of sewage-based surveillance for Public Health hazards. This virus is too small to be filtered out via conventional methods, and the monitoring for and detection of it may provide an early warning system as to whether SARS-CoV-2 can be identified in communities more quickly:

> With limitations on COVID-19 testing making it hard to know how many people actually have the disease, turning to the sewer systems for a fast snapshot seems to be promising and a useful method to provide complementary and additional information to the clinical testing. The SARS-CoV-2's faecal signature could actually turn out to be very useful, helping track how and where disease is spreading among the population.
>
> **(Michael-Kordatou et al., 2020, p. 1)[7]**

8.2.4 Italian Utility Drains Lake to Send More Water to Rome

In 2017, the local water utility started draining Lake Bracciano, northwest of Rome, to continue potable water service into Rome. Demand was up – over time – and this year had a severe drought:

> Residents of the surrounding villages were able to stop the water company from draining the lake, but the damage was done. Habitat and ecosystems were lost, and the lake has never fully recovered. Now, they are bringing the company to court for environmental crimes.
>
> The trial could set a precedent throughout Europe, by creating more public awareness about environmental crimes that involve water use and drainage, a long-overlooked issue, in addition to water pollution. It could inspire other small communities to fight to safeguard their natural havens and force big cities to find new solutions to deal with their water needs.
>
> The Lazio region, where Rome and Lake Bracciano are located, is rich in lakes, rivers, and pristine springs, but frequent droughts, climate change, and terrible water management are creating a recurring state of crisis.
>
> **(di Donfrancesco, 2023, p. 1)[8]**

Italy's national health minister then warned that the public health of the residents and visitors to Rome would be compromised. The Vatican shut off its fountains that year. Back-to-back years of lower-than-average rainfall had a significant financial impact on Italy's agricultural industry, at an estimated cost that year of nearly €2 billion ($2.3 billion; £1.8 billion). (BBC, 2017)[9]

8.2.5 Ultrafiltration of Rainwater and Chlorination Can Produce Potable Water

As part of a decentralized potable water supply construct, for small buildings and remote areas without access to a centralized water supply, collection of rainwater which is then put through an ultrafiltration process and then chlorinated, can produce small amounts of potable water – enough for the personal consumption use by the building, for example. That ultrafiltration process can be powered by photoelectric (PV) generation and is required since there are a multitude of hazards in raw rainwater or rainwater collected from rooftops and/or stored in open vessels.[10] This has applicability in the off-the-grid parts of the continental United States, islands, and U.S. territories, in addition to developing countries.

8.3 IMPACTS TO THE DISASTER PHASE CYCLES

As shown in Table 8.1, there are specific impacts to all four of the disaster cycle phases, from human-made hazards to water source systems:

8.4 ADVERSE IMPACTS TO THE INCIDENT COMMAND SYSTEM

There may be unique hazards directly impacting Responses and Responders, which should be considered adverse impacts to the Incident Command System (ICS). For Natural Water Source Contaminations, Table 8.2 details what may be occurring.

TABLE 8.1
Impacts to the Disaster Phase Cycles from Human-Made Hazards to Water Sources

Preparedness/Protection/Prevention

There are steps and courses of action that emergency managers can take to protect water sources, including public messaging about watershed protection, alerting officials when threats are noticed, etc. Communities – especially geographic areas which have high socially vulnerable individuals (SVIs) populations – need preparedness messaging and training to respond themselves to natural water source contaminations. This includes having potable water on-hand as part of their sheltering-in-place kits and pre-identifying alternative locations to obtain potable water. Diverting to alternative dialysis centers for treatment, if necessary and other personal health issues should be reminders to the public that they have the ultimate responsibility for their own life safety, first. Proper use of filters, based on historic and geographic elements, **plus *additional* filtration** for events and planned changes, such as waterway dredging, strip mining, fracking, etc. is something that emergency managers should be aware of and monitor for their communities.

Response

Emergency Management's response courses of action for natural water source contamination are similar to those for a human-made hazard to the local water source, includes providing alternative sources of potable water, crisis communications about the status of the water, and public health syndromic surveillance. This is not a comprehensive list, but a reminder that there are significant response actions needed for this hazard, to be managed through the local emergency management agency. There may be more notice – more time – to advise the public and plan for alternatives, as drought conditions (generally the initial causal factor) or systemic/chronic conditions (such as naturally occurring contaminants, which can adversely impact potable water systems) are part of the equation for the water system contamination. Do not rely on the water systems authority to provide all of the resources and response capability – in fact, their representative should be a liaison officer (or possibly part of unified command) to your jurisdiction's incident command system, not the other way around. An example of this is having independent testing, including that of higher governmental levels, for water contaminants coming into water treatment systems **and** the water coming out of those systems, as well.

Recovery

There will be cleanup and remediation work, possible infrastructure repair work, financial reimbursement to government, possibly by higher level government (see case study section in this chapter, for example) and other entities, continued health monitoring for the public, continued crisis communications, cooperation with investigators, including state/tribal/territorial and even federal level officials, and more. And most likely, in this longer-term time frame for recovery, another incident of scale will occur and consume the time and staffing of emergency managers. Self-care and wellness are critical for not only first responders, but everyone involved in the disaster phase cycle – which will be longer than other threats and hazards.

Mitigation

If there was a Declared Emergency or Disaster for this hazard, in this same jurisdiction or any jurisdiction which impacted the water sources from this natural hazard, it is possible that mitigation work[11] to help prevent and protect against the adverse impacts of that natural hazard can be obtained. This may also help prevent and protect against the adverse impacts of any future human-made hazards to those same systems.

Source: Author.

TABLE 8.2
Adverse Impacts to the Incident Command System

Command (including SO, PIO, and LNOs)

Safety Officers (SOs) need emergency management intelligence about this threat, as soon as possible. As the water supply may become contaminated, potable water will be needed for responders, emergency operations center staff, etc. This may extend beyond just drinking water into handwashing, bathing, etc. These workforce safety elements are paramount. While there are always pathogens and waterborne diseases which do not get filtered out of a water treatment process – extra attention – and advanced notice to deploying response teams – of this **hardship** should be made.

Public Information Officers (PIOs) need to urgently activate accurate templated messages to the public which provide critical crisis communications in the appropriate languages and modalities for oversaturation. If an "avoid use" warning is issued, for example, it needs to be communicated far and wide, and quickly. See the Shelter-in-Place section of Chapter 2 and also the Additional References/Reading section links for more specifics. Also recall that this hazard will impact agricultural uses of the water, too.

Liaison Officers (LNOs) are needed from both the water supply companies/entities and potentially any organizations aligned to human-made actions which may have exacerbated the natural contamination – or are involved in the remediation and recovery aspects, for example, the U.S. Army Corps of Engineers (USACE). These LNOs will need to be prepared to financially commit staffing and resources to the Incident/Unified Command for their support quickly, in order to effect solutions on a timely basis. In the United States, there may also be LNOs from federal agencies (such as the U.S. Environmental Protection Agency), who need to represent the full capabilities and capacities of their organizations, regardless of the status of a Presidentially Declared Disaster or Emergency. If a disaster or emergency is Presidentially Declared, then a unified Federal Coordinating Officer is assigned to support the State (or Tribal) Coordinating Officer.

Intelligence

Sources of Emergency Management Intelligence for incidents where hazards are from natural water source contamination will need to include more than just water systems officials. Additional subject-matter expertise of public health, geologists, ecologists, and other experts from this jurisdiction, higher levels of government and possibly derived from mutual-aid compacts (i.e., EMAC), may be needed.

Finance/Administration

- Budgets for alternative water storage, delivery, and distribution may not be adequate. The same may be true for additional filtration processes at water treatment plans, cost-share match portions of USACE work, etc. Emergency declarations at only the local jurisdiction level may not be sufficient to support emergency contracts, etc. Overtime and other staffing costs may exceed what is covered in normal fiscal budgets and/or emergency assistance reimbursements from higher governmental entities.
- If it is determined that a private entity is financially responsible in any part, for the natural water source contamination, the costs for remediation and cleanup may be borne in part or whole by that entity. Governmental Finance/Administration officials will need to manage costs and expenses in such a way as to obtain reimbursement on a timely basis. Additional elements of penalties and fines are not normally the responsibility of emergency management, but the organizations who are support partners to a jurisdiction's emergency management organization (such as the governmental environmental protection agency) may have dual responsibilities of both supporting the response/recovery work and issuing penalties and/or fines to the entity responsible for the adverse impacts caused by them. Emergency Management staff at those agencies must clearly divest themselves from any direct missions involved in the investigation or prosecution of malfeasance or criminal activity.

TABLE 8.2 *(Continued)*
Adverse Impacts to the Incident Command System

Logistics
With adverse impacts to raw water sources and/or water supply systems, command may designate points of distribution (PODs) for potable water to be donated to the public. This can take many forms (water from a truck where the public must bring their own containers, individual size bottled water cases which have weight and are usually distributed to residents via vehicle-based, gallon bottles of water, etc.). How this water is ordered/requisitioned is critical for logistics to align to the needs of the operation. Many times, there are also donations of water – from both corporations and the general public – which have to be managed as in-kind donations, probably through a non-governmental organization or consortium of organizations. In the event the incident is Stafford Act qualified for Public Assistance reimbursement of costs – these material donations, as well as hours of donated labor from volunteers, can be applied to public assistance projects.[12]
Operations
• Firefighting may be impacted – for other threats/hazards, for example, responding to a fire in the same jurisdiction as the impacted water supply system, if low or no water pressure adversely affects the water available through fire hydrants. • The operational distribution of water to the public – and for agricultural use, such as livestock watering – may be a required incident mission.
Planning
There are no specific adverse impacts to the Planning section, from this hazard. There are, however, additional planning activities, aligned planners from other incident commands, etc. associated with this hazard. Full-cycle planning through Mitigation work will most likely involve large infrastructure projects which may require Planning elements maintain the operational tempo/battle rhythm beyond shorter-term incidents, such as winter storms, minor river flooding, etc.

Source: Author.

8.5 POETE PROCESS ELEMENTS FOR THIS HAZARD

Table 8.3 shows the specifics for **Natural Water Source Contamination**.

8.6 CHAPTER SUMMARY/KEY TAKEAWAYS

As with human-made water quality issues, natural water source contamination has significant impacts on hospitals, dialysis centers, nursing homes/assisted living centers, daycare sites, etc. Boil alerts mean massive protocol changes at hospitals – longer than the alert itself. Community preparedness messaging can be very effective in minimizing adverse impacts from incidents involving this hazard. This should be accomplished through a whole-of-government/whole-community effort. A community cannot overcommunicate hazards associated with threats to their potable water supply systems, so messaging via public media, community/civic/religious groups, libraries, social media, etc. – all should be maximized using consistent messaging before, during, and after any incidents. There will be a need to monitor public health

TABLE 8.3
POETE Process Elements for This Hazard

Planning	While there will most likely **not** be a separate annex or appendix to any jurisdiction's emergency operations plan for this specific hazard, there should be elements of other main plan portions and annexes/appendices which cover the planning needed for a natural water source contamination to water supply systems, including individual's systems (wells, cisterns, etc.). From this chapter, there are a number of adverse impacts, historical references of past incidents to learn from, and elements of exercises which can be performed, in order to facilitate positive changes and modifications to preparedness, response, recovery, mitigation, and continuity of operations plans.
Organization	The **trained** staff needed to effect plans, using proper equipment should be exercised and evaluated on a regular basis each year. Actual incident response and recovery can help determine elements of POETE, including how these types of incidents are organized for staffing. Also, as part of both deployments and exercises, staff involved should be evaluated for both further training needed and also the possibility of advancement in their disaster state roles to leadership positions.
Equipment	Equipment needed for this hazard specifically includes anything needed to remediate/repair the water systems plant. Filtration and monitoring equipment is key. Also, any special equipment (traffic controls, pallet jacks, water tanker trucks, etc.) is needed at – and in support of – emergency potable water distribution points. This can also include at warehouses and supply depots.
Training	In addition to standardized Incident Command System training, specific training on adverse impacts to water supply and wastewater systems[13] should be taken by emergency management officials, from an all-hazards perspective.
Exercises and evaluation	A full cycle of discussion and performance exercises should be designed for natural water source contamination to Water Supply Systems – both on their own and as part of various cascading incident scenarios. These exercises should involve the whole community in exercise planner's invitations for both additional exercise planners/evaluators and actual players, especially since many water supply system operators are separate legal entities from the local governments they serve. In the same way, there are adversely impacted ESFs, RSFs, and CLs as noted in this chapter, those elements should be exercised and evaluated so that the elements of the POETE cycle are reviewed and revised as needed (i.e., agreed upon for change, as part of the Improvement Plan components).

Source: Author.

impacts for years after the incident was discovered. If the water supply system was damaged itself, longer-term remediation, repair, restoration, and additional elements which protect it from harm/prevent adverse impacts in the future should be incorporated into Mitigation work. On U.S. Presidentially declared emergencies and disasters, the partial (usually 75%) Federal funding for this work may be available. There may also be other governmental funding[14] in the United States related to natural hazard mitigation.

8.6.1 What to Read Next

If you did not read the previous chapter on Human-Made Water Source Contamination, please do so as well. While there are many repeated elements between the two chapters, there are some specifics found in one, which may be applicable to the other. The next chapter will continue the theme of Water Source Contamination but cover some of the other possible hazards from a water quality threat. Also, be prepared for the possibility of a complex or cascading incident: It may start out as a human-made hazard to a water source system, but then may produce different hazards, including natural ones, somewhere else. As noted in the introduction to this chapter, water sources can be adversely impacted by the actions to remediate and correct hazards in other geographies, other jurisdictions, even other nations. A fix in one country can generate a problem in another. Historically, this aspect has generated complex or cascading incidents.

NOTES

1 See https://www.hsph.harvard.edu/nutritionsource/water/#:~:text=General%20recommendations,exposed%20to%20very%20warm%20climates

2 See www.un.org/waterforlifedecade/pdf/human_right_to_water_and_sanitation_media_brief.pdf

3 Chakraborti, D., Rahman, M. M., Mukherjee, A., Alauddin, M., Hassan, M., Dutta, R. N., Pati, S., Mukherjee, S. C., Roy, S., Quamruzzman, Q., Rahman, M., Morshed, S., Islam, T., Sorif, S., Selim, M., Islam, M. R., & Hossain, M. M. (2015). Groundwater Arsenic Contamination in Bangladesh: 21 Years of Research. *Journal of Trace Elements in Medicine and Biology*, *31*, 237–248. https://doi.org/10.1016/j.jtemb.2015.01.003

4 Li, P., Karunanidhi, D., Subramani, T., & Srinivasamoorthy, K. (2021). Sources and Consequences of Groundwater Contamination. *Archives of Environmental Contamination and Toxicology*, *80*(1), 1–10. https://doi.org/10.1007/s00244-020-00805-z

5 Missimer, T. M., Teaf, C. M., Beeson, W. T., Maliva, R. G., Woolschlager, J., & Covert, D. J. (2018). Natural Background and Anthropogenic Arsenic Enrichment in Florida Soils, Surface Water, and Groundwater: A Review with a Discussion on Public Health Risk. *International Journal of Environmental Research and Public Health*, *15*(10), 2278.

6 Smith, A. H., Lingas, E. O., & Rahman, M. (2000). Contamination of Drinking-Water by Arsenic in Bangladesh: A Public Health Emergency. *Bulletin of the World Health Organization*, *78*(9), 1093–1103.

7 Michael-Kordatou, I., Karaolia, P., & Fatta-Kassinos, D. (2020). Sewage Analysis as a Tool for the COVID-19 Pandemic Response and Management: The Urgent Need for Optimised Protocols for SARS-CoV-2 Detection and Quantification. *Journal of Environmental Chemical Engineering*, *8*(5), 104306–104306. https://doi.org/10.1016/j.jece.2020.104306

8 di Donfrancesco, G. (2023). Italian Utility in Hot Water for Draining a Picturesque Lake to Send Water to Rome. *Bulletin of the Atomic Scientists*. Accessed on December 11, 2023. See https://thebulletin.org/2023/12/italian-utility-in-hot-water-for-draining-a-picturesque-lake-to-send-water-to-rome/

9 BBC. (2017). Italy Drought: Water Cuts Pose Rome "Health Risk". *BBC*. Accessed on December 11, 2023. See www.bbc.com/news/world-europe-40733218

10 Baú, S. R. C., Bevegnu, M., Giubel, G., Gamba, V., Cadore, J. S., Brião, V. B., & Shaheed, M. H. (2022). Development and Economic Viability Analysis of Photovoltaic (PV) Energy Powered Decentralized Ultrafiltration of Rainwater for Potable Use. *Journal of Water Process Engineering*, *50*, 103228. https://doi.org/10.1016/j.jwpe.2022.103228
11 FEMA. (2023). *Hazard Mitigation Policy Aid Drought Mitigation.* FEMA. See www.fema.gov/sites/default/files/documents/fema_hma_drought-mitigation-policy-aid_09282023.pdf
12 See www.govinfo.gov/content/pkg/FR-2013-11-01/pdf/2013-26018.pdf
13 FEMA. (2019). *IS-1024: Water and Wastewater Treatment System Considerations.* FEMA EMI. See https://training.fema.gov/is/courseoverview.aspx?code=IS-1024&lang=en
14 See www.fema.gov/grants/mitigation/building-resilient-infrastructure-communities

REFERENCES/ADDITIONAL READING

Camarillo, M. K., Jain, R., & Stringfellow, W. T. (2014). *Drinking Water Security for Engineers, Planners, and Managers.* Elsevier/Butterworth-Heinemann.

Howard, G., & Bartram, J. (2010). *Vision 2030: The Resilience of Water Supply and Sanitation in the Face of Climate Change Technical Report.* World Health Organization. See https://iris.who.int/bitstream/handle/10665/70462/WHO_HSE_WSH_10.01_eng.pdf?sequence=1

U.S. Environmental Protection Agency Office of Water. (2017). *Incident Action Checklist – Harmful Algal Blooms.* EPA. See www.epa.gov/sites/default/files/2017-11/documents/171030-incidentactionchecklist-hab-form_508c.pdf

World Health Organization. (n.d.). *Water Sanitation and Health Humanitarian Emergencies.* WHO. Accessed December 31, 2023. See www.who.int/teams/environment-climate-change-and-health/water-sanitation-and-health/environmental-health-in-emergencies/humanitarian-emergencies

Zulkifli, S. N., Rahim, H. A., & Lau, W. J. (2018). Detection of Contaminants in Water Supply: A Review on State-of-the-Art Monitoring Technologies and Their Applications. *Sensors and Actuators B: Chemical*, *255*, 2657–2689. https://doi.org/10.1016/j.snb.2017.09.078

BIBLIOGRAPHY

Baú, S. R. C., Bevegnu, M., Giubel, G., Gamba, V., Cadore, J. S., Brião, V. B., & Shaheed, M. H. (2022). Development and Economic Viability Analysis of Photovoltaic (PV) Energy Powered Decentralized Ultrafiltration of Rainwater for Potable Use. *Journal of Water Process Engineering*, *50*, 103228. https://doi.org/10.1016/j.jwpe.2022.103228

BBC. (2017). Italy Drought: Water Cuts Pose Rome 'Health Risk'. *BBC.* Accessed December 11, 2023. See www.bbc.com/news/world-europe-40733218

Chakraborti, D., Rahman, M. M., Mukherjee, A., Alauddin, M., Hassan, M., Dutta, R. N., Pati, S., Mukherjee, S. C., Roy, S., Quamruzzman, Q., Rahman, M., Morshed, S., Islam, T., Sorif, S., Selim, M., Islam, M. R., & Hossain, M. M. (2015). Groundwater Arsenic Contamination in Bangladesh-21 Years of Research. *Journal of Trace Elements in Medicine and Biology*, *31*, 237–248. https://doi.org/10.1016/j.jtemb.2015.01.003

di Donfrancesco, G. (2023). Italian Utility in Hot Water for Draining a Picturesque Lake to Send Water to Rome. *Bulletin of the Atomic Scientists.* Accessed December 11, 2023. See https://thebulletin.org/2023/12/italian-utility-in-hot-water-for-draining-a-picturesque-lake-to-send-water-to-rome/

FEMA. (2019). *IS-1024: Water and Wastewater Treatment System Considerations.* FEMA EMI. See https://training.fema.gov/is/courseoverview.aspx?code=IS-1024&lang=en

FEMA. (2023). *Hazard Mitigation Policy Aid Drought Mitigation.* FEMA. See www.fema.gov/sites/default/files/documents/fema_hma_drought-mitigation-policy-aid_09282023.pdf

Li, P., Karunanidhi, D., Subramani, T., & Srinivasamoorthy, K. (2021). Sources and Consequences of Groundwater Contamination. *Archives of Environmental Contamination and Toxicology, 80*(1), 1–10. https://doi.org/10.1007/s00244-020-00805-z

Michael-Kordatou, I., Karaolia, P., & Fatta-Kassinos, D. (2020). Sewage Analysis as a Tool for the COVID-19 Pandemic Response and Management: The Urgent Need for Optimised Protocols for SARS-CoV-2 Detection and Quantification. *Journal of Environmental Chemical Engineering, 8*(5), 104306–104306. https://doi.org/10.1016/j.jece.2020.104306

Missimer, T. M., Teaf, C. M., Beeson, W. T., Maliva, R. G., Woolschlager, J., & Covert, D. J. (2018). Natural Background and Anthropogenic Arsenic Enrichment in Florida Soils, Surface Water, and Groundwater: A Review with a Discussion on Public Health Risk. *International Journal of Environmental Research and Public Health, 15*(10), 2278.

Smith, A. H., Lingas, E. O., & Rahman, M. (2000). Contamination of Drinking-Water by Arsenic in Bangladesh: A Public Health Emergency. *Bulletin of the World Health Organization, 78*(9), 1093–1103.

Zulkifli, S. N., Rahim, H. A., & Lau, W. J. (2018). Detection of Contaminants in Water Supply: A Review on State-of-the-Art Monitoring Technologies and Their Applications. *Sensors and Actuators B: Chemical, 255*, 2657–2689. https://doi.org/10.1016/j.snb.2017.09.078

9 Water Source Quality Hazards from Other Incidents

> As we reimagine and shape a post-COVID world, making sure we are sending children and mothers to places of care equipped with adequate water, sanitation and hygiene (WASH) services is not merely something we can and should do. It is an absolute must.
>
> **UNICEF Executive Director Henrietta Fore[1]**

This chapter covers how there can be other water source quality hazards, from incidents other than human-made water source contamination (Chapter 7) or natural water source contamination (Chapter 8). This chapter is designed as a catch-all for the other hazards which do not align with the previous two chapters and will include more details on Waterborne Diseases and Parasites, as well as direct technological hazards, including cyberattacks. Adverse impacts to human and agricultural potable water sources, as well as navigable freshwater waterways, will most likely have the same hazards as either of those other water source quality chapters on hazard types.

This chapter is one of the Water Quality Hazard Part. Each chapter in every Hazard part will have (1) an overview of the specific hazard and its adverse impacts to communities, (2) case examples for historic reference, many of these will be what not to do, as lessons learned, (3) impacts to the disaster phase cycles (the before, during, and after incidents occur), (4) adverse impacts of this specific hazard to the Response's Incident Command System itself, (5) a summary of the Planning, Organization, Equipping, Training, and Exercising (POETE) process for deliberative planning and problem-solving, and finally (6) chapter-specific references/reading material.

Besides local jurisdictional emergency management groups, this hazard will need to be managed by water management districts, hospitals and other healthcare facilities, colleges/universities, etc. in a coordinated and collaborative way. All will have unmet needs which they will consider as a priority – and some of those priorities may conflict and compete with each other for Response and Recovery resource allocation. Unified Command should always prioritize the LIPER (see Chapter 2), when making decisions.

9.1 OVERVIEW

As with other hazards, the question of causality is immaterial to Emergency Management's Response and Recovery operations, except that there may be law

DOI: 10.1201/9781003474685-12

enforcement intelligence and investigation elements imbedded within the Incident Command System (ICS) established, specifically, to respond to and recover from the adverse impacts of the incident. If this is the case, then actions of the other Command and General Staff branches/sections may need to be adjusted and/or expanded to support law enforcement intelligence and investigation elements. This could be investigators for environmental protection laws, for example. There may be a parallel incident command system established by law enforcement officials and there may or may not be a liaison officer (LNO) linking the two – or more – ICSs. Each incident, emergency, disaster, etc. is different. Emergency managers should be involved in the full disaster cycle planning, organization, equipping, training, and exercising for freshwater, potable water, drinking water supply systems, and/or navigable waterways in their respective communities (including those which provide support or services to their jurisdiction's water system from another jurisdiction – such as a neighboring community's shared water treatment plant, watershed areas, dams, etc.).

9.1.1 Waterborne Diseases and Parasites

Regardless of method of introduction (naturally occurring or human-induced accidental/terrorism), water quality can be adversely affected by waterborne diseases and parasites. There are a number of these which survive past water treatment plants/systems and others can be introduced into water supply systems at points post-filtration and possibly post-testing/surveillance. View these hazards in this chapter, as connected to the hazards in Part 5.

Also, halting or channeling natural water flow can increase adverse ecological changes in the water – some of which can be hazardous to humans. For example, the Aswan High Dam in Egypt[2] and dam sites in Nigeria[3] have seen increases in snail hosts for both urinary and intestinal schistosomiasis, caused by parasitic worms. While not found in the United States, the impact of this hazard worldwide is "second only to malaria as the most devastating parasitic disease".[4]

9.1.2 Sewage Treatment and Waste Issues

Freshwater sources are part of a water processing system cycle, which includes sewage treatment. In many communities, the water moves through the delivery system and gets recycled. It can also move through rivers, lakes, and oceans downstream.

During heavy flow incidents[5] (flooding, dam breeches, storm surge, etc.), tons of debris can get caught up in waterways. Not only can this cause damage to infrastructure (bridges, roadways, etc.) but can also impact the oceans, as all freshwater sources flow into them.[6]

9.1.3 Water Resources and Armed Conflict

War is the worst type of disaster, and while not managed by Emergency Management, any adverse water source or supply impacts to civilian populations will become concerns for Emergency Management, in most cases.

While listed under this book's Part for Quality issues, hazards from armed conflict/war to water resources can also impact water quantity. Purposely damaging civilian water supplies in armed conflict is most likely a violation of International Humanitarian Law (IHL) (Zemmali, 1998).[7] Regardless of the legal specifics, for any emergency manager to be involved in the purposeful adverse impacts to a civilian water source or supply system would certainly be unethical activity. That may sound like a problem for only parts of the globe where there is armed conflict regularly, but historically this has occurred *everywhere*, even in the United States. The emergency manager, who is actively participating in the reduction or elimination of the water sources to a civilian population anywhere, has made the decision to conduct themselves in a manner against generally accepted emergency management principles worldwide. This section is effectively a call to action on what ***not to do***.

When it comes to armed conflict within a nation – or between nations – the lines between governmental response and military actions can become convoluted. In some cases, International Humanitarian Law (IHL) has applicability to human-made adverse impacts to water and wastewater supply systems. The rights of citizens to have access to clean/potable water resources (water reservoirs, water tanks, water supply systems, water delivery systems, etc.) should not be abridged or destroyed by armed conflict. And with this conflict can come adverse impacts to humanitarian personnel and their missions to support civilian populations:

- Water used as a target:
  - Intentional targeting of water systems is a violation of IHL Additional Protocol 1, Article 54.[8]
  - Water sources may be shared between civilian and military populations (especially in urban areas) but should be considered a civilian resource when considering whether it is allowable during an armed conflict between nations to be attacked.
- Water used as a weapon:
  - See Chapter 12, for emergency management impacts of threats/hazards such as dam breeches, flooding, etc.
  - See Chapter 19, for emergency management impacts of threats/hazards such as hacking, ransomware, and attacks on ports.
- Water can be a tool for peace. United Nations' peacekeeping forces can be utilized to protect water resources.[9]
- There are specific international water resources laws and rules, established through extensions of the Helsinki Accords, Armed Annex IV – Berlin Rules on Water Resources.[10]

9.2 CASE EXAMPLES

Below, please find some case examples from recent history, which exhibit both best practices for disaster readiness and how there can be significant adverse impacts to the LIPER (life safety, incident stabilization, property/asset protection, environmental/economic impacts, and recovery operations) from Water Source Quality Hazards from other Incidents.

Water Systems operators – including their Emergency Management teams – will need a strong understanding of the monitoring, filtration methods, process flow, contingency measures, countermeasures, etc. needed to help keep their systems safe and secure. Figure 8.1 has an example of the history of elemental monitoring and filtration flow.

9.2.1 COVID-19 Pandemic Healthcare Impacts

In 2021, residents of Orlando, Florida, were asked to conserve potable water usage, due to the prioritization of using liquid oxygen for treating COVID-19 patients, over its routine use in ozonation in their water treatment systems. Ozonation is used to remove some contaminants and the naturally occurring sulfur smell from freshwater in this part of Florida.[11]

9.2.2 New Jersey's Superstorm Sandy and the Sayreville Pump Station

Just days before Superstorm Sandy struck the U.S. East Coast in 2012, the Middlesex County Utility Authority in New Jersey made temporary fixes and replaced three of the sluice gates at the Sayreville pump station, with temporary metal plates. Sandy brought 15 ft (4.6 m) of stormwater into the building, flooding the basement with raw sewage and sending millions of gallons of untreated waste into the Raritan River, nearby.

Working around the clock for days to fix the problem, and finally finding a new sluice gate to put into place – 30 ft (9.1 m) underneath the sewage waterline – divers had to manipulate the gate by hand to get it into place.[12]

9.2.3 *Legionella*: A Compounding/Cascading Hazard for Flint, Michigan

If the water supply contamination concerns – based on a change in water sources – in Flint, Michigan (a case example mentioned in multiple chapters in this book), were not enough on their own, that community's water source disaster generated another compounding or cascading hazard: cases of *Legionella* in 2014 as well.[13]

> From 2000 to 2014, the rate of reported cases of legionellosis, which includes Legionnaires' disease and a milder flu-like illness called Pontiac fever, increased nearly fourfold, from .42 to 1.62 cases per 100,000 persons, according to a CDC report released Tuesday.[14] Please see Figure 9.1 for specifics.

9.2.4 IRGC-Affiliated Cyber Actors Exploit PLCs in Multiple Sectors, including U.S. Water and Wastewater Systems Facilities

In 2023, a number of U.S. Federal agencies, as well as the Israel National Cyber Directorate, issued a warning that malicious cyber actors – organized by the Iranian national government's Islamic Revolutionary Guard Corps (IRGC), designated by the United States as a foreign terrorist organization – was targeting and compromising

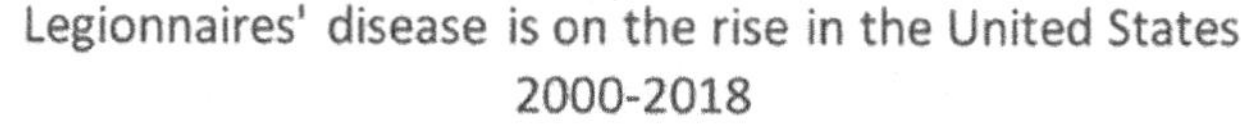

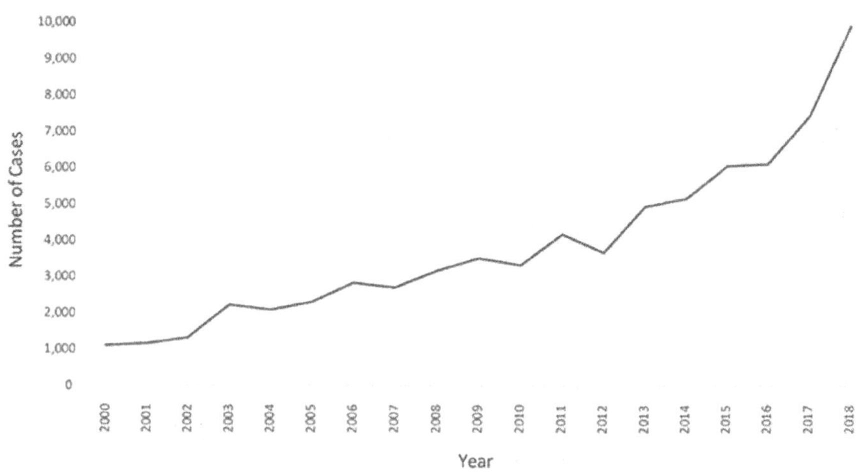

FIGURE 9.1 Legionnaires disease is on the rise in the United States, 2000–2018.

Source: CDC. (2021). *Legionella (Legionnaires' Disease and Pontiac Fever).* Accessed December 14, 2023. See www.cdc.gov/legionella/qa-media.html

programmable logic controllers (PLCs), commonly found in the water and wastewater systems sector in the United States[15] and other countries.

> These PLC and related controllers are often exposed to outside internet connectivity due to the remote nature of their control and monitoring functionalities. The compromise is centered around defacing the controller's user interface and may render the PLC inoperative. With this type of access, deeper device and network level accesses are available and could render additional, more profound cyber physical effects on processes and equipment. It is not known if additional cyber activities deeper into these PLCs or related control networks and components were intended or achieved. Organizations should consider and evaluate their systems for these possibilities.
>
> **(CISA, 2023, p. 1)[16]**

9.2.5 Water Rights Challenges in the United States

There have been violent challenges with water rights in the United States, one example is the 1927s fights between the Owens Valley ranchers and residents, and the City of Los Angeles, in California. Infrastructure was sabotaged, and local law enforcement sided with the detractors. And while Los Angeles paid for the land, the Owens Valley ranchers and farmers did not fully realize what the loss of the water rights would mean for their future:[17]

> Los Angeles gets its water by reason of one of the costliest, crookedest, most unscrupulous deals ever perpetrated, plus one of the greatest pieces of engineering folly ever heard of. Owens Valley is there for anybody to see. The city of Los Angeles moved through this valley like a devastating plague. It was ruthless, stupid, cruel and crooked. It stole the waters of the Owens River. It drove the people of Owens Valley from their home, a home which they had built from the desert. For no sound reason, for no sane reason, it destroyed a helpless agricultural section and a dozen towns. It was an obscene enterprise from beginning to end.
>
> **(Mayo, 1933, p. 246)**[18]

9.2.6 Manzanar, California: Forced Relocations Twice in History

Water rights, land rights, racism, and big city politics were all in play in the history of supporting the ever-growing water needs for Los Angeles, California. First there was the forced relocation of the Paiute from the Owens Valley, then the ranchers from that same land in order to take the water for Los Angeles (see the case above), and then the forced relocation/internment of Japanese Americans during World War II – all on the same land in California. The memorialization and recognition of the wrongs committed then is still controversial and challenging today.[19]

9.3 IMPACTS TO THE DISASTER PHASE CYCLES

As shown in Table 9.1, there are specific impacts to all four of the disaster cycle phases, from Water Supply Quality Hazards from other Incidents:

TABLE 9.1
Impacts to the Disaster Phase Cycles from Water Supply Quality Hazards from Other Incidents

Preparedness/Protection/Prevention
There are steps and courses of action that emergency managers can take to protect water sources, including public messaging about watershed protection, alerting officials when threats are noticed, etc. Communities – especially geographic areas which have high socially vulnerable individuals (SVIs) populations – need preparedness messaging and training to respond themselves to any type of water source contaminations. This includes having potable water on-hand as part of their sheltering-in-place kits and pre-identifying alternative locations to obtain potable water. Diverting to alternative dialysis centers for treatment, if necessary, and other personal health issues should be reminders to the public that they have the ultimate responsibility for their own life safety, first. Proper use of filters, based on historic and geographic elements, **plus *additional* filtration** for events and planned changes, such as waterway dredging, strip mining, fracking, etc. is something that emergency managers should be aware of and monitor for their communities.

(Continued)

TABLE 9.1 *(Continued)*
Impacts to the Disaster Phase Cycles from Water Supply Quality Hazards from Other Incidents

Response

Emergency Management's response courses of action for natural water source contamination are similar to those for a human-made hazard to raw water sources, and includes providing alternative sources of potable water, crisis communications about the status of the water, and public health syndromic surveillance. This is not a comprehensive list, but a reminder that there are significant response actions needed for this hazard, to be managed through the local emergency management agency. There may be more notice – more time – to advise the public and plan for alternatives, as drought conditions (generally one initial causal factor) or systemic/chronic conditions (such as naturally occurring contaminants or biologics, which can adversely impact water sources) are part of the equation for the overall water system contamination. Do not rely on the water systems authority to provide all of the resources and response capability – in fact, their representative should be a liaison officer (or possibly part of unified command) to your jurisdiction's incident command system, not the other way around. An example of this is having independent testing, including that of higher governmental levels, for water contaminants coming into water treatment systems **and** the water coming out of those systems, as well.

Recovery

There will be cleanup and remediation work, possible infrastructure repair work, financial reimbursement to government, possibly by higher level government (see case study section in this chapter, for example) and other entities, continued health monitoring for the public, continued crisis communications, cooperation with investigators, including state/tribal/territorial and even federal level officials, and more. And most likely, in this longer-term time frame for recovery, another incident of scale will occur and consume the time and staffing of emergency managers. Self-care and wellness are critical for not only first responders but also everyone involved in the disaster phase cycle – which will be longer than other threats and hazards.

Mitigation

If there was a Declared Emergency or Disaster for this hazard, in this same jurisdiction or any jurisdiction which impacted the water supply system from this hazard, it is possible that mitigation work[20] to help prevent and protect against the adverse impacts of the hazard can be obtained. This may also help prevent and protect against the adverse impacts of any future human-made hazards to those same systems.

Source: Author.

9.4 ADVERSE IMPACTS TO THE INCIDENT COMMAND SYSTEM

There may be unique hazards directly impacting Responses and Responders, which should be considered adverse impacts to the Incident Command System (ICS). For Water Supply Quality Hazards from other Incidents, Table 9.2 details what may be occurring.

9.5 POETE PROCESS ELEMENTS FOR THIS HAZARD

Table 9.3 shows the specifics for **Water Source Quality Hazards from Other Incidents**.

TABLE 9.2
Adverse Impacts to the Incident Command System

Command (including SO, PIO, and LNOs)

Safety Officers (SOs) need emergency management intelligence about this threat, as soon as possible. As the water supply may become contaminated, potable water will be needed for responders, emergency operations center staff, etc. This may extend beyond just drinking water into handwashing, bathing, etc. These workforce safety elements are paramount. While there are always pathogens and waterborne diseases which do not get filtered out of a water treatment process – extra attention – and advanced notice to deploying response teams – of this **hardship** should be made.

Public Information Officers (PIOs) need to urgently activate accurate templated messages to the public which provide critical crisis communications in the appropriate languages and modalities for oversaturation. If an "avoid use" warning is needed, for example, it needs to be communicated far and wide, and quickly. See the Shelter-in-Place section of Chapter 2 and also References/Additional Reading section links for more specifics. Also recall that this hazard will also impact agricultural uses of the water.

Liaison Officers (LNOs) are needed from both the water supply companies/entities and potentially any organizations aligned to human-made actions which may have exacerbated the water source contamination – or are involved in the remediation and recovery aspects, for example, the U.S. Army Corps of Engineers (USACE). These LNOs will need to be prepared to financially commit staffing and resources to the Incident/Unified Command for their support quickly, in order to effect solutions on a timely basis. In the United States, there may also be LNOs from federal agencies (such as the U.S. Environmental Protection Agency), who need to represent the full capabilities and capacities of their organizations, regardless of the status of a Presidentially Declared Disaster or Emergency. If a disaster or emergency is Presidentially Declared, then a unified Federal Coordinating Officer is assigned to support the State (or Tribal) Coordinating Officer.

Intelligence

Sources of Emergency Management Intelligence for incidents where hazards are from any type of water source contamination will need to include more than just water systems officials. Additional subject-matter expertise of public health, geologists, ecologists, and other experts from this jurisdiction, higher levels of government and possibly derived from mutual-aid compacts (i.e., EMAC), may be needed.

Finance/Administration

- Budgets for alternative water storage, delivery, and distribution may not be adequate. The same may be true for additional filtration processes at water treatment plans, cost-share match portions of USACE work, etc. Emergency declarations at the local jurisdiction may not be sufficient to support emergency contracts, etc. Overtime and other staffing costs may exceed what is covered in normal fiscal budgets and/or emergency assistance reimbursements from higher governmental entities.
- If it is determined that a private entity is financially responsible in any part for the water source contamination, the costs for remediation and cleanup may be borne in part or whole, by that entity. Governmental Finance/Administration officials will need to manage costs and expenses in such a way as to obtain reimbursement on a timely basis. Additional elements of penalties and fines are not normally the responsibility of emergency management, but the organizations who are support partners to a jurisdiction's emergency management organization (such as the governmental environmental protection agency) may have dual responsibilities of both supporting the response/recovery work and issuing penalties and/or fines to the entity responsible for the adverse impacts caused by them. Emergency Management staff at those agencies must clearly divest themselves from any direct missions involved in the investigation or prosecution of malfeasance or criminal activity.

(Continued)

TABLE 9.2 *(Continued)*

Adverse Impacts to the Incident Command System

Logistics

With adverse impacts to water sources and then the community's water supply systems, command may designate points of distribution (PODs) for potable water to be donated to the public. This can take many forms (water from a truck where the public must bring their own containers, individual size bottled water cases which have weight and are usually distributed to residents via vehicle-based, gallon bottles of water, etc.). How this water is ordered/requisitioned is critical for logistics to align to the needs of the operation. Many times, there are also donations of water – from both corporations and the general public – which have to be managed as in-kind donations, probably through a non-governmental organization or consortium of organizations. In the event the incident is Stafford Act qualified for Public Assistance reimbursement of costs – these material donations, as well as hours of donated labor from volunteers, can be applied to public assistance projects.[21]

Operations

- Firefighting may be impacted – for other threats/hazards, for example, responding to a fire in the same jurisdiction as the impacted water supply system, if low or no water pressure adversely affects the water available through fire hydrants.
- The operational distribution of water to the public – and for agricultural use, such as livestock watering – may be a required incident mission.

Planning

There are no specific adverse impacts to the Planning section, from this hazard. There are, however, additional planning activities, aligned planners from other incident commands, etc. associated with this hazard. Full-cycle planning through Mitigation work will most likely involve large infrastructure projects which may require Planning elements maintain the operational tempo/battle rhythm beyond shorter-term incidents, such as winter storms, minor river flooding, etc.

Source: Author.

TABLE 9.3

POETE Process Elements for This Hazard

Planning	While there will most likely **not** be a separate annex or appendix to any jurisdiction's emergency operations plan for this specific hazard, there should be elements of other main plan portions and annexes/appendices which cover the planning needed for any type of water source contamination to that community's water supply systems, including individual's systems (wells, cisterns, etc.). From this chapter, there are a number of adverse impacts, historical references of past incidents to learn from, and elements of exercises which can be performed, in order to facilitate positive changes and modifications to preparedness, response, recovery, mitigation, and continuity of operations plans.
Organization	The **trained** staff needed to effect plans, using proper equipment should be exercised and evaluated on a regular basis each year. Actual incident response and recovery can help determine elements of POETE, including how these types of incidents are organized for staffing. Also, as part of both deployments and exercises, staff involved should be evaluated for both further training needed and the possibility of advancement in their disaster state roles to leadership positions.

TABLE 9.3 *(Continued)*
POETE Process Elements for This Hazard

Equipment	Equipment needed for this hazard specifically includes anything needed to remediate/repair the water systems plant. Filtration and monitoring equipment is key. Also needed is any special equipment (traffic controls, pallet jacks, water tanker trucks, etc.) needed at – and in support of – emergency potable water distribution points. This can also include at warehouses and supply depots.
Training	In addition to standardized Incident Command System training, specific training on adverse impacts to water supply and wastewater systems[22] should be taken by emergency management officials, from an all-hazards perspective.
Exercises and evaluation	A full cycle of discussion and performance exercises should be designed for any type of water source contamination to Water Supply and Wastewater Management Systems – both on their own and as part of various cascading incident scenarios. These exercises should involve the whole community in exercise planner's invitations for both additional exercise planners/evaluators and actual players, especially since many water supply system operators are separate legal entities from the local governments they serve. In the same way, there are adversely impacted ESFs, RSFs, and CLs as noted in this chapter, those elements should be exercised and evaluated so that the elements of the POETE cycle are reviewed, and revised as needed (i.e., agreed upon for change, as part of the Improvement Plan components).

Source: Author.

9.6 CHAPTER SUMMARY/KEY TAKEAWAYS

As with both human-made water quality issues and natural water source contamination, Water Source Quality Hazards from other Incidents have significant impacts on hospitals, dialysis centers, nursing homes/assisted living centers, daycare sites, etc. Boil alerts mean massive protocol changes at hospitals – longer than the alert itself. Community preparedness messaging can be very effective in minimizing adverse impacts from incidents involving this hazard. This should be accomplished through a whole-of-government/whole-community effort. A community cannot overcommunicate hazards associated with threats to their potable water supply systems, so messaging via public media, community/civic/religious groups, libraries, social media, etc. all should be maximized using consistent messaging before, during, and after any incidents. There will be a need to monitor public health impacts for years after the incident was discovered. If the water supply system was damaged itself, longer-term remediation, repair, restoration, **and additional** elements which protect it from harm/prevent adverse impacts in the future, should be incorporated into Mitigation work. On U.S. Presidentially Declared emergencies and disasters, the partial (usually 75%) Federal funding for this work may be available.

9.6.1 What to Read Next

If you did not read the previous chapters on Human-Made Water Source Contamination and Natural Water Source Contamination, please do so as well. While there are many repeated elements between this chapter and the other two chapters, there are some specifics found in one, which may be applicable to the other. The next chapter is the first one in the Quantity Hazards with Water – Too Little Potable Water. Also, be prepared for the possibility of a complex or cascading incident: It may start out as a human-made hazard to a water supply system, but then may produce different hazards including natural ones impacting the raw water sources, somewhere else. As noted in the introduction to this chapter, water sources can be impacted by the actions to remediate and correct hazards in other geographies, other jurisdictions, even other nations. Historically, this aspect has generated complex or cascading incidents, just from the jurisdictional concerns themselves.

NOTES

1 See www.unicef.org/tajikistan/press-releases/almost-2-billion-people-depend-health-care-facilities-without-basic-water-services

2 Malek, E. A. (1975). Effect of the Aswan High Dam on Prevalence of Schistosomiasis in Egypt. *Tropical and Geographical Medicine*, *27*(4), 359–364.

3 Betterton, C., Ndifon, G. T., Bassey, S. E., Tan, R. M., & Oyeyi, T. (1988). Schistosomiasis in Kano State, Nigeria. I. Human Infections Near Dam Sites and the Distribution and Habitat Preferences of Potential Snail Intermediate Hosts. *Annals of Tropical Medicine and Parasitology*, *82*(6), 561–570.

4 CDC. (2018). *Parasites – Schistosomiasis*. Accessed November 28, 2023. See www.cdc.gov/parasites/schistosomiasis/index.html

5 Lam, A. (2023). *What Stops Rain from Flooding Your City*? YouTube Video. Accessed December 14, 2023. See www.youtube.com/watch?v=coXe8_xnAOs

6 The Ocean Cleanup. (n.d.). *The Largest Cleanup in History*. The Ocean Cleanup Technologies B.V. Accessed December 14, 2023. See https://theoceancleanup.com/

7 Zemmali, A. (1998). Dying for Water. *Forum, War and Water*. ICRC. See https://casebook.icrc.org/case-study/water-and-armed-conflicts

8 See www.un.org/en/genocideprevention/documents/atrocity-crimes/Doc.34_AP-I-EN.pdf

9 Tignino, M., & Irmakkesen, Ö. (2020). Water in Peace Operations: The Case of Haiti. *RECIEL*, *29*. https://doi.org/10.1111/reel.12333

10 International Law Association. (2004). *Berlin Conference (2004) Water Resources Law*. United Nations Economic Commission for Europe. See https://unece.org/fileadmin/DAM/env/water/meetings/legal_board/2010/annexes_groundwater_paper/Annex_IV_Berlin_Rules_on_Water_Resources_ILA.pdf

11 Fox, G. (2021). *Orlando Mayor Asks Residents to Start Conserving Water Immediately Due to Liquid Oxygen Shortage*. WESH. Accessed December 14, 2023. See www.wesh.com/article/orlando-mayor-to-hold-newser-on-unprecedented-event-requiring-community-assistance/37359831

12 Chesler, C. (2014). *Down the Drain: NJ's Sewage System*. NJ Monthly. Accessed on December 14, 2023. See https://njmonthly.com/articles/jersey-living/down-the-drain-njs-sewage-system/

13 Goodnough, A. (2016). Legionnaires' Outbreak in Flint Was Met with Silence. *The New York Times*. Accessed November 21, 2023. See www.nytimes.com/2016/02/23/us/legionnaires-outbreak-in-flint-was-met-with-silence.html

14 Sun, L. H. (2016). Legionnaires' Outbreaks: Cases Nearly Quadrupled in 15 Years. *The Washington Post*. Accessed November 21, 2023. See www.washingtonpost.com/news/to-your-health/wp/2016/06/07/legionnaires-outbreaks-cases-nearly-quadrupled-in-15-years/

15 See www.cisa.gov/water

16 CISA. (2023). *IRGC-Affiliated Cyber Actors Exploit PLCs in Multiple Sectors, Including U.S. Water and Wastewater Systems Facilities*. Accessed December 14, 2023. See www.cisa.gov/news-events/cybersecurity-advisories/aa23-335a?fbclid=IwAR2Q2SglxjAfzgGtcI7vzRi84hQP1hA1r-C0gN4fgnisYf-1eEneEb7L2DU

17 Owens Valley History. (n.d.). *Owen's Valley's – Los Angeles Aqueduct*. Accessed December 14, 2023. See www.owensvalleyhistory.com/ov_aqueduct/page20c.html

18 Mayo, M. (1933). *Los Angeles*. A. A. Knopf.

19 Linnarz, R. (2021). *Documentary Review: Manzanar, Diverted: When Water Becomes Dust (2021) by Ann Kaneko*. Accessed December 14, 2023. See https://asianmoviepulse.com/2021/10/documentary-review-manzanar-diverted-when-water-becomes-dust-2021-by-ann-kaneko/

20 FEMA. (2023). *Hazard Mitigation Policy Aid Drought Mitigation*. FEMA. See www.fema.gov/sites/default/files/documents/fema_hma_drought-mitigation-policy-aid_09282023.pdf

21 See www.govinfo.gov/content/pkg/FR-2013-11-01/pdf/2013-26018.pdf

22 FEMA. (2019). *IS-1024: Water and Wastewater Treatment System Considerations*. FEMA EMI. See https://training.fema.gov/is/courseoverview.aspx?code=IS-1024&lang=en

REFERENCES/ADDITIONAL READING

Amann, D. M., & Sellers, M. N. S. (2002). The United States of America and the International Criminal Court. *The American Journal of Comparative Law*. https://doi.org/10.2307/840883

Bos, R., Roaf, V., Payen, G., Rousse, M. J., Latorre, C., McCleod, N., & Alves, D. (2016). *Manual on the Human Rights to Safe Drinking Water and Sanitation for Practitioners. Water Intelligence Online* (1st ed.). IWA Publishing. https://doi.org/10.2166/9781780407449

Collier, S. A., Deng, L., Adam, E. A., et al. (2021). Estimate of Burden and Direct Healthcare Cost of Infectious Waterborne Disease in the United States. *Emerging Infectious Diseases*, *27*(1), 140–149. https://doi.org/10.3201/eid2701.190676

International Committee of the Red Cross (ICRC). (2017). *The Additional Protocols at 40*. ICRC. Accessed December 14, 2023. See www.icrc.org/en/document/the-additional-protocols-at-40

International Criminal Court. (1998). *Rome Statute of the International Criminal Court*. United Nations. See www.icc-cpi.int/sites/default/files/RS-Eng.pdf

Lorenz, F. M. (2003). *The Protection of Water Facilities under International Law*. UNESCO. See https://unesdoc.unesco.org/ark:/48223/pf0000132464

Russell, A. (2019). The Human Right to Water in a Transboundary Context. In *Research Handbook on International Water Law*. Edward Elgar Publishing.

Tignino, M. (2011). The Right to Water and Sanitation in Post-Conflict Peacebuilding. *Water International, 36*(2), 242–249. https://doi.org/10.1080/02508060.2011.561523

Watson, G. R. (1993). Constitutionalism, Judicial Review, and the World Court. *Harvard International Law Journal, 34*(1). See https://scholarship.law.edu/cgi/viewcontent.cgi?referer=&httpsredir=1&article=1409&context=scholar

BIBLIOGRAPHY

Betterton, C., Ndifon, G. T., Bassey, S. E., Tan, R. M., & Oyeyi, T. (1988). Schistosomiasis in Kano State, Nigeria. I. Human Infections Near Dam Sites and the Distribution and Habitat Preferences of Potential Snail Intermediate Hosts. *Annals of Tropical Medicine and Parasitology, 82*(6), 561–570.

CDC. (2018). *Parasites – Schistosomiasis.* Accessed November 28, 2023. See www.cdc.gov/parasites/schistosomiasis/index.html

CDC. (2021). *Legionella (Legionnaires' Disease and Pontiac Fever).* Accessed December 14, 2023. See www.cdc.gov/legionella/qa-media.html

Chesler, C. (2014). *Down the Drain: NJ's Sewage System.* NJ Monthly. Accessed December 14, 2023. See https://njmonthly.com/articles/jersey-living/down-the-drain-njs-sewage-system/

CISA. (2023). *IRGC-Affiliated Cyber Actors Exploit PLCs in Multiple Sectors, Including U.S. Water and Wastewater Systems Facilities.* Accessed December 14, 2023. See www.cisa.gov/news-events/cybersecurity-advisories/aa23-335a?fbclid=IwAR2Q2Sglx jAfzgGtcI7vzRi84hQP1hA1r-C0gN4fgnisYf-1eEneEb7L2DU

FEMA. (2019). *IS-1024: Water and Wastewater Treatment System Considerations.* FEMA EMI. See https://training.fema.gov/is/courseoverview.aspx?code=IS-1024&lang=en

FEMA. (2023). *Hazard Mitigation Policy Aid Drought Mitigation.* FEMA. See www.fema.gov/sites/default/files/documents/fema_hma_drought-mitigation-policy-aid_09282023.pdf

Fox, G. (2021). *Orlando Mayor Asks Residents to Start Conserving Water Immediately Due to Liquid Oxygen Shortage.* WESH. Accessed December 14, 2023. See www.wesh.com/article/orlando-mayor-to-hold-newser-on-unprecedented-event-requiring-community-assistance/37359831

Goodnough, A. (2016). Legionnaires' Outbreak in Flint Was Met With Silence. *The New York Times.* Accessed November 21, 2023. See www.nytimes.com/2016/02/23/us/legionnaires-outbreak-in-flint-was-met-with-silence.html

International Law Association. (2004). *Berlin Conference (2004) Water Resources Law.* United Nations Economic Commission for Europe. See https://unece.org/fileadmin/DAM/env/water/meetings/legal_board/2010/annexes_groundwater_paper/Annex_IV_Berlin_Rules_on_Water_Resources_ILA.pdf

Lam, A. (2023). *What Stops Rain from Flooding Your City*? YouTube Video. Accessed December 14, 2023. See www.youtube.com/watch?v=coXe8_xnAOs

Linnarz, R. (2021). *Documentary Review: Manzanar, Diverted: When Water Becomes Dust (2021) by Ann Kaneko.* Accessed December 14, 2023. See https://asianmoviepulse.com/2021/10/documentary-review-manzanar-diverted-when-water-becomes-dust-2021-by-ann-kaneko/

Malek, E. A. (1975). Effect of the Aswan High Dam on Prevalence of Schistosomiasis in Egypt. *Tropical and Geographical Medicine, 27*(4), 359–364.

Mayo, M. (1933). *Los Angeles.* A. A. Knopf.

The Ocean Cleanup. (n.d.). *The Largest Cleanup in History.* The Ocean Cleanup Technologies B.V. Accessed December 14, 2023. See https://theoceancleanup.com/

Owens Valley History. (n.d.). *Owen's Valley's – Los Angeles Aqueduct.* Accessed December 14, 2023. See www.owensvalleyhistory.com/ov_aqueduct/page20c.html

Sun, L. H. (2016). Legionnaires' Outbreaks: Cases Nearly Quadrupled in 15 Years. *The Washington Post.* Accessed November 21, 2023. See www.washingtonpost.com/news/to-your-health/wp/2016/06/07/legionnaires-outbreaks-cases-nearly-quadrupled-in-15-years/

Tignino, M., & Irmakkesen, Ö. (2020). Water in Peace Operations: The Case of Haiti. *RECIEL, 29.* https://doi.org/10.1111/reel.12333

Zemmali, A. (1998). *Dying for Water. Forum, War and Water.* ICRC. See https://casebook.icrc.org/case-study/water-and-armed-conflicts

Part 4

Quantity Hazards with Water

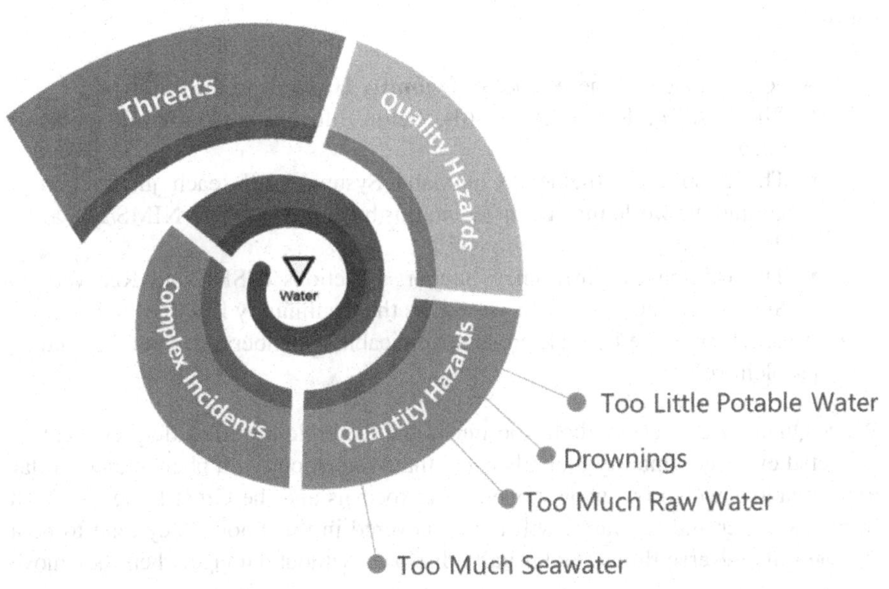

This part covers quantity hazards with water-related threats. It has four chapters:

- Too Little Potable Water (Chapter 10)
- Drownings (Chapter 11)
- Too Much Raw Water (Chapter 12)
- Too Much Seawater (Chapter 13)

DOI: 10.1201/9781003474685-13

Chapter 10 has the hazard impacts from all types of threats, including droughts and water-main breaks. Chapter 11 will have their own hazards as they have some unique Emergency Management impacts. Chapters 12 and 13 will have some overlapping information, but there are unique elements to each as well. Complex incidents which involve both excess raw water and seawater at the same time will be covered in Part 5. Regardless of the water quantity **threat**, the actions for preparedness/prevention/protection, response, recovery, and mitigation will most likely be similar, when it comes to specific quantity **hazards** with water.

Each chapter in this Quantity Hazards part – and all the other chapters in the Quality Hazards and Complex Incidents parts – will contain the following:

- An overview of the specific hazard and its adverse impacts to communities
- Case examples for historic reference (many of these will hopefully guide officials as to what ***not*** to do, as lessons learned from other's missteps and misfortunes)
- References and Additional Reading, more specific to the hazard

This overview section under Part 4 will also have a common set of the following:

- Adverse impacts of these types of **Quantity Hazards** to the following:
  - The Disaster Phase Cycles (the before, during, and after incidents occur)
  - The Response's Incident Command System itself (each jurisdiction around the world may be different, this book uses the U.S. NIMS/ICS as the baseline)
  - The U.S.-based Emergency Support Functions (ESFs) and Recovery Support Functions (RSFs), as well as the Community Lifelines (CLs)
- A summary of the POETE process, applicable to deliberative planning and problem-solving

Water Quantity concerns – both too much and too little are life safety elements – impacted every community globally. Note that Waterspouts[1] – a phenomena similar to a tornado – which occur over open-water (oceans and the Great Lakes, so both from raw water and seawater) will not be covered in this book. They tend to be a marine-only adverse threat, and usually dissipate without damage when they move over land.[2]

ADVERSE IMPACTS TO THE EMERGENCY AND RECOVERY SUPPORT FUNCTIONS AND COMMUNITY LIFELINES

Following table shows some of the adverse impacts of water quantity hazards on the U.S.-based Emergency Support Functions. There are similar tables in Parts 3 and 5 for the other hazard sections of this book.

Water-Related Threats and Hazards to the ESFs

Emergency Support Function (ESF)	Example Supporting Actions or Capabilities Related to Water-Related Threats or Hazards
#1 – Transportation	Adverse impacts to water quantity – especially flooding – generally will generate targeted hazards directly to the Transportation ESF. The agencies, departments, staff, etc. who support it will have additional responsibilities and courses of action related logistics support needed to restore Community Lifelines for the public. For example, transportation of bulk water (via tankers, on pallets carried on trucks, etc.) will impact this ESF. Also, public transportation routes (bus, paratransit, etc.) will need to be restored and maintained during the Response phase.
#2 – Communications	There should be no direct adverse impacts to this ESF (in terms of Communications Systems) from a water quantity hazard, if the infrastructure has been adequately hazard mitigated. However, the level of raw and seawater expectations may be outdated, especially when considering climate change. Also, if the potable water systems are damaged due to a water quantity threat, additional crisis communications to the public about alternative sources for potable water, water quality alerts and warnings, etc. will need to be made. Agencies, departments, staff, etc. should be on alert for false information about restoration and remediation work (i.e., predatory contractors, insurance adjustors, price gouging, etc.) and counter against them, if found. A higher volume of calls into PSAPs will occur and need to be managed and triaged. As some hazard Response/Recovery time frames may be longer-term, rerouting some call volume to 2–1–1 should be considered. Details about internal (for non-public use) responder sheltering, points-of-distribution of water, etc. will need to be communicated internally within the operation, possibly through the Safety Officer
#3 – Public Works and Engineering	**This ESF is significantly impacted by this hazard. Subject-matter experts from the water system authority, treatment plant management, etc. may be in unified command and/or acting as liaison officers to support the restoration of full capability of the water supply system, if damaged by either a low-quantity or high-quantity incident. See the cases in chapters in this part and other chapters for examples where drought threats can generate hazards for potable water treatment plants.**
#4 – Firefighting	Fire trucks cannot pass through flooded roadways. In some communities, fire departments are responsible for swift-water rescue support teams. Firefighters may be deployed to flooded roadways to help prevent vehicles from traveling through floodwaters. If the incident shuts down a water treatment plant, reducing or eliminating water pressure to the community's fire hydrant system, significant adverse impacts to that community's firefighting capability will occur.
#5 – Information and Planning	Essential Elements of Intelligence/Information about the impacts to all Community Lifelines is critical for planning the next operational period for water quantity incidents. This can include capabilities status of all of the other ESFs, such as alternative water supply sources (both fixed and mobile), water quality testing results, distribution status, restoration timelines, and more. All of this Emergency Management Intelligence is needed by ESF#5 to work through the operational planning cycle (see Chapter 2 for details).

(Continued)

Water-Related Threats and Hazards to the ESFs *(Continued)*

Emergency Support Function (ESF)	Example Supporting Actions or Capabilities Related to Water-Related Threats or Hazards
#6 – Mass Care, Emergency Assistance, Temporary Housing, and Human Services	While sheltering, feeding, and family reunification will most likely occur during an excess water quantity event such as flooding, those ESF#6 items can also occur for drought threats, too. Distribution of emergency supplies, such as cleanup kits will also occur during flooding incidents. That distribution may also include potable water if that community's water quality is impacted as well. See this table in Part 3 for more specifics.
#7 – Logistics	Flooding incidents require many of the same logistics support elements as other large-scale incidents (tornados, earthquakes, etc.). Potable water-related logistical support can include the temporary supply of bottled water, bulk potable water, etc. This may also involve distribution at PODs, door-to-door, or both – and flooded roadways may impact supply-chain integrity.
#8 – Public Health and Medical Services	As previously noted, every incident, emergency, or disaster has a public health and medical services impact. Water quantity incidents – especially excess water ones – can generate deadly mold in homes, sewage backups, and other public health hazards. Hospitals may become overwhelmed with injuries associated with people traveling through floodwaters, medical patients rerouted from clinics and services in flooded areas, and backup other hospitals which are in the impacted geography. Drowning incidents – especially those with multiple casualties and/or high media attention – may adversely impact hospitals and emergency rooms with patient surge.
#9 – Search and Rescue	For excess water incidents (i.e., flooding), this ESF will be activated. FEMA has specific resource typing associated with swift water/flood search and rescue teams.[3] And this is a resource which a U.S. community can request help from a higher-level government within their state, territory, or tribal nation. And those state or territorial governors can request teams from other jurisdictions – even in advance of a flooding incident – to fill gaps in local capacity. See the EMAC section in Chapter 2 for details.
#10 – Oil and Hazardous Materials Response	As noted in the case studies in the chapters in this Quantity Part (also in the Quality Part), there can be incidents where oil and hazardous materials make their way into the streams, rivers, watershed, etc.; and therefore, adversely impact downstream raw water and water sources. The adverse health impacts from those human-made hazards – as well as heavy metals, elements occurring naturally, and microorganisms/parasites – may require missions for cleanup and remediation from this ESF. The high priority of this cleanup and remediation of the oil and/or hazardous materials is both a life safety and an incident stabilization course of action.
#11 – Agriculture and Natural Resources	Crop irrigation, livestock watering, parks and recreation areas all can be adversely impacted by droughts, for example. Also, flooding will impact parks, cultural resources, etc. and may include adverse impacts to boating, fishing, swimming, etc.

Water-Related Threats and Hazards to the ESFs *(Continued)*

Emergency Support Function (ESF)	Example Supporting Actions or Capabilities Related to Water-Related Threats or Hazards
#12 – Energy	While excess quantity may not be a concern – if protective elements of the infrastructure are sufficient, if the water supply is reduced or cut off to a nuclear power plant, that can have significant impact on the Energy ESF. The same is true for hydroelectric power generation.
#13 – Public Safety and Security	In addition to new incident-related Response missions for law enforcement officials, there may be interim Recovery missions as well, such as curfews, no entry zones, etc. There may be investigations related to fatalities attributed to the incident. Public Safety and Security may also be needed to support ESF#6 and ESF#7 in mass care support missions. And of course, Public Safety is never confined to only one incident at a time, so staffing will need to be extended or expanded to cover multiple missions simultaneously. Staff wellness and health is important as well – since some may be impacted by the incident directly themselves.
#14 – Cross-Sector Business and Infrastructure	There can be homeland security implications to water quantity hazards. For example, if a refinery is flooded, it could impact fuel production and other economic aspects. Part 5 covers the complex threats of when a water quality and/or quantity hazard impacts a water or wastewater treatment plant.
#15 – External Affairs	Coordinate with ESF#2 on both internal and external messaging regarding water quantity status, restoration of systems status, evacuation shelters, alternative sources of potable water, etc. Note that there may be annual memorial services or markers for fatalities, and thank-you ceremonies, when survivors of drowning incidents exist. The PIO needs to activate templated messages to the public which provide the LIPER actions they need to take, in the languages and access to crisis communication that the impacted community needs. Emergency Management's public crisis communications messages for this hazard of too much water, can be summarized as "run from the water, hide from the wind".[4]

Source: Barton Dunant.

Recovery Support Functions

Similarly, following table shows some of the adverse impacts of water quantity hazards on the U.S.-based Recovery Support Functions (RSF). There are similar tables in Parts 3 and 5 for the other hazard sections of this book.

Recovery Support Functions

Federal Recovery Support Functions (RSFs)

Economic

Most water quantity incidents will have an economic impact across a community: From droughts impacting agriculture to floods impacting everyone. Federal mitigation funding, through their Hazard Mitigation Assistance program – which is a competitive grant program – including early warning systems and stabilization, may be of financial assistance to local communities, for droughts.[5]

Recovery Support Functions *(Continued)*

Federal Recovery Support Functions (RSFs)

Health and Social Services

Flooding can have significant adverse impacts to the health of individuals and families, due to hazards associated with contaminated water, mold found in homes after floodwaters recede, etc. Individuals with compromised immune systems, pulmonary issues, and other medical concerns can be significantly harmed after a flooding incident. Social Services programs.

Community Assistance

One aspect that FEMA has for Community Assistance in Recovery[6] is to help achieve **equitable** recovery, as incident survivors work toward rebuilding their community. Incorporating community partners, stakeholders, faith-based organizations, etc. is one way to achieve this.

Infrastructure Systems

As Infrastructure Systems can be damaged through either too little or too much water (i.e., drought conditions or flooding), the recovery aspects of restoration of the critical infrastructure/key resources.[7] This will include both public and private sector organizations.

Housing

If Housing is adversely impacted – for example, due to flooding – recovery efforts will need to transition from the ESF#6 – Mass Care/Sheltering through the Recovery phase into temporary housing, interim housing, and possibly permanent housing replacement solutions. In many communities, there is an affordable housing shortage now – before any disasters occur of any kind. The U.S. Department of Housing and Urban Development (HUD)[8] – upon a request from the state or territorial governor – may provide Community Development Block Grant Disaster Recovery (CDBG-DR) funds to help cities, counties, and states recover from disasters, through housing vouchers for Section 8 approved housing units.

Natural and Cultural Resources

The above-noted economic benefits from the Hazard Mitigation Assistance program can be applied toward nature-based solutions, floodplain and stream restoration, flood diversion and storage, and aquifer recharge, storage, and recovery. Smart land use practices, including best practices for floodplain management, should be tools in an all-hazards whole cycle emergency manager's toolbox.[9]

Mitigation work – which starts from projects identified through the RSFs – should also be considered by emergency managers, for children-related sites such as daycare, camps, etc., in addition to what has been traditionally focused on government-run K12 schools. The remediation, repair, and future hazard protection work for some NGOs (such as a magnet school or a VOAD's site) can count toward Public Assistance[10] projects in the United States.

Source: FEMA.

Generally, adverse impacts from water quantity threats and hazards target all of the RSFs, but there can be focused attention by Emergency Management for the adverse impacts to Infrastructure Systems, including potable water systems, such as water and wastewater treatment plants. The lack of potable water can also impact all the RSFs.

Community Lifelines

As noted in Chapter 2, the Community Lifelines (CLs) are a highly useful list of EEIs, which can be visualized as a dashboard. Water Quantity hazards will have implications for every community lifeline. If roads and power stations and other critical infrastructure are flooded, almost every CL will become quickly adversely impacted. There are life safety impacts to each CL:

Safety and Security (Law Enforcement/Security, Fire Service, Search and Rescue, Government Service, Community Safety): Too much water, from raw or ocean water flooding, human-made water main breaks, etc. can block roads, generate additional security and safety missions for law enforcement, and all the other sub-elements within this CL, such as Search and Rescue.

Food, Hydration, Shelter (Food, Hydration, Shelter, Agriculture): Obvious impacts to this Community Lifeline include evacuation missions (including sheltering and feeding) associated with too much water, but threats such as droughts can also create hazards such as too little potable water (see Part 3 for those elements). People may evacuate their homes if the quantity of potable water for drinking, bathing, toileting, etc. at their homes is adversely impacted. For example, after a flooding incident subsides, if that community's water treatment plant was damaged, residents may not return to their homes – and choose to stay at shelters, for feeding and hydration assistance.

Health and Medical (Medical Care, Public Health, Patient Movement, Medical Supply Chain, Fatality Management): Aligns to both Emergency Support Function ESF#8) Public Health missions and the Health and Social Services RSF missions. As noted in the ESFs and RSFs, this Community Lifeline is significantly adversely impacted by any disruption to either site access (roads blocked due to flooding), critical infrastructure damage to health or medical facilities, or damage to their potable water supply systems, which may happen with water quantity hazards.

Energy (Power Grid, Fuel): Water and Energy have many nexus points, all of which need emergency management intelligence. If a hydroelectric dam were to breach, not only would it flood the land below it, but also the electrical generation capabilities of the plant would be severely impacted.

Communications (Infrastructure, Responder Communications, Alerts Warnings and Messages, Finance, 911 and Dispatch): One aspect of crisis communications for Public Information Officers, related to water quantity threats is to pre-plan (and exercise) both the notice and no-notice aspects of incidents, utilizing templated public messaging in the languages and delivery methods which have the greatest outreach capabilities for that community. Drought hazards can be forecasted, and crisis communications messages delivered before, during, and after the lack of potable water threat occurs. Flooding may be "flash flooding" and have little notice to first responders and the impacted jurisdiction, but their adverse impacts – even down to specific streets and buildings – should be known in advance. Chapter 2 has several online tools that can help predict areas which can flood, where sea level rise will occur, etc.

Transportation (Highway/Roadway/Motor Vehicle, Mass Transit, Railway, Aviation, Maritime): Water quantity issues can have an adverse impact on this CL, and the negative status of this CL can impact the others as it relates to providing any CL support to a community. If water quantity becomes a hazard in a community, the full capabilities of the Transportation CL will be needed for emergency services Response (police, fire, EMS, etc.), electrical and other utility restoration, and potentially transporting potable water to Points of Distribution, and in bulk transport to critical infrastructure/key resource sites, such as hospitals and dialysis centers. If navigable waterways are blocked or hampered by a water quantity hazard, this CL will be impacted as well. Critical repair parts, filters, staff, etc. need to be transported to water and wastewater treatment sites, to make repairs and work toward restoration of the **Water Systems** CL, if they are adversely impacted by a water quantity threat, such as flooding.

Hazardous Material (Facilities, HAZMAT, Pollutants, Contaminants): Impacts to this Community Lifeline can be for Pollutants and/or Contaminants and the HAZMAT response to the incident, if those commercial and industrial sites are adversely impacted by too much water (flooding). This could be from natural or human-made hazards. There may be longer-term impacts to facilities, if a contaminated waterway supports those Critical Infrastructure/Key Resources (CIKR), from a needed clean water supply basis. Considerations on how natural freshwater supplies are utilized in dialysis centers and even nuclear power plants, for example.

Water Systems (Potable Water Infrastructure, Wastewater Management):[11] Understanding how drought and flooding conditions both can impact a potable water supply system (and the wastewater treatment, as well) is critical for Emergency Management.

There may also be longer-term impacts to the food supply if agriculturally used land is flooded, or experiences severe drought conditions for long periods of time.

MITIGATION PARADOXES FOR COMMUNITIES

Many communities are constantly faced with a series of paradoxes – amplified by climate change, in many cases – where they must make long-term urban and master planning decisions, driven by disasters:

- Droughts can occur slowly and have long-lasting adverse impacts to communities anywhere.[12]
- Recurrent river and stream flooding is expensive to correct.
- Yearly coastal flooding in certain areas require response and cleanup work – which many times is not covered by insurance,[13] state, or federal funding.

- "Blue Acres" programs save recovery costs, but take parcels off the tax rolls.
- Coastal Inundation requires individuals to pay more in flood insurance – or become self-insured.
- The highest valued properties are also the riskiest, in terms of excess water from the ocean.
- Long-time residents quickly become outpriced for flood insurance, and then sell properties to developers who build multi-million-dollar homes, etc.

There can be significant cost-savings to communities when proper Mitigation work is completed. Estimates between $6 and $8 (€5–7) saved in Response and Recovery costs per dollar (Euro) spent on Mitigation work for excess water threats and hazards are well established.[14]

National Institute of BUILDING SCIENCES

	ADOPT CODE	ABOVE CODE	BUILDING RETROFIT	LIFELINE RETROFIT	FEDERAL GRANTS
Overall Benefit-Cost Ratio	11:1	4:1	4:1	4:1	6:1
Cost ($ billion)	$1/year	$4/year	$520	$0.6	$27
Benefit ($ billion)	$13/year	$16/year	$2200	$2.5	$160
Riverine Flood	6:1	5:1	6:1	8:1	7:1
Hurricane Surge	not applicable	7:1	not applicable	not applicable	not applicable
Wind	10:1	5:1	6:1	7:1	5:1
Earthquake	12:1	4:1	13:1	3:1	3:1
Wildland-Urban Interface Fire	not applicable	4:1	2:1	not applicable	3:1

U.S. Nationwide average cost-benefit ratio by hazard and mitigation measure. BCRs can vary geographically and can be much higher in some places.

Source: National Institute of Building Sciences. (2020). *Mitigation Saves: Mitigation Saves Up to $13 Per $1 Invested.* NIBS. See www.nibs.org/files/pdfs/ms_v4_overview.pdf. Used with permission.

Also, as to Mitigation, efforts for Whole Community work should be collaborative and cooperative with the work that NGOs are doing, such as the American Red Cross. In 2023, the Red Cross started an initiative they call their "Community Adaptation Program" in the United States. They are partnering with local nonprofits to help build more disaster resilient communities in areas prone to flooding and other natural hazards, especially in communities which are already facing social inequities. Their goal is long-term coordination, and collaboration with those local partners to provide services such as access to healthcare, nutritious food, and housing for underserved local families. At the same time, this will build the capacity for those partners to work with the Red Cross during times of disaster to help those same families recover and mitigate disaster-caused poverty. In 2023, their Community

Adaptation Program was introduced in ten targeted areas across eight states. In FY 2024, they plan to expand the program to an additional five communities.

The American Red Cross also

> helped its global Red Cross and Red Crescent network partners in 10 countries develop strategies to enable their climate adaptation investments. The Red Cross also supported Red Cross partners in Indonesia, Bangladesh, Honduras, Nepal and Tanzania and their local governments in adjusting to coastal risks and/or extreme heat in 14 cities and towns. Through the Red Cross Global Disaster Preparedness Center, the Red Cross funded 15 research teams across 13 countries to fill critical gaps in extreme heat research. The Red Cross is dedicated to scaling investments in pre-disaster anticipatory action. In FY 2023, we held a workshop for nine global Red Cross and Red Crescent network partners in the Pacific region to learn about anticipatory action approaches and begin developing early action protocols.
>
> **(American Red Cross, 2023, p. 17)**[15]

THE AMERICAN RED CROSS

Adapting Our Response Services to Address the Increasing Impacts of Extreme Weather on Vulnerable Communities

In FY 2023, to address increasing disaster caused humanitarian needs and help bridge the gap between immediate disaster relief and long-term recovery, the Red Cross launched casework follow-up pilots for our highest-need clients from Hurricane Ida and the severe California floods. Red Cross caseworkers work with families for up to one year after their stay at a Red Cross shelter to address any gaps in their recovery. Assistance is focused on improving access to healthcare, nutritious food and housing. The Red Cross is also improving the shelter experience for our clients by providing more nutritious snacks and meals, offering additional personal assistance care, and partnering with animal welfare organizations to ensure that household pets are welcome and cared for in our shelters. The Red Cross is committed to building climate leadership capacity by providing opportunities for youth to lead local climate action.

In FY 2023, we partnered with the Solomon Islands Red Cross and its partners to develop and test a youth climate leadership curriculum intended for global scaling. It includes peer coaching and communication strategies. These skills empower young people as local leaders to raise awareness and help alleviate climate-related risks for their communities. Working with schools, universities, local organizations, and other partners, the Red Cross will facilitate these types of actions in our priority countries to amplify youth voices and empower youth as climate champions (American Red Cross, 2023, p. 18).[16]

NOTES

1 See www.weather.gov/apx/waterspout
2 See www.weather.gov/mlb/waterspout_threat
3 See www.fema.gov/sites/default/files/2020-08/fema_swiftwater-flood-sar-team_definition_08-03-2020.pdf
4 See https://oceantoday.noaa.gov/hurricanestormsurge/
5 See www.fema.gov/sites/default/files/documents/fema_hma_drought-mitigation-policy-aid_09282023.pdf
6 See www.fema.gov/sites/default/files/documents/fema_equitable-recovery-post-disaster-guide-local-officials-leaders.pdf
7 See www.fema.gov/pdf/emergency/nrf/nrf-support-cikr.pdf
8 See https://www.hud.gov/disaster_resources
9 See https://planning.org/blog/9283634/mitigating-flood-risk-through-smart-land-use-practice/
10 See www.fema.gov/assistance/public
11 See www.fema.gov/sites/default/files/documents/fema_p-2181-fact-sheet-4-1-drinking-water-systems.pdf
12 See www.fema.gov/sites/default/files/documents/fema_drought-planning-fact-sheet_092021.pdf
13 See https://www.bloomberg.com/news/articles/2024-04-30/uk-s-nationwide-pulls-mortgage-offers-to-homes-at-flood-risk?embedded-checkout=true
14 Ganderton, P. T. (2005). 'Benefit – Cost Analysis' of Disaster Mitigation: Application as a Policy and Decision-Making Tool. *Mitigation and Adaptation Strategies for Global Change*, *10*, 445–465. https://doi.org/10.1007/s11027-005-0055-6
15 American Red Cross. (2023). *Environmental, Social and Governance Report 2023*. See www.redcross.org/content/dam/redcross/about-us/publications/2023-publications/2023_ESG_Report_FINAL.pdf
16 American Red Cross. (2023). *Environmental, Social and Governance Report 2023*. See www.redcross.org/content/dam/redcross/about-us/publications/2023-publications/2023_ESG_Report_FINAL.pdf

BIBLIOGRAPHY

American Red Cross. (2023). *Environmental, Social and Governance Report 2023*. See www.redcross.org/content/dam/redcross/about-us/publications/2023-publications/2023_ESG_Report_FINAL.pdf

Ganderton, P. T. (2005). 'Benefit – Cost Analysis' of Disaster Mitigation: Application as a Policy and Decision-Making Tool. *Mitigation and Adaptation Strategies for Global Change*, *10*, 445–465. https://doi.org/10.1007/s11027-005-0055-6

National Institute of Building Sciences. (2020). *Mitigation Saves: Mitigation Saves Up to $13 Per $1 Invested*. NIBS. See www.nibs.org/files/pdfs/ms_v4_overview.pdf

10 Too Little Potable Water

> Drought is one of the most complex hazards. Its drivers stem from many factors, with impacts that are wide and cascading, from worsening food security to triggering displacement.
>
> If we are successful in achieving universal early warning coverage, then all countries will have a valuable tool for the early detection and mitigation of drought.
>
> **UNDRR Head Mami Mizutori, speaking at COP28[1]**

10.1 OVERVIEW

This chapter covers how there can be water quantity hazards from incidents which generate a reduction or break in the potable water supply for any community or communities. This chapter will include droughts, low water pressure, and health issues associated with lacking access to enough potable water. Adverse impacts to human and agricultural potable water supplies, as well as navigable freshwater waterways, will most likely require Emergency Management courses of action, across the full disaster phase cycle.

As with other hazards, the question of causality is immaterial to Emergency Management's Response and Recovery operations, except that there may be law enforcement intelligence and investigation elements imbedded within the Incident Command System (ICS) established, specifically, for responding to and recovery from the adverse impacts of the incident. If this is the case, then actions of the other Command and General Staff branches/sections may need to be adjusted and/or expanded to support law enforcement intelligence and investigation elements. This could be investigators for environmental protection laws, for example. There may be a parallel incident command system established by law enforcement officials and there may or may not be a liaison officer (LNO) linking the two – or more – ICSs. Each incident, emergency, disaster, etc. is different. Emergency managers should be involved in the full disaster cycle planning, organization, equipping, training, and exercising for freshwater, potable water, drinking water supply systems, and/or navigable waterways in their respective communities (including those which provide support or services to their jurisdiction's water system from another jurisdiction – such as a neighboring community's shared water treatment plant, watershed areas, dams, etc.).

Besides local jurisdictional emergency management groups, this hazard will need to be managed by water management districts, hospitals and other healthcare facilities, colleges/universities, etc. in a coordinated and collaborative way. All these groups and organizations will have unmet needs which they will consider as a priority – and some of those priorities may conflict and compete with each other for

DOI: 10.1201/9781003474685-14

Response and Recovery resource allocation. Unified Command should always prioritize the LIPER (see Chapter 2) when making decisions.

AMERICAN RED CROSS LOWERS ITS WATER CONSUMPTION IN THE UNITED STATES

The Red Cross is committed to reducing their water consumption by at least 20% in the next four years. To achieve this goal, they are focusing on three types of investments:

- Low-flow fixtures and aerators
- Leak detection and rain sensors
- Landscaping

In January 2023, the Red Cross partnered with Waste Reduction Partners to analyze the water usage in one of their larger facilities. Waste Reduction Partners is a unique technical assistance program that engages the expertise of retired and volunteer engineers to serve businesses and institutions in North Carolina. Using their findings, the Red Cross identified several improvements to make across all their facilities, specifically focusing on upgrades to older fixtures in their restrooms and kitchen areas. In addition to that effort, the Red Cross has investigated water usage at their top ten sites by water intensity (gallons per square foot/liters per square meter). At five of those sites, they identified and implemented interventions and are actively monitoring those sites to measure the reductions.

(American Red Cross, 2023, p. 11)[2]

10.1.1 Droughts

Emergency managers worldwide need to consider the insidiousness of this hazard. It is a "Grey Rhino"[3] hazard: a life-threatening problem we see coming at us, and yet we do not get out of its way. While in the United States, droughts may be considered only part of the agricultural sector's purview, worldwide drought conditions are an ongoing crisis of extinction-level proportions. Droughts can adversely impact a nation's water security.[4]

> Freshwater shortages, once considered a local issue, are increasingly a global risk. In every annual risk report since 2012, the World Economic Forum[5] has included water crisis as one of the top-five risks to the global economy. Half of the global population – almost 4 billion people – live in areas with severe water scarcity for at least one month of the year, while half a billion people face severe water scarcity all year round.
>
> **(Smedley, 2023, p. 1)**[6]

Globally, the United Nations Office for Disaster Risk Reduction[7] monitors and reports on droughts and drought conditions. And drought conditions in one country can exacerbate climate change adverse impacts elsewhere, which in turn add to more drought conditions everywhere. For example, drought conditions in the Amazon rainforest are lowering river levels – and adversely impacting local economies – as well as killing off vegetation and parts of the canopies which absorb huge amounts of worldwide carbon production. This deforestation is growing exponentially and will have impacts in other parts of the world, as well.[8] Emergency managers everywhere need to see and act upon threats globally today, even though the hazards in their jurisdictions may not appear until tomorrow.

In the United States, the National Oceanic and Atmospheric Administration (NOAA) has the National Integrated Drought Information System (NIDIS) and other emergency management intelligence found at www.drought.gov.

This can include:

- **Flash Droughts**
  - Droughts are usually slow-developing and long- lasting climate-driven hazards, whose onsets are difficult to detect. On the contrary, flash droughts rapidly evolve, often with strong impacts. Flash droughts are generally driven by precipitation deficits, extremely high temperatures and a rapid increase in evaporative demand.
- **Megadroughts**
  - Megadroughts are defined as multi-decadal events, referring to long and abnormally dry periods, more severe than multi-year droughts registered since the 1880s with the onset of regular meteorological measurement.
- **Cold Region Droughts**
  - Different processes play a role in the development of droughts in cold climates compared to droughts in warmer climates. For example, temperature is a highly significant variable because it determines whether precipitation falls as rain or snow and whether water is available for use or locked up in frozen form. In regions with seasonal snow cover, the amount of snow accumulation is crucial. A below- normal snow accumulation (or snow drought) depresses the tourism sector, constrains downstream water use and weakens ecosystems dependent on snow-melt.

(UNDRR, 2021, pp. 35–36)[9]

As noted above, droughts can have economic impacts to any jurisdiction. Drought conditions can cause river levels, including navigable rivers, to lower significantly – the point where they can become unnavigable, as well as expose additional hazards (silt with hazardous elements, sandbars, rocks, trees, etc.) for swimmers, boaters, and others.

Droughts can occur at the same time as flooding.[10] Areas which are normally drought-stricken – prone to limited normal water absorption, such as deserts, playas, and salt flats – can become hazardous when heavy, sudden rain impacts them.

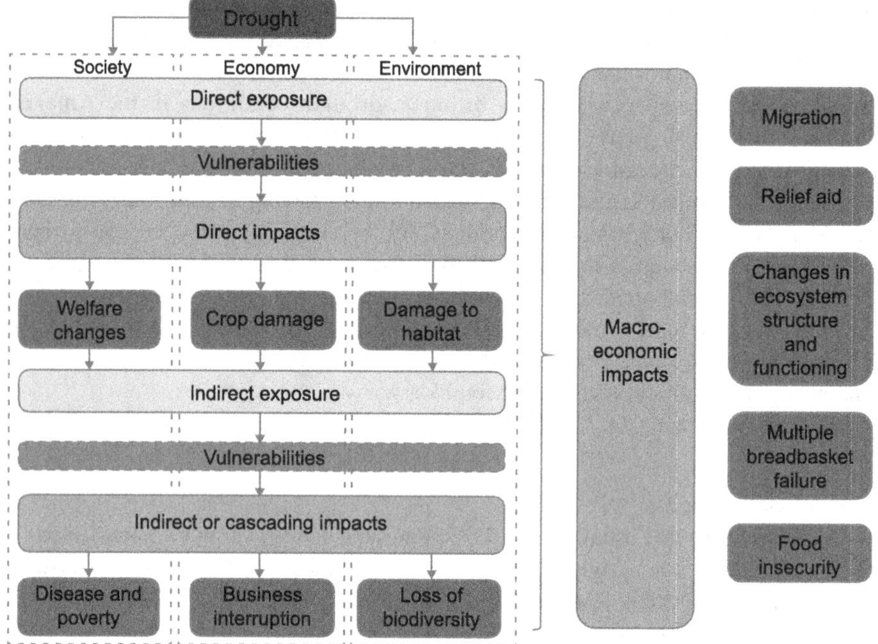

FIGURE 10.1 Schematic representation of direct and indirect drought impacts and their interrelations.

Source: UNDRR – CC BY-NC IGO.

Figure 10.1 shows various direct and indirect impacts of droughts to communities, which cross Emergency Support Functions, Recovery Support Functions, and Community Lifelines.

Overconsumption from growing communities – in addition to potential climate change impacts – can significantly reduce lake levels.[11] Lakes which are dammed for hydroelectric production can be adversely affected by both this overconsumption, as well as drought conditions and heatwaves. All together or separately, those threats produce hazards which impact communities.

10.1.2 Low Water Pressure in Water Supply Systems

The weight of water, plus height and gravity through pipes from its source to its destination, is how water pressure is generated.[12] Getting potable water to high-rise buildings and other locations where the calculation does not quite work (uphill, rural irrigation systems, etc.) requires pumps and additional equipment to maintain water pressure throughout the system. The concerns for Emergency Management when there is low water pressure include not enough pressure for fire suppression (for sprinklers, hydrants, etc.), a building's steam generation equipment, nuclear power plant cooling, dialysis center operations, and more.

10.1.3 Health Issues Associated with Lacking Access to Potable Water

The global concern for universal access to clean water is primarily associated with public health. Water for sanitation and hygiene (WASH)[13] is a critical factor for children's health, welfare, and mortality.

The United Nations Global Assessment Report on Disaster Risk Reduction (GAR) Special Report on Drought 2021 notes there are broad categories of the health impacts, from droughts:

- Malnutrition (including micronutrient malnutrition and antinutrient consumption)
- Waterborne diseases (including algal bloom, cholera, and *Escherichia coli*)
- Vector-borne diseases (including dengue, malaria, and West Nile virus)
- Airborne diseases (including coccidioidomycosis, Covid-19, and silo gas exposure under reduced water availability for sanitation)
- Mental health, including distress (UNDRR, 2021, p. 52).[1]

Worldwide, the majority (~58%) of the deaths attributed to extreme weather incidents from 1900 to 2008 were caused by droughts (UNDRR, 2021).[14]

10.2 CASE EXAMPLES

Below, please find some case examples from recent history, which exhibit both best practices for disaster readiness and how there can be significant adverse impacts to the LIPER (life safety, incident stabilization, property/asset protection, environmental/economic impacts, and recovery operations) from too little potable water hazards from other incidents.

10.2.1 Mississippi River Drought Conditions in 2022

> In 2022, abnormal drought conditions combined with record heat waves caused the Mississippi River's water levels to recede, nearly approaching the record low set in 1988.[15] The river receded to levels low enough for barges to get stuck in mud and sandbars, jeopardizing millions of tons of critical supplies that are transported via water from states in the upper and lower Mississippi River watershed. Ultimately, the U.S. Army Corps of Engineers had to conduct emergency dredging at points where as many as 2,000 barges plateaued.
>
> **(FEMA, 2023b, p. 24)**[16]

10.2.2 Desert Flooding at the "Burning Man" Music Festival in 2023

The annual "Burning Man" outdoor festival each year in the Black Rock Desert of Nevada is a logistical challenge and an Emergency Management outlier (70,000+ people assemble for a nine-day concert with no infrastructure in place). In 2023, heavy rain from Hurricane Hilary[17] on the West Coast dumped rain and created flooding and impassible mud. There was only one road into and out of the area where the festival was held, which quickly became impassible. This delayed logistical support (fuel for generators, equipment to cleanout port-a-potties, etc.), even as organizers kept the party going as best they could.[18]

There was one death that year at the festival, but it was attributed to a drug overdose[19] and not considered storm related. The same Emergency Management concerns for bringing in logistics, would have applied for evacuating a medical patient. The concern that sudden weather – such as heavy rain, hail,[20] etc. at outdoor events with large populations – and not nearby emergency medical and support services – can be exacerbated by dry/drought conditions of the ground beforehand.

10.2.3 Quantitative Research on Reduced Water Consumption Correlation with Health Concerns

In 2015, Stelmach and Clasen performed a literature view analysis of prior water consumption studies to validate that there were health risks associated with reduced water consumption. Their

> results showed a positive association between water quantity and health outcomes, but the effect depended on how the water was used. Increased water usage for personal hygiene was generally associated with improved trachoma[21] outcomes, while increased water consumption was generally associated with reduced gastrointestinal infection and diarrheal disease and improved growth outcomes. In high-income countries, increased water consumption was associated with higher rates of renal cell carcinoma and bladder cancer but not associated with type II diabetes, cardiac-related mortality, or all-cause mortality.
>
> **(Stelmach and Clasen, 2015, p. 5954)**[22]

The study which also showed that an *increase* in water consumption generated higher rates of renal cell carcinoma was related to a study done in Canada – and the question of whether water *quality* was a contributing factor should be considered. Either way, the premise that both water quality and water quantity issues can adversely affect public health can be validated.

10.2.4 2023 El Nino Flooding in Nairobi, Kenya

In November 2023, the severe rain which occurred this year – and followed a decades-high drought season – left 90,000 families homeless and killed 120 people. The flooding in neighboring Somalia from this same weather pattern displaced 700,000 people and killed more than 95.

> Climate change is causing more intense and more frequent extreme weather events, according to scientists. In response, African leaders have proposed new global taxes and changes to international financial institutions to help fund climate change action.
>
> **(Holland, 2023, p. 1)**[23]

10.2.5 Restoring Utrecht's Catharijnesingel Canal, in the Netherlands

Utrecht's canals when built were major focuses of commerce and transportation. Over time – and reduction in the amount of water flowing through them – their usefulness subsided. That community's canal was covered with a roadway in 1969, and now restored to its original state for recreation and ecological purposes. This 6 km (3.7 mile) long stretch of the Catharijnesingel canal adds biodiversity, more public space, and enhances the city's historical heritage.[24]

10.2.6 Mexico City's Lack of Potable Water

In 2023, after years of low rainfall and centuries of poor urban planning (dating back to the 16th century), Mexico City may be on the verge of depleting all its potable water supply.

> Wetlands and rivers have been replaced with concrete and asphalt. In the rainy season, it floods. In the dry season, it's parched.
>
> Around 60% of Mexico City's water comes from its underground aquifer, but this has been so over-extracted that the city is sinking at a frightening rate – around 20 inches a year, according to recent research. And the aquifer is not being replenished anywhere near fast enough. The rainwater rolls off the city's hard, impermeable surfaces, rather than sinking into the ground.
>
> The rest of the city's water is pumped vast distances uphill from sources outside the city, in an incredibly inefficient process, during which around 40% of the water is lost through leaks.
>
> The Cutzamala water system, a network of reservoirs, pumping stations, canals and tunnels, supplies about 25% of the water used by the Valley of Mexico, which includes Mexico City. But severe drought has taken its toll. Currently, at around 39% of capacity, it's been languishing at a historic low.
>
> **(Paddison et al., 2024, p. 1)**[25]

10.3 IMPACTS TO THE DISASTER PHASE CYCLES

As shown in Table 10.1, there are specific impacts to all four of the disaster cycle phases, from **Too Little Potable Water Quantity Hazards**.

10.4 ADVERSE IMPACTS TO THE INCIDENT COMMAND SYSTEM

There may be unique hazards directly impacting Responses and Responders, which should be considered adverse impacts to the Incident Command System (ICS). For **Too Little Potable Water Quantity Hazards**, Table 10.2 details what may be occurring.

10.5 ADVERSE IMPACTS OF THIS HAZARD TO THE ESFS, RSFS, AND CLS

10.5.1 Emergency Support Functions (ESFs)

The full list of possible ESF impacts to any threat can be found in Chapter 2. Table 10.3 shows the specifics for **Too Little Potable Water Quantity Hazards**.

TABLE 10.1
Impacts to the Disaster Phase Cycles from Water Supply Quality Hazards from Other Incidents

Preparedness/Protection/Prevention
There are steps and courses of action that emergency managers can take to protect water supply systems (dikes, levees, canals, aqueducts, etc.), including public messaging about watershed protection, alerting officials when threats are noticed, etc. Communities – especially geographic areas which have high socially vulnerable individuals (SVIs) and medically fragile populations – need preparedness messaging and training to respond themselves to the hazards from too little potable water. This may include having extra potable water on-hand as part of their sheltering-in-place kits and pre-identifying alternative locations to obtain potable water. Diverting to alternative dialysis centers for treatment, if necessary and other personal health issues should be part of the reminders to the public that they have the ultimate responsibility for their own life safety first. Proper use of filters, based on historic and geographic elements, **plus *additional* filtration** for low water incidents, such as droughts, is something that emergency managers should be aware of and monitor for their communities.
Response
Emergency Management's response courses of action for low water pressure and other hazards with too little potable water are like those for water quality hazards noted in Part 3. There are also significant impacts to other response actions (such as firefighting) still needed in the community, which need to be managed through the local emergency management agency. There may be more notice – more time – to advise the public and plan for alternatives, as drought conditions (generally the initial causal factor) or systemic/chronic conditions (such as naturally occurring contaminants, which can adversely impact potable water systems) are part of the equation for the water system contamination. Do not rely on the water systems authority to provide all the resources and response capability – in fact, their representative should be a liaison officer (or possibly part of unified command) to your jurisdiction's incident command system, not the other way around. An example of this is having independent testing – including pressure testing – to be conducted through higher governmental levels. This is also for water contaminants coming into water treatment systems **and** the water coming out of those systems as well.
Recovery
There may be cleanup and remediation work, possible infrastructure repair work, financial reimbursement to government, possibly by higher level government and other entities, continued health monitoring for the public, continued crisis communications, cooperation with investigators, including state/tribal/territorial and even federal level officials, and more. And most likely, in this longer-term time frame for recovery, another incident of scale will occur and consume the time and staffing of emergency managers. Self-care and wellness are critical for not only first responders, but everyone involved in the disaster phase cycle – which may be longer than other threats and hazards.
Mitigation
If there was a Declared Emergency or Disaster for this hazard, in this same jurisdiction or any jurisdiction which impacted the water supply system from this hazard, it is possible that mitigation work[26] to help prevent and protect against the adverse impacts of that hazard can be obtained. This may also help prevent and protect against the adverse impacts of any future hazards to those same systems.

Source: Author.

TABLE 10.2
Adverse Impacts to the Incident Command System

Command (including SO, PIO, and LNOs)

Safety Officers (SOs) need emergency management intelligence about this threat, as soon as possible. As the pressure in the water supply may become diminished, potable water will be needed for firefighting. If that were to happen, while a fire was being responded to, the loss in pressure could harm firefighters themselves. Plus, prolonged loss of water from the primary supply will require backup water sourced for responders, emergency operations center staff, etc. This may extend beyond just drinking water into handwashing, bathing, etc. These workforce safety elements are paramount. Extra attention – and advanced notice to deploying response teams – of this **hardship** should be made.

Public Information Officers (PIOs) need to urgently activate accurate templated messages to the public which provide critical crisis communications in the appropriate languages and modalities for oversaturation. If an "avoid use" warning is needed (when a low-pressure situation is resolved), it may kick up sediment in the pipes and generate a boil water alert, for example, it needs to be communicated far and wide, and quickly. See the Shelter-in-Place section in Chapter 2 and also References/Additional Reading links for more specifics. Also recall that this hazard will also impact agricultural, healthcare, and other economic uses of the water.

Liaison Officers (LNOs) are needed from both the water supply companies/entities and potentially any organizations aligned to human-made actions which may have exacerbated the low/no water situation – or are involved in the remediation and recovery aspects, for example, the U.S. Army Corps of Engineers (USACE). These LNOs will need to be prepared to financially commit their staffing and resources to the Incident/Unified Command for their support quickly, in order to effect solutions on a timely basis. In the United States, there may also be LNOs from federal agencies (such as the U.S. Environmental Protection Agency), who need to represent the full capabilities and capacities of their organizations, regardless of the status of a Presidentially Declared Disaster or Emergency. If a disaster or emergency is Presidentially Declared, then a unified Federal Coordinating Officer is assigned to support the State (or Tribal) Coordinating Officer.

Intelligence

Sources of Emergency Management Intelligence for incidents where hazards are from natural water supply contamination will need to include more than just water systems officials. Additional subject-matter expertise of public health, geologists, ecologists, and other experts from this jurisdiction, higher levels of government and possibly derived from mutual-aid compacts (i.e., EMAC), may be needed.

Finance/Administration

- Budgets for alternative water storage, delivery, and distribution may not be adequate. The same may be true for additional filtration processes at water treatment plans, cost-share match portions of USACE work, etc. Emergency declarations at the local jurisdiction may not be sufficient to support emergency contracts, etc. Overtime and other staffing costs may exceed what is covered in normal fiscal budgets and/or emergency assistance reimbursements from higher governmental entities.
- If it is determined that a private entity is financially responsible in any part, for the reduction/halting of potable water through the normal water supply system, the costs for remediation and cleanup may be borne in part or whole, by that entity. Governmental Finance/Administration officials will need to manage costs and expenses in such a way as to obtain reimbursement on a timely basis. Private entities may need to utilize insurance coverage for business interruption coverage. Additional elements of penalties and fines are not normally the responsibility of emergency management, but the organizations who are support partners to a jurisdiction's emergency management organization (such as the governmental environmental protection agency) may have dual responsibilities of both supporting the response/recovery work and issuing penalties and/or fines to the entity responsible for the adverse impacts caused by them. Emergency Management staff at those agencies must clearly divest themselves from any direct missions involved in the investigation or prosecution of malfeasance or criminal activity.

(Continued)

TABLE 10.2 *(Continued)*
Adverse Impacts to the Incident Command System

Logistics

With any adverse impacts to water supply systems, command may designate points of distribution (PODs) for potable water to be donated to the public. This can take many forms (water from a truck where the public must bring their own containers, individual size bottled water cases which have weight and are usually distributed to residents via vehicle-based, gallon bottles of water, etc.). Dialysis Centers, for example, need water trucks and pumping equipment to provide equivalent water pressure to their filtration systems for renal support of patients. How this water is ordered/requisitioned is critical for logistics to align to the varying constituent needs of the operation. Many times, there are also donations of water – from both corporations and the general public – which have to be managed as in-kind donations, probably through a non-governmental organization or consortium of organizations. In the event the incident is Stafford Act qualified for Public Assistance reimbursement of costs – these material donations, as well as hours of donated labor from volunteers, can be applied to public assistance projects.[27]

Operations

- Firefighting will be impacted – for other threats/hazards, for example, responding to a fire in the same jurisdiction as the impacted water supply system, if low or no water pressure adversely affects the water available through fire hydrants.
- The operational distribution of water to the public – and for agricultural use, such as livestock watering – may be a required incident mission.

Planning

There are no specific adverse impacts to the Planning section from this hazard. There are, however, additional planning activities, aligned planners from other incident commands, etc. associated with this hazard. Full-cycle planning through Mitigation work will most likely involve large infrastructure projects which may require Planning elements maintain the operational tempo/battle rhythm beyond shorter-term incidents within a single operational period, such as winter storms, minor river flooding, etc.

Source: Author.

TABLE 10.3
Adverse Impacts to the Emergency Support Functions

ESF #1: Transportation	While an adverse impact of low/no water pressure to the water supply system will not generate hazards directly to the Transportation ESF, this ESF and the agencies, departments, staff, etc. who support it will have additional responsibilities and courses of action related to the emergency distribution of potable water to the public. Transportation of bulk water (via tankers, on pallets carried on trucks, etc.) will impact this ESF. Also, public transportation routes (bus, paratransit, etc.) will need to be examined, to pass to the PODs. Water may have to be transported for agricultural use. And threats such as a water-main break will generate both too much water in one place (as well as road closures, detours, etc.) and too little water, elsewhere.

TABLE 10.3 *(Continued)*
Adverse Impacts to the Emergency Support Functions

ESF #2: Communications	There are generally no direct adverse impacts to this ESF (in terms of Communications Systems) from a water supply hazard, but the additional crisis communications to the public about alternative sources for potable water, water quality alerts and warnings, etc. will need to be made. Agencies, departments, staff, etc. should be on alert for false information about the water quality and points-of-distribution, and counter against them, if found. A higher volume of calls into PSAPs will occur and need to be managed and triaged. As this may be longer-term, re-routing call volume to 2–1–1 should be considered. Details about internal (for non-public use) points-of-distribution of water will need to be communicated internally within the operation, possibly through the Safety Officer.
ESF #3: Public Works and Engineering	**This ESF is significantly impacted by this hazard. Subject-matter experts from the water system authority, treatment plant management, etc. may be in unified command and/or acting as liaison officers to support the restoration of full capability of water pressure and the safe/sanitary water supply system.**
ESF #4: Firefighting	If the incident shuts down a water treatment plant thereby reducing or eliminating water pressure to the community's fire hydrant system, significant adverse impacts to that community's firefighting capability will occur.
ESF #5: Information and Planning	Essential Elements of Information about alternative water supply sources (both fixed and mobile), water quality testing results, distribution status, restoration timelines, and more, all are needed by ESF#5 to work through the operational planning cycle (see Chapter 2 for details).
ESF #6: Mass Care, Emergency Assistance, Temporary Housing, and Human Services	While sheltering, feeding, and family reunification will most likely not be adversely impacted by a natural water supply contamination to a community's water supply system, the fourth element of ESF#6: distribution of emergency supplies, is definitely a course of action/response mission, which will most likely be implemented if the incident lasts more than a pre-planned time, MOUs/MOAs with the water authorities should have triggers, or leadership orders emergency distribution of water to the public. Note that if there are concurrent disasters needing bottled water (low/no pressure generating PODs) *and* a sheltering incident such as a wildfire or tornado – the prioritization of water distribution must be made by Unified Command.
ESF #7: Logistics	Temporary supply of bottled water, potable water, etc. are all needed. This may involve distribution at PODs, door-to-door, or both. Logistical supply chain integrity and security may be needed, cross-walked with ESF#13.
ESF #8: Public Health and Medical Services	As previously noted, every incident, emergency, or disaster has a public health and medical services impact. Water Supply System adverse impacts include hospitals, clinics, assisted living facilities, nursing homes, **dialysis centers,**[28] blood/plasma collection sites, and more. While many of these licensed facilities have mandates for contingency planning – covering water supply systems concerns which should include a lack of water, as well as others – those plans are generally short-term (incidents lasting hours or days) and not long-term (weeks into months).
ESF #9: Search and Rescue	This ESF will most likely not be adversely impacted by this hazard.

(Continued)

TABLE 10.3 ***(Continued)***
Adverse Impacts to the Emergency Support Functions

ESF #10: Oil and Hazardous Materials Response	As noted in the case studies, the adverse health impacts from disturbed heavy metals, elements occurring naturally, and microorganisms/parasites, may require missions for cleanup from this ESF. The high priority of cleanup and remediation of the hazardous substances/materials is both a life safety and an incident stabilization course of action.
ESF #11: Agriculture and Natural Resources Annex	Crop irrigation, livestock watering, parks and recreation areas all can be adversely impacted by this hazard. If parks cannot be watered – and have working bathrooms and drinking fountains, for example, they may need to be temporarily closed.
ESF #12: Energy	If the water supply is reduced or cut off to a nuclear power plant, that can have significant impact on the Energy ESF. The same is true for hydroelectric power generation.
ESF #13: Public Safety and Security	May need to support ESF#6 and ESF#7 in the distribution of emergency supplies (potable water). Also, first responders and other operational groups will need their own secure point-of-distribution for water, separate from the public one.
ESF #14: Cross-Sector Business and Infrastructure	Water Management Districts may be a separate legal entity from the local government, and have their own emergency management staff, protocols, legal requirements, etc. How problems are caused – and solved – may depend on how staff at water treatment plants, wastewater treatment plants, water supply systems, etc. collaborate, coordinate, cooperate, and communicate day-to-day with local emergency management officials in the jurisdiction. The same is true for exercising the emergency management scenarios, including reductions/depletions of potable water and its impact to water supply systems. Recognizing in advance that there are differences in the actions of for-profit organizations from nonprofit organizations and governmental organizations, is key. For example, in many jurisdictions if after some period of time (usually days) the water supply remains low or unhealthy, alternative processes for the rerouting of water systems and/or purchase, delivery and distribution of potable water will most likely be necessary. Who is responsible for these missions needs to be established in advance and memorialized in formal plan documents.
ESF #15: External Affairs	Coordinate with ESF#2 on both internal and external messaging regarding water quality status, restoration of systems status, alternative sources of potable water, etc.

Source: Author.

10.5.2 Recovery Support Functions (RSFs)

Table 10.4 shows the specifics for **Too Little Potable Water Quantity Hazards**.

10.5.3 Community Lifelines (CLs)

Community Lifelines enable the operation of critical governmental and non-governmental functions and are essential to life safety or economic security. Table 10.5 shows the specifics for **Too Little Potable Water Quantity Hazards**.

TABLE 10.4
Adverse Impacts to the Recovery Support Functions

Cultural and Natural Resources	If the lack of potable water at cultural resources and sites continues beyond the short-term (hours, possibly days) or take extended time and financial resources to remediate and restore capability to the cultural resource, Emergency Management support may be needed. Even more complex would be cultural or natural resource sites with well water which has been depleted or compromised. In these cases, sites may have to be closed or temporarily restricted in access (i.e., potable water brought in, in addition to port-a-potties, if freshwater systems are compromised). This can involve long-term remediation and mitigation – even complete reworking of water systems for those sites.
Economic	The sooner that full capacity restoration of the water supply system of a community is completed, the sooner the adverse impacts to the economy of that community will be alleviated. If cascading incidents occur (a warehouse fire is unable to be extinguished quickly, because of low/no water pressure in nearby fire hydrants), the economic loss of that business – and jobs – could be massive to a community. There is also reputational risk to communities with repeating natural water supply pressure/capacity incidents, as the public may view these incidents as chronic or systemic, and therefore unable to be mitigated against.
Health and Social Services	Hospitals and Healthcare facilities – including walk-in facilities such as clinics and dialysis centers – all are adversely impacted by a lack of potable water. Recovery for this support function is critical.
Housing	Long-term housing is not possible without potable water systems. Generally, the housing RSF is activated when housing is damaged or destroyed by an incident (such as a hurricane or typhoon), but it may be necessary to call upon additional state/territorial/tribal nation and federal resources to support alternative housing – especially for socially vulnerable populations – due to a long-term lack of potable water.
Infrastructure Systems	As noted in ESF#3, recovery of this infrastructure element, to support the restoration of full capability of the water supply system is also critical. Most of the recovery focus of emergency management – and the community leadership – will be on this RSF.

Source: Author.

10.5.4 POETE Process Elements for This Hazard

Table 10.6 shows the specifics for **Too Little Potable Water Quantity Hazards**.

10.6 CHAPTER SUMMARY/KEY TAKEAWAYS

As with all the water **quality** hazards noted in Part 3, both too much water and too little water **quantity** hazards from incident threats, have significant impacts on hospitals, dialysis centers, nursing homes/assisted living centers, daycare sites, etc. As noted, a lack of sufficient potable water – and distributed through the normal water supply

TABLE 10.5
Adverse Impacts to Community Lifelines

Safety and Security	Food, Hydration, Shelter	Health and Medical	Energy (Power & Fuel)
Aligns to ESF #13 missions. The status of this CL should not be easily impacted by specific hazards from this incident.	Aligns to ESF#6 missions, and the distribution of emergency supplies (potable water) is a lifeline which must be maintained and/or restored.	Aligns to both ESF#8 missions and the Health and Social Services RSF missions. As noted in the ESFs and RSFs, this CL is significantly adversely impacted by any disruption to the potable water supply systems.	Aligns to ESF#12 missions, as well as the overall Infrastructure Systems RSF – since energy production and distribution is an infrastructure. To the extent which water is utilized in the production of energy (hydroelectric power, nuclear power, etc.) and is adversely impacted by hazards from this incident, the status of this CL may be impacted.
Communications	Transportation	Hazardous Materials	Water Systems
Aligns to ESF #2 and ESF #15 missions. The status of this CL should not be easily impacted by specific hazards from this incident.	Aligns to ESF #1 Missions. As noted in this chapter's ESF Table 2.5, there may be impacts to this CL from any disruption in the potable water supply systems and there is a need to move potable water via transport.	Aligns to ESF #10 missions. The status of this CL should not be easily impacted by specific hazards from this incident, **unless** there is hazardous material cleanup work **beyond the capability and capacity** of the collaborative response and recovery teams assigned to the hazardous material work needed to remediate and restore the water supply system.	Aligns to ESF#3 missions and the Infrastructure RSF. The two major groupings within this CL are Potable Water Infrastructure and Wastewater Management – both of which can be adversely impacted by too little potable water in a community's water supply systems.

Source: Author.

systems – will shut down a dialysis center, which could cause patients to be redirected in a surge, to area hospitals and other dialysis centers. This can create a new disaster for a community. Community preparedness messaging can be very effective in minimizing adverse impacts from incidents involving this hazard. This should be accomplished through a whole-of-government/whole-community effort. A community cannot over-communicate hazards associated with threats to their potable water supply systems, so

TABLE 10.6
POETE Process Elements for This Hazard

Planning	While there will most likely **not** be a separate annex or appendix to any jurisdiction's emergency operations plan for this specific hazard, there should be elements of other main plan portions and annexes/appendices which cover the planning needed for too little potable water, impacting a community's water supply systems, including individual's systems (wells, cisterns, etc.). From this chapter, there are a number of adverse impacts, historical references of past incidents to learn from, and elements of exercises which can be performed, in order to facilitate positive changes and modifications to preparedness, response, recovery, mitigation, and continuity of operations plans.
Organization	The **trained** staff needed to effect plans, using proper equipment should be exercised and evaluated on a regular basis each year. Actual incident response and recovery can help determine elements of POETE, including how these types of incidents are organized for staffing. Also, as part of both deployments and exercises, staff involved should be evaluated for both further training needed and also the possibility of advancement in their disaster state roles to leadership positions
Equipment	Equipment needed for this hazard specifically includes anything needed to remediate/repair the water systems plant. Filtration and monitoring equipment is key. Also needed is any special equipment (traffic controls, pallet jacks, water tanker trucks, etc.) needed at – and in support of – emergency potable water distribution points. This can also include equipment needed at warehouses and supply depots
Training	In addition to standardized Incident Command System training, specific training on adverse impacts to water supply and wastewater systems[29] should be taken by emergency management officials, from an all-hazards perspective
Exercises and evaluation	A full cycle of discussion and performance exercises should be designed for natural water supply contamination to Water Supply Systems – both on their own and as part of various cascading incident scenarios. These exercises should involve the whole community in exercise planner's invitations for both additional exercise planners/evaluators and actual players, especially since many water supply system operators are separate legal entities from the local governments they serve. In the same way, there are adversely impacted ESFs, RSFs, and CLs as noted in this chapter, those elements should be exercised and evaluated so that the elements of the POETE cycle are reviewed, and revised as needed (i.e., agreed upon for change, as part of the Improvement Plan components)

Source: Author.

messaging via public media, community/civic/religious groups, libraries, social media, etc. – all should be maximized using consistent messaging before, during, and after any incidents. There will be a need to monitor public health impacts in the short term once the sufficient water quantity levels are restored. Water supply systems – including aqueducts, spillways, etc. – can be damaged due to low or no water pressure. The longer-term remediation, repair, restoration, **and additional** elements which protect it from harm/prevent adverse impacts in the future, should be incorporated into Mitigation work. On U.S. Presidentially declared emergencies and disasters, the partial (usually 75%) Federal funding for this work, may be available.

10.6.1 What to Read Next

If you did not read the previous part on Water Quality hazards, please do so as well. While there are many repeated elements between this chapter and the other chapters in this Quantity part, there are some specifics found in one, which may be applicable to the other. The next chapter is the second one in the Quantity Hazards with Water – Drownings. While that chapter could have been placed elsewhere, it is a critical one for emergency managers to become more familiar with – and for communities, even one drowning death is a disaster itself.

Also, be prepared for the possibility of a complex or cascading incident: It may start out as a drought impacting a river, but then become saltwater inundation from the ocean, adversely impacting a water supply system, upstream. As noted in the introduction to this chapter, water supply systems can be impacted by the actions to remediate and correct hazards in other geographies, other jurisdictions, even other nations. Historically, this aspect has generated complex or cascading incidents, just from the jurisdictional concerns themselves.

NOTES

1 See www.youtube.com/watch?v=TfH3M6WiQeg
2 American Red Cross. (2023). *Environmental, Social and Governance Report 2023.* See www.redcross.org/content/dam/redcross/about-us/publications/2023-publications/2023_ESG_Report_FINAL.pdf
3 Wucker, M. (2016). *The Gray Rhino: How to Recognize and Act on the Obvious Dangers We Ignore.* Macmillan.
4 See www.homelandsecuritynewswire.com/dr20231010-are-we-running-out-of-water-water-security-threatened-by-droughts-and-heatwaves-worldwide
5 See https://www3.weforum.org/docs/WEF_Global_Risks_Report_2023.pdf
6 Smedley, T. (2023). 'Drought Is on the Verge of Becoming the Next Pandemic'. *The Guardian.* Accessed December 16, 2023. See www.theguardian.com/news/2023/jun/15/drought-is-on-the-verge-of-becoming-the-next-pandemic
7 See www.undrr.org/water-risks-and-resilience
8 Rodrigues, M. (2023). The Amazon's Record-Setting Drought: How Bad Will It Be? *Nature.* Accessed December 28, 2023. See www.nature.com/articles/d41586-023-03469-6
9 United Nations Office for Disaster Risk Reduction. (2021). *GAR Special Report on Drought 2021.* Geneva. See www.undrr.org/media/49386/download
10 See www.wyomingpublicmedia.org/natural-resources-energy/2022-07-11/experts-say-flooding-and-drought-can-happen-simultaneously-as-they-are-two-different-concepts
11 See www.science.org/content/article/utah-s-great-salt-lake-has-lost-half-its-water-thanks-thirsty-humans
12 Scientific American. (1857). *The Pressure of Water.* Scientific American. Accessed December 16, 2023. See www.scientificamerican.com/article/the-pressure-of-water/
13 See www.unicef.org/wash
14 United Nations Office for Disaster Risk Reduction. (2021). *GAR Special Report on Drought 2021.* Geneva. See www.undrr.org/media/49386/download
15 McFall-Johnsen, M., & Rosa-Aquino, P. *Photos Show the Mississippi River Is So Low that It's Grounding Barges, Disrupting the Supply Chain, and Revealing a 19th-Century Shipwreck.* Business Insider. See www.businessinsider.com/photos-drymississippi-river-waters-ground-barges-reveal-old-shipwreck-2022-10

16 FEMA. (2023). *National Preparedness Report*. FEMA. See www.fema.gov/sites/default/files/documents/fema_2023-npr.pdf
17 See www.independent.co.uk/climate-change/news/hurricane-hilary-2023-california-b2395549.html
18 See www.independent.co.uk/news/world/americas/what-is-burning-man-festival-b2404911.html
19 See https://people.com/burning-man-victim-suspected-died-drug-intoxication-7966029
20 See Too Much Freshwater Chapter for case example
21 See www.who.int/news-room/fact-sheets/detail/trachoma
22 Stelmach, R. D., & Clasen, T. (2015). Household Water Quantity and Health: A Systematic Review. *International Journal of Environmental Research and Public Health*, *12*(6), 5954–5974. https://doi.org/10.3390/ijerph120605954
23 Holland, H. (2023). *Death Toll from Kenya's El Nino Floods Jumps to 120*. Reuters. Accessed December 16, 2023. See www.reuters.com/world/africa/death-toll-kenyas-el-nino-floods-jumps-120-2023-11-28/
24 Arquitectura Viva. (2020). *Restoration Utrecht's Catharijnesingel Canal*. Accessed December 16, 2023. See https://arquitecturaviva.com/works/recuperacion-del-canal-catharijnesingel-en-utrecht
25 Paddison, L., Guy, J., & Gutiérrez, F. (2024). One of the World's Biggest Cities May Be Just Months Away from Running Out of Water. *CNN*. See www.cnn.com/2024/02/25/climate/mexico-city-water-crisis-climate-intl/index.html
26 FEMA. (2023). *Hazard Mitigation Policy Aid Drought Mitigation*. FEMA. See www.fema.gov/sites/default/files/documents/fema_hma_drought-mitigation-policy-aid_09282023.pdf
27 See www.govinfo.gov/content/pkg/FR-2013-11-01/pdf/2013-26018.pdf
28 Centers for Medicare and Medicaid Services. (n.d.). *Emergency Preparedness for Dialysis Facilities: A Guide for Chronic Dialysis Facilities*. U.S. Centers for Medicare & Medicaid Services. Accessed November 20, 2023. See www.cms.gov/medicare/end-stage-renal-disease/esrdnetworkorganizations/downloads/emergencypreparednessforfacilities2.pdf
29 FEMA. (2019). *IS-1024: Water and Wastewater Treatment System Considerations*. FEMA EMI. See https://training.fema.gov/is/courseoverview.aspx?code=IS-1024&lang=en

REFERENCES/ADDITIONAL READING

American Planning Association. (2019). *Falling Dominoes: A Planner's Guide to Drought and Cascading Impacts*. APA. See https://planning.org/publications/document/9188906/

Fu, X., Tang, Z., Wu, J., & McMillan, K. (2013). Drought Planning Research in the United States: An Overview and Outlook. *International Journal of Disaster Risk Science*, *4*(2), 51–58. https://doi.org/10.1007/s13753-013-0006-x

McRoberts, D. B., & Nielsen-Gammon, J. W. (2012). The Use of a High-Resolution Standardized Precipitation Index for Drought Monitoring and Assessment. *Journal of Applied Meteorology and Climatology*, *51*(1), 68–83. https://doi.org/10.1175/JAMC-D-10-05015.1

Ramos, M. (2019). *Global Review of Water, Sanitation and Hygiene (WASH) Components in Rapid Response Mechanisms and Rapid Response Teams in Cholera Outbreak Settings – Haiti, Nigeria, South Sudan and Yemen*. UNICEF. See www.unicef.org/media/73121/file/UNICEF-WASH-Global-Review-Rapid-Response-Teams.pdf

Shroder, J. F. (Ed.). (2013). *Treatise on Geomorphology*. Academic Press. https://doi.org/10.1016/B978-0-12-374739-6.09021-7

United Nations Office for Disaster Risk Reduction. (2021). *GAR Special Report on Drought 2021*. Geneva. See www.undrr.org/media/49386/download

U.S. EPA. (n.d.). *Drought Resilience and Water Conservation.* Accessed January 27, 2024. See www.epa.gov/water-research/drought-resilience-and-water-conservation

World Economic Forum. (2023). *The Global Risks Report 2023* (18th ed.). World Economic Forum. www3.weforum.org/docs/WEF_Global_Risks_Report_2023.pdf

Wucker, M. (2016). *The Gray Rhino: How to Recognize and Act on the Obvious Dangers We Ignore.* Macmillan.

BIBLIOGRAPHY

American Red Cross. (2023). *Environmental, Social and Governance Report 2023.* See www.redcross.org/content/dam/redcross/about-us/publications/2023-publications/2023_ESG_Report_FINAL.pdf

Arquitectura Viva. (2020). *Restoration Utrecht's Catharijnesingel Canal.* Accessed December 16, 2023. See https://arquitecturaviva.com/works/recuperacion-del-canal-catharijnesingel-en-utrecht

Centers for Medicare and Medicaid Services. (n.d.). *Emergency Preparedness for Dialysis Facilities: A Guide for Chronic Dialysis Facilities.* U.S. Centers for Medicare & Medicaid Services. Accessed November 20, 2023. See www.cms.gov/medicare/end-stage-renal-disease/esrdnetworkorganizations/downloads/emergencypreparednessforfacilities2.pdf

FEMA. (2019). *IS-1024: Water and Wastewater Treatment System Considerations.* FEMA EMI. See https://training.fema.gov/is/courseoverview.aspx?code=IS-1024&lang=en

FEMA. (2023a). *Hazard Mitigation Policy Aid Drought Mitigation.* FEMA. See www.fema.gov/sites/default/files/documents/fema_hma_drought-mitigation-policy-aid_09282023.pdf

FEMA. (2023b). *National Preparedness Report.* FEMA. See www.fema.gov/sites/default/files/documents/fema_2023-npr.pdf

Holland, H. (2023). *Death Toll from Kenya's El Nino Floods Jumps to 120.* Reuters. Accessed December 16, 2023. See www.reuters.com/world/africa/death-toll-kenyas-el-nino-floods-jumps-120-2023-11-28/

McFall-Johnsen, M., & Rosa-Aquino, P. *Photos Show the Mississippi River Is So Low That It's Grounding Barges, Disrupting the Supply Chain, and Revealing a 19th-Century Shipwreck.* Business Insider. See www.businessinsider.com/photos-drymississippi-river-waters-ground-barges-reveal-old-shipwreck-2022-10

Paddison, L., Guy, J., & Gutiérrez, F. (2024). One of the World's Biggest Cities May Be Just Months Away from Running Out of Water. *CNN.* See www.cnn.com/2024/02/25/climate/mexico-city-water-crisis-climate-intl/index.html

Rodrigues, M. (2023). The Amazon's Record-Setting Drought: How Bad Will It Be? *Nature.* Accessed December 28, 2023. See www.nature.com/articles/d41586-023-03469-6

Scientific American. (1857). *The Pressure of Water.* Scientific American. Accessed December 16, 2023. See www.scientificamerican.com/article/the-pressure-of-water/

Smedley, T. (2023). Drought Is on the Verge of Becoming the Next Pandemic. *The Guardian.* Accessed December 16, 2023. See www.theguardian.com/news/2023/jun/15/drought-is-on-the-verge-of-becoming-the-next-pandemic

Stelmach, R. D., & Clasen, T. (2015). Household Water Quantity and Health: A Systematic Review. *International Journal of Environmental Research and Public Health, 12*(6), 5954–5974. https://doi.org/10.3390/ijerph120605954

United Nations Office for Disaster Risk Reduction. (2021). *GAR Special Report on Drought 2021.* Geneva. See www.undrr.org/media/49386/download

Wucker, M. (2016). *The Gray Rhino: How to Recognize and Act on the Obvious Dangers We Ignore.* Macmillan.

11 Drownings

> Too much of water hast thou, poor Ophelia,
> And therefore I forbid my tears.
>
> **– William Shakespeare[1]**

11.1 OVERVIEW

Whether it is from the dangers of a 5-gallon paint bucket[2] or an entire ocean – accidentally drownings can be tragic for a community – and they are almost always preventable hazards. Some people may not view drownings as a disaster or rising to the level of Emergency Management and Emergency Management practitioners. The World Health Organization describes drownings as "a neglected Public Health issue" (WHO, 2014, p. viii)[3] globally. Prevention and Protection measures and crisis communications to the public should be made to help protect life safety and prevent drowning deaths from occurring. Not only will these efforts save lives, but they provide a positive way to continue dialogue with the public on overall all-hazards disaster readiness initiatives and crisis communications in general – again, using an all-hazards approach. Emergency managers need to be a trusted voice in their communities, and providing critical information on how to prevent accidental drownings is one way to do this.

> Drowning happens when a person is submerged in liquid, inhales water, and cannot breathe air. The signs of drowning happen immediately. The person will have trouble breathing, excessive coughing, or foam in the mouth. Some people's breathing difficulty may resolve rapidly, but for some, breathing difficulty can progressively get worse over the next few hours (up to 5–8 hours after having had trouble in the water).
>
> **(Hawkins, 2018, p. 1)[4]**

Encourage adults and near-adults to learn CPR and take lifesaving courses – which in turn, can increase the qualified candidates for trained/qualified lifeguards for community. This could also lead to people becoming Emergency Medical Technicians (EMTs)[5] or other first responders, as well. Encourage safe boating practices as well and reinforce both the need to learn to swim for boaters and the need for properly fitting and approved life vests for everyone on a boat, in any type of water.

Emergency managers should understand that there has been historic racism in the United States and in other countries,[6] when it comes to segregation and inaccessibility to pools, beaches, etc. Preparedness crisis communications should follow IDEA=B concepts (see Chapter 2) in their distribution to the public.

DOI. 10.1201/9781003474685-15

As with other hazards, the question of causality for a drowning is immaterial to Emergency Management's Response and Recovery operations, except that there may be law enforcement intelligence and investigation elements imbedded within the Incident Command System (ICS) established, specifically, for responding to and recovery from the adverse impacts of the incident. If this is the case, then actions of the other Command and General Staff branches/sections may need to be adjusted and/or expanded to support law enforcement intelligence and investigation elements. This could include investigators for criminal conduct laws, for example. There may be a parallel incident command system established by law enforcement officials and there may or may not be a liaison officer (LNO) linking the two – or more – ICS's. Each incident, emergency, disaster, etc. is different. Emergency managers should be involved in the full disaster cycle planning, organization, equipping, training, and exercising for drownings anywhere in their respective communities (including adjacent jurisdictions – such as a neighboring community's shared oceanfront/beach area or lake/river/waterway).

Besides local jurisdictional emergency management groups, this hazard may need to be managed by hospitals and other healthcare facilities, corporations, colleges/universities, etc. in a coordinated and collaborative way. All may have different unmet needs which they will consider as a priority (risk management, liability, surge capacity, etc.) – and some of those priorities may conflict and compete with each other for overall Response and Recovery resource allocation. Unified Command should always prioritize the LIPER (see Chapter 2), when making decisions.

11.1.1 Drownings in Locations without a Lifeguard

The U.S. Consumer Products and Safety Commission (CPSC) reports on fatal child drownings and nonfatal drownings each year. As previously noted, child drownings account for the highest deadly hazard for children between 1 and 4 years old in the United States. Emergency managers can echo and amplify public safety messaging, including Prevention and Protection items, such as follows:

> Never leave a child unattended in or near water, and always designate an adult Water Watcher.[7] This person should not be reading, texting, using a phone or being otherwise distracted. In addition to pools and spas, this warning includes bathtubs, buckets, decorative ponds, and fountains.
>
> If you own a pool or spa, install layers of protection, including barriers to prevent an unsupervised child from accessing the water. Homes can use door alarms, pool covers, and self-closing, self-latching devices on fence gates and doors that access pools.
>
> Learn how to perform CPR on children and adults. Many communities offer online CPR training.
>
> Learn how to swim and teach your child how to swim.
>
> Keep children away from pool drains, pipes, and other openings to avoid entrapments.
>
> Ensure any pool and spa you use has drain covers that comply with federal safety standards. If you do not know, ask your pool service provider about safer drain covers.
>
> **(CPSC, 2023, p. 1)**[8]

Encouraging people to learn to swim – and to make sure their children know how to swim is critical. As is reminding residents who are considering swimming in open water to do so, only in areas where there are lifeguards present. Home pools, spas, etc. should have lockable covers, alarms for unattended access, and fences/gates to prevent unauthorized use.

And reminding people about the drowning hazards associated with flooding is critical as well. Public Information Officers need to echo and amplify crisis communications messages warning people not to drive through floodwater and flooded roadways.[9]

11.1.2 Drownings in Locations with a Lifeguard

The two best practices for Prevention and Protection Emergency Management activities against drownings are having professional, trained lifeguards at swimming areas, and restricting access to areas where there are no lifeguards present. Support for better rescue equipment,[10] detection techniques,[11] and how lifeguards manage critical patients[12] – all are areas where Emergency Management can aid in the training, education, and Preparedness work, to improve life safety in their communities.

One factor which needs further development is the research on and practical support for the mental health of lifeguards. Their continual responses to traumatic incidents can take a toll on their mental health and wellness, in the same ways it does for other emergency and first responders.[13]

11.1.3 Ocean Drownings

Emergency managers need the Emergency Management Intelligence that ocean-based drowning incidents have additional factors, which are different from river, lake, pool, etc. drowning incidents. Snorkeling and scuba diving can be factors in drownings.[14] As can surfing and bodyboarding.[15] Rip Currents, as noted in the Chapter 4, are a major threat to life safety.[16]

11.2 CASE EXAMPLES

Below, please find some case examples from recent history, which exhibit both best practices for disaster readiness and how there can be significant adverse impacts to the LIPER (life safety, incident stabilization, property/asset protection, environmental/economic impacts, and recovery operations) from Drownings.

11.2.1 River Drowning Reductions via Combined Prevention Strategies

Researchers Peden, Franklin, and Leggat in 2020, using a modified Delphi process, identified 11 prevention strategies – when used together as much as possible – to reduce river drownings.[17] Table 11.1 shows those strategies found to be effective, as determined by a group of worldwide experts, who were surveyed as to best practices for river drowning prevention strategies.

TABLE 11.1
Categories of River Drowning Prevention Strategies

Categories of river drowning prevention strategies proposed by Delphi participants at phase 1 ($n = 11$)

Category	Strategies (n)	Example Strategy
Life jackets	9	Lifejacket wear for children.
Personal behaviors	17	Do not engage in water recreation in a river alone.
Knowledge	10	Strategies to survive cold water immersion.
Public awareness and advocacy	12	Raise awareness of the dangers of submerged obstacles.
Cardiopulmonary resuscitation (CPR) and rescue	7	Training for all boat personnel in CPR, calling for rescue, and search and rescue.
Personal skills	9	Teach self-rescue skills to enable unaided movement to water's edge.
Signage	13	Highly visible signs warning of local hazards at popular swimming destinations.
Engineering	19	Safe and accessible infrastructure, such as bridges, for crossing rivers.
Flooding	18	Establish effective early warning systems for notifying at-risk citizens when rivers are flooded.
Alcohol	11	Restriction of alcohol usage around hire and drive vessels, such as houseboats and party boats.
Other	14	Include river drowning prevention in national and local water safety plans.
Total	139	

Source: © 2020 Peden, Franklin, and Leggat, p. 241 CC BY-NC.

11.2.1.1 Programs for Reducing Drowning

Maryland's Department of Health has a Center for Recreation and Community Environmental Health Services[18] which provides proactive water safety information for its residents and visitors. The state also maintains high standards for youth camp aquatic programs, in requiring certified lifeguards, water watchers, and aquatic directors.[19] In 2010, a faith-based group[20] in the Baltimore area and a school district in Prince George's County[21] embarked on utilizing the U.S. Swimming Foundation's "Make a Splash" swimming education program.

> Through the Make a Splash initiative, the USA Swimming Foundation provides the opportunity for every child in America to learn to swim – regardless of race, gender or financial circumstances. The USA Swimming Foundation partners with learn-to-swim providers, community-based water safety advocates, and national organizations to provide swimming lessons and educate children and their families on the importance of learning how to swim.
>
> As the philanthropic arm of USA Swimming, the USA Swimming Foundation works to strengthen the sport by raising funds to support programs that save lives and build champions in the pool and in life. Thus far, the USA Swimming Foundation has

> provided over $4.5 million in grants to Make a Splash Local Partners for free and low-cost swimming lessons nationwide to spread national awareness on the importance of learning to swim and bring together strategic partners to end drowning.
>
> **(USA Swimming Foundation, 2017, p. 1)[22]**

If cost and access to swimming lessons/facilities are barriers to increasing youth swimming capabilities – and therefore their life safety around open water – emergency managers should assist in communicating and providing resources to increase the Protective elements found in knowing how to swim.

11.2.2 Drowning: A Public Health Issue

Researchers at the World Health Organization (Meddings et al., 2021) have recognized that drownings are a leading cause of death worldwide, especially for children and young adults. Drowning disproportionately impacts low-income and middle-income counties. Meddings et al. noted that drownings can occur anywhere, and that drowning deaths – including those of fishermen and people who collect water, work/travel on water, or recreate on any body of water. Drowning is a universal hazard – wherever there is water, there is a high probability of a drowning hazard.[23]

Also noted by Meddings et al.:

> Flood-related disasters can also lead to drowning. Both the number of people exposed to flood hazards and the severity of flood-related disasters are increasing, and are projected to grow further, as a result of climate change.[24] Floods are the most common type of natural disaster and the leading cause of mortality in disasters, and drowning is the leading cause of death during floods.[25]

11.2.3 Water Safety Ireland

Water Safety Ireland[26] is the organization created by legislation to provide national-level standards and training for lifeguards, and also promote water safety, swimming, and rescue/resuscitation. Dating back to 1945 and with a start in the Irish Red Cross (and some help from the water safety experts over at the American Red Cross), this organization was aligned with fire and road safety, to become part of a National Safety Council.[27]

> Any swimming pool in Ireland, that is open to the public, to residents or members, must be managed in accordance with the Safety, Health and Welfare at Work Act, 2005, and the accompanying regulations. There is a duty of care under this Act to ensure that the quality of swimming & spa pools is managed at safe levels and that bathers can expect competent staff and a safe hygienic bathing environment.
>
> **(Water Safety Ireland, 2021, p. 7)[28]**

On average, over 100 people die from drowning each year in Ireland. The majority of these deaths occur in lakes and rivers, on farms, and at private home pools. Ireland's water safety program is organized through 30 water safety area committees and people can volunteer with any of them, as well as swim instructors can be affiliated with Water Safety Ireland and take advantage of insurance coverage for their work.[29]

11.2.4 Repurposing Quarries – and Reducing Drownings

A number of communities are viewing abandoned quarries[30] as potential sites for recreation, redevelopment, ecological restoration, and revenue sources. At the same time, eliminating (or at least reducing) the hazard of drowning can be accomplished as well. Quarries have very deep areas, and very cold water. Also, the walls are very steep and hard to climb out of, and there may be machinery, abandoned vehicles, and other hazards in the water.[31]

Rehabilitation[32] and repurposing quarries as secure recreational areas – with lifeguards while operating and guards when closed – will have long-term benefits to communities and reduce drownings. See the Butchart Gardens in Canada, the Bellwood Quary in the United States,[33] and the Shimao Shanghai Wonderland Hotel in China, for examples.

11.2.5 A New Possible Drowning Prevention Tool

Researchers at Barton Dunant,[34] an Emergency Management training and consulting firm, are proposing new tools be researched and implemented to help prevent drowning worldwide, specifically for non-ocean locations without a lifeguard. It is predominantly these locations – such as home swimming pools, rivers, private lakes, etc. – which the CPSC has indicated have yearly, *preventable* fatalities. Included in these tragedies are a high number of children who drown because they are not being properly watched in the water. Their concept, a *Smarter Water Watcher*™ paradigm shift for people who swim in areas without a lifeguard, could help responsible adults be less distracted and provide simple instructions to people in the water to help themselves, warn for threats and hazards, and increase the wayfinding of first responders to a potential drowning incident's location, especially in rural or remote areas.[35]

11.2.6 When a Lifeguard Is Needed

The question of whether a lifeguard is required at a private pool is a local jurisdictional one. Generally speaking (and this book claims no legal advice status in any jurisdiction), one article indicated the need for a qualified lifeguard – or lifeguards – was dependent on pool size and number of people who could use the pool at the same time. Also, if the pool has a diving board, regardless of size or capacity, may dictate whether a lifeguard is required by law. Organizations (both governmental and non-governmental/commercial) with pools (hotels, clubs, universities, etc.) must also have signage, safety equipment, and some jurisdictions may require a landline telephone to call for first aid.[36]

There are many countries which have private organizations which can provide – for a fee – a lifeguard on a short-term basis for private pools, spas, etc. Some companies are specifically focused on lifeguarding only, while others have this as an aspect of security services. A quick internet search can find entities in almost every jurisdiction. It would be prudent to integrate these groups and companies into your Emergency Management POETE process for all-hazards planning, so at the very least they are better educated on the emergency services protocols and procedures for your jurisdiction.

11.3 IMPACTS TO THE DISASTER PHASE CYCLES

As shown in Table 11.2, there are specific impacts to all four of the disaster cycle phases, from Drownings:

TABLE 11.2
Impacts to the Disaster Phase Cycles from Water Supply Quality Hazards from Other Incidents

Preparedness/Protection/Prevention

There are steps and courses of action that emergency managers can take to protect their constituents – and visitors to their jurisdiction – from drowning hazards, including public messaging about water and boating safety, learning to swim, following directions as to where not to swim, etc. Communities – especially geographic areas which have high socially vulnerable individuals (SVIs) populations – need this preparedness messaging and opportunities for free/low-cost training (i.e., swimming lessons) for their own Protective and Prevention capabilities, for life safety. Emergency managers can integrate other groups for coordination on Response and Recovery protocols and procedures as well.

Response

Emergency Management's response courses of action for drowning incidents will depend on the capacity and capability of the local Emergency Services group or groups to respond. Ocean drowning incidents may require more equipment and staff, incidents where multiple people have drowned will require more EMS, more hospital support, etc. There may be a need to coordinate law enforcement investigation missions; and to help coordinate and transition into Recovery operations. Large mass casualty drowning incidents – especially involving children and/or high media coverage – may require a Friends and Family Reception Center to be established, and possibly a Family Assistance Center, if there are fatalities. This is not a comprehensive list, but a reminder that there are significant Response phase actions needed for this hazard, which should be managed and coordinated locally, through the local Emergency Management agency. This hazard will most likely be a no-notice incident. There will not be time to advise the public to protect them against this specific hazard at this location, with the exception of forecasted severe weather impacting open-water sources (i.e., tropical storms generate great waves on the ocean – and also posing greater risks of drownings – which should be part of the crisis communications made in advance of said storm).

Recovery

There may be Recovery operations, if the Search and Rescue work does not find all the people in the water. There may be infrastructure repair work, if the drownings – or responder rescue/recovery work – involved any damage to facilities such as storm drains, protective fencing, or water inlets. There needs to be mental health and wellness support provided to all the Response and Recovery teams. Self-care and wellness are critical for not only first responders, but everyone involved in the disaster phase cycle – which may be longer than other threats and hazards, particularly if children are involved.

(Continued)

TABLE 11.2 *(Continued)*
Impacts to the Disaster Phase Cycles from Water Supply Quality Hazards from Other Incidents

Mitigation
There will most likely **not** be any SLTT or Federal Declarations for a drowning incident, so there will not be funding for any Mitigation work, related to drownings. SLTTs need to capitalize (not profit, of course) on any high media and political attention associated with a recent drowning – being most cognizant of the potential negative impacts to the family of the injured/deceased – in the most tactful ways possible. It is worth a conversation with the family to see what they would prefer the crisis communications to the public be, immediately after their injury or death of one of their family members. One mother whose child drowned in a home pool, turned her grief into advocacy for water safety awareness.[37] In 1989, Stew Leonard III drowned at 22 months old. His parents, Kim and Stew Leonard, Jr. created a water safety foundation in his memory, which trains lifeguards and provides free/low-cost swimming lessons to 10,000 children a year.[38]

Source: Author.

11.4 ADVERSE IMPACTS TO THE INCIDENT COMMAND SYSTEM

There may be unique hazards directly impacting Responses and Responders, which should be considered adverse impacts to the Incident Command System (ICS). For Drownings, Table 11.3 details what may be occurring:

11.4.1 POETE Process Elements for This Hazard

Table 11.4 shows the specifics for **Drownings**.

11.5 CHAPTER SUMMARY/KEY TAKEAWAYS

Community preparedness messaging can be very effective in minimizing the threats from incidents involving this hazard. This should be accomplished through a whole-of-government/whole-community effort.

> Drowning is a leading cause of death in childhood, especially for children 1 to 4 years of age. Drowning risks and prevention measures change as children grow, but at all ages, multiple layers of prevention are needed. Key prevention measures include 4-sided fencing of all residential pools, close supervision of young children whenever they are in or near water, and swim lessons once children are ready to learn to swim.
>
> **(American Academy of Pediatrics, 2021, p. 1)[42]**

A community cannot **overcommunicate** hazards associated with life safety adverse impacts associated with drownings – especially for children. Messaging via public media, community/civic/religious groups, libraries, social media,

TABLE 11.3
Adverse Impacts to the Incident Command System

Command (including SO, PIO, and LNOs)

Safety Officers (SOs) need emergency management intelligence about this threat, as soon as possible – especially for hazardous open-water situations (swift water, rip tides, near utility inlets, etc.). These workforce safety elements are paramount. While there are always risks to responders (including pathogens which can transfer from victim to responder), there may be additional hazards such as oil and hazardous materials in the water, and uncharted risks for search and rescue, as well as recovery work. Dive teams, for example, will need counseling after recovering any drowning victims – so their own life safety is sustained.

Public Information Officers (PIOs) need to urgently activate accurate templated messages to the public which provide critical crisis communications in the appropriate languages and modalities for oversaturation. If an "avoid the area" or "Turn Around, Don't Drown®"[39] warning is needed, for example, it needs to be communicated far and wide, and quickly. Consult with the impacted family or families before issuing any Mitigation communications about water safety, so that messaging does not come across as victim-blaming.

Liaison Officers (LNOs) may be in place on this type of hazard, again if any water utility or other private-sector infrastructure is involved, such as dams, quarries, and reservoirs. The same may be true if there are any higher-level investigations (USCG, EPA, FBI, etc.)

Intelligence

Sources of Emergency Management Intelligence for incidents where hazards are from drowning will need to include the topography of the water source (river, lake, etc.), especially if there are natural underwater hazards (wildlife, coral reefs, debris, tree branches, etc.) as well as human-made hazards (water system inlets, sunken vehicles, drains, etc.). Current flow, temperature, and water depth (quarries, as noted in the Case Examples, can be hundreds of ft/m deep). Knowledge of mutual-aid support capabilities for both search and rescue, as well as recovery is needed.

Finance/Administration

Generally, there are no adverse impacts to the Finance/Administration branch, specifically associated with a drowning hazard. Overtime and other staffing costs may exceed what is covered in normal fiscal budgets and/or emergency assistance reimbursements from higher governmental entities on Declared Emergencies or Disasters (Search and Rescue efforts during a Tropical Storm, for example).

Logistics

Generally, there are no adverse impacts to the Logistics branch from this hazard.

Operations

Generally, there are no adverse impacts to the Operations branch – if it is properly staffed and equipped for this Response and Recovery work.

Planning

There are no specific adverse impacts to the Planning section, from this hazard.

Source: Author.

etc. – all should be maximized using consistent messaging before, during, and after any incidents.

There may be a need to monitor public mental health and wellness impacts for years after the incident occurred. A drowning incident will have a very long Recovery phase timeline.

TABLE 11.4
POETE Process Elements for This Hazard

Planning	While there will most likely **not** be a separate annex or appendix to any jurisdiction's emergency operations plan for this specific hazard, there should be elements of other main plan portions and annexes/appendices which cover the planning needed for a mass casualty incident, which can include drownings. They can occur in all types of weather, indoors or outdoors, etc. From this chapter, there are adverse impacts of death and near-death for both the public and responders, historical references of past incidents to learn from, and elements of exercises which can be performed, in order to facilitate positive changes and modifications to preparedness, response, recovery, mitigation, and continuity of operations plans.
Organization	The **trained** staff needed to effect plans, using proper equipment – especially first responder life safety equipment[40] – should be exercised and evaluated on a regular basis each year. Actual incident response and recovery can help determine elements of POETE, including how these types of incidents are organized for staffing. Also, as part of both deployments and exercises, staff involved should be evaluated for both further training needed and also the possibility of advancement in their disaster state roles to leadership positions
Equipment	Equipment needed for this hazard specifically includes anything needed to rescue someone from any type of water source (ocean, floodwater, quarry, frozen lake, etc.). This should also include decontamination equipment.
Training	In addition to standardized Incident Command System training, specific training on adverse impacts from drownings[41] should be taken by emergency management officials, from an all-hazards perspective.
Exercises and evaluation	A full cycle of discussion and performance exercises should be designed for drownings – from the single child in a swimming pool, to 30 people on a pontoon boat in a lake. These exercises should stand on their own and be part of various cascading incident scenarios. Exercises should involve the whole community in exercise planner's invitations for both additional exercise planners/evaluators and actual players, especially since many private organizations (swim clubs, private lakes, etc.) are separate legal entities from the local governments they support. In the same way, there are adversely impacted ESFs, RSFs, and CLs as noted in the Part 4 overview, those elements should be exercised and evaluated so that the elements of the POETE cycle are reviewed, and revised as needed (i.e., agreed upon for change, as part of the Improvement Plan components).

Source: Author.

NOTES

1 Shakespeare, W. (2016). *Hamlet, Prince of Denmark*. B. Mowat & P. Werstine (Eds.). See www.folgerdigitaltexts.org/html/Ham.html#line-4.7.211 (Original work published 1599).

2 See www.cpsc.gov/Newsroom/News-Releases/1989/Large-Buckets-Are-Drowning-Hazards-For-Young-Children

3 World Health Organization. (2014). *Global Report on Drowning: Preventing a Leading Killer*. See www.who.int/publications/i/item/global-report-on-drowning-preventing-a-leading-killer

4 Hawkins, S. C. (2018). *Dry Drowning, Delayed Drowning, Secondary Drowning Are Myths*. Water Safety USA. Accessed December 20, 2023. See www.watersafetyusa.org/mythical-drowning-terms.html

5 See www.sandiego.gov/lifeguards/about/respons/emt

6 See https://thebsa.co.uk/

7 See https://bartondunant.com/smarter-water-watcher

8 U.S. Consumer Product Safety Commission. (2023). *CPSC Issues New Drowning Report with Child Fatalities; Reminder for Extra Water Safety Vigilance*. U.S. Consumer Product Safety Commission. Accessed December 21, 2023. See www.cpsc.gov/Newsroom/News-Releases/2023/CPSC-Issues-New-Drowning-Report-with-Child-Fatalities-Reminder-for-Extra-Water-Safety-Vigilance

9 See www.weather.gov/safety/flood-turn-around-dont-drown

10 Barcala-Furelos, R., Szpilman, D., Palacios-Aguilar, J., Costas-Veiga, J., Abelairas-Gomez, C., Bores-Cerezal, A., López-García, S., & Rodríguez-Nuñez, A. (2016). Assessing the Efficacy of Rescue Equipment in Lifeguard Resuscitation Efforts for Drowning. *The American Journal of Emergency Medicine*, *34*(3), 480–485. https://doi.org/10.1016/j.ajem.2015.12.006

11 Laxton, V., & Crundall, D. (2018). The Effect of Lifeguard Experience upon the Detection of Drowning Victims in a Realistic Dynamic Visual Search Task. *Applied Cognitive Psychology*, *32*(1), 14–23. https://doi.org/10.1002/acp.3374

12 Fernández-Méndez, F., Otero-Agra, M., Abelairas-Gómez, C., Sáez-Gallego, N. M., Rodríguez-Núñez, A., & Barcala-Furelos, R. (2019). ABCDE Approach to Victims by Lifeguards: How Do They Manage a Critical Patient? A Cross Sectional Simulation Study. *PLoS ONE*, *14*(4), e0212080–e0212080. https://doi.org/10.1371/journal.pone.0212080

13 Fien, S., Lawes, J. C., Ledger, J., Drummond, M., Simon, P., Joseph, N., Daw, S., Best, T., Stanton, R., & de Terte, I. (2023). A Preliminary Study Investigating the Neglected Domain of Mental Health in Australian Lifesavers and Lifeguards. *BMC Public Health*, *23*, 1–13. https://doi.org/10.1186/s12889-023-15741-5

14 Dunne, C. L., Madill, J., Peden, A. E., Valesco, B., Lippmann, J., Szpilman, D., & Queiroga, A. C. (2021). An Underappreciated Cause of Ocean-Related Fatalities: A Systematic Review on the Epidemiology, Risk Factors, and Treatment of Snorkelling-Related Drowning. *Resuscitation Plus*, *6*, 100103. https://doi.org/10.1016/j.resplu.2021.100103

15 Lawes, J. C., Koon, W., Berg, I., van de Schoot, D., & Peden, A. E. (2023). The Epidemiology, Risk Factors and Impact of Exposure on Unintentional Surfer and Bodyboarder Deaths. *PLoS ONE*, *18*(5), e0285928. https://doi.org/10.1371/journal.pone.0285928

16 Houser, C., Barrett, G., & Labude, D. (2011). Alongshore Variation in the Rip Current Hazard at Pensacola Beach, Florida. *Natural Hazards (Dordrecht)*, *57*(2), 501–523. https://doi.org/10.1007/s11069-010-9636-0

17 Peden, A. E., Franklin, R. C., & Leggat, P. A. (2020). Developing Drowning Prevention Strategies for Rivers through the Use of a Modified Delphi Process. *Injury Prevention:*

Journal of the International Society for Child and Adolescent Injury Prevention, 26(3), 240–247. https://doi.org/10.1136/injuryprev-2019-043156

18 See https://health.maryland.gov/phpa/OEHFP/CHS/Pages/HealthySwimming.aspx

19 Stokes, R. (2007). *Drowning Prevention*. Maryland Department of Health. See https://health.maryland.gov/phpa/OEHFP/CHS/Shared%20Documents/Youth%20Camps/drwnngprevtn%20ycoptrs.pdf

20 Monteiro, E. H. (2010). *Factors Influencing Childhood Swimming Instruction: Evaluation of the Safe Water Initiative: Maryland Program for Drowning Prevention*. ProQuest Dissertations Publishing.

21 See www.nrpa.org/success-stories/making-a-splash-in-schools/

22 USA Swimming Foundation. (2017). *USA Swimming Foundation Surpasses 5 Million Children Served with Swim Lessons through Make a Splash*. Accessed December 22, 2023. See www.usaswimming.org/news/2017/11/14/usa-swimming-foundation-surpasses-5-million-children-served-with-swim-lessons-through-make-a-splash

23 Meddings, D. R., Scarr, J.-P., Larson, K., Vaughan, J., & Krug, E. G. (2021). Drowning Prevention: Turning the Tide on a Leading Killer. *The Lancet, 6*(9), e692–e695. https://doi.org/10.1016/S2468-2667(21)00165-1

24 See www.ipcc.ch/report/managing-the-risks-of-extreme-events-and-disasters-to-advance-climate-change-adaptation/

25 Doocy, S., Daniels, A., Murray, S., & Kirsch, T. D. (2013). The Human Impact of Floods: A Historical Review of Events 1980–2009 and Systematic Literature Review. *PLoS Currents, 5*. See www.researchgate.net/publication/249647995_The_Human_Impact_of_Floods_A_Historical_Review_of_Events_1980-2009_and_Systematic_Literature_Review

26 See https://watersafety.ie/

27 See https://watersafety.ie/about/#

28 Water Safety Ireland. (2021). *Swimming Pool Safety Guidelines*. Government of Ireland. See https://watersafety.ie/wp-content/uploads/2021/05/Swimming-Pool-Safety-Guidelines-2021-Screen-Read.pdf

29 See https://watersafety.ie/members/

30 McCandless, C. (2013). *No Longer Just a Hole in the Ground: The Adaptive Re-use of Resource Depleted Quarries*. Massachusetts Institute of Technology. See https://web.mit.edu/people/spirn/Public/Ulises-11-308/Quarrying.pdf

31 Seewww.nebosh.org.uk/our-news-and-events/our-news/how-do-you-get-the-public-to-stay-safe-and-out-of-quarries/

32 See https://gravelwatch.org/wp-content/uploads/2016/02/Rehabilitation-of-Pits-and-Quarries.pdf

33 See www.archpaper.com/2017/09/atlanta-beltline-quarry-park/

34 See https://bartondunant.com/smarter-water-watcher

35 Mell, H. K., Mumma, S. N., Hiestand, B., Carr, B. G., Holland, T., & Stopyra, J. (2017). Emergency Medical Services Response Times in Rural, Suburban, and Urban Areas. *JAMA Surgery, 152*(10), 983–984. https://doi.org/10.1001/jamasurg.2017.2230

36 Gentile, M. (2019). *Risk Management: Do I Need a Lifeguard*? Hospitalitylawyer.com. Accessed December 22, 2023. See https://hospitalitylawyer.com/wp-content/uploads/2019/01/lifeguard.pdf

37 Koopmans, K. (2022). *Mother Pours Grief from Her Child's Drowning into Creating Nonprofit, Film*. KOMO. Accessed December 22, 2023. See https://wpde.com/news/nation-world/mother-pours-grief-from-her-childs-drowning-into-creating-nonprofit-film-centers-for-disease-control-and-prevention-kid-deaths-drowning-in-silence-no-more-under-water-safety-education

38 See https://stewietheduck.org/foundation/
39 Turn Around, Don't Drown® Is a Registered Trademark of the National Weather Service. Used with permission. See www.weather.gov/safety/flood-turn-around-dont-drown
40 See www.fema.gov/sites/default/files/2020-08/fema_swiftwater-flood-sar-team_definition_08-03-2020.pdf
41 See www.ang.af.mil/Media/Article-Display/Article/3140557/hawaii-california-nevada-airmen-practice-water-rescue/
42 American Academy of Pediatrics. (2021). *Drowning Prevention and Water Safety*. American Academy of Pediatrics. See www.aap.org/en/patient-care/drowning-prevention-and-water-safety/

REFERENCES/ADDITIONAL READING

American Academy of Pediatrics. (2021). *Drowning Prevention and Water Safety*. American Academy of Pediatrics. See www.aap.org/en/patient-care/drowning-prevention-and-water-safety/

American Red Cross. (n.d.). *Drowning Prevention & Facts*. American National Red Cross. Accessed January 31, 2024. See www.redcross.org/get-help/how-to-prepare-for-emergencies/types-of-emergencies/water-safety/drowning-prevention-and-facts.html

Cortés, L. M., Hargarten, S. W., & Hennes, H. M. (2006). Recommendations for Water Safety and Drowning Prevention for Travelers. *Journal of Travel Medicine*, *13*(1), 21–34. https://doi.org/10.1111/j.1708-8305.2006.00002.x

Dean, C. (2021). *Wade in the Water: Drowning in Racism* (Documentary Film). BlackCat Media.

National Safe Boating Council. (n.d.). *Boating Safety Tips and Resources*. National Weather Service. Accessed January 31, 2024. See www.weather.gov/safety/safeboating

National Water Safety Forum. (2015). *A Future Without Drowning: The UK National Drowning Prevention Strategy 2016–2026*. National Water Safety Forum (UK). See www.nationalwatersafety.org.uk/media/1005/uk-drowning-prevention-strategy.pdf

National Weather Service. (n.d.a). *Beach Safety*. NOAA. Accessed January 31, 2024. See www.weather.gov/safety/beach

National Weather Service. (n.d.b). *How to Avoid Getting Caught in a Rip Current*. NOAA. Accessed January 31, 2024. See www.weather.gov/safety/ripcurrent

Tushemereirwe, R., Tuhebwe, D., Cooper, M. A., & D'ujanga, F. M. (2017). The Most Effective Methods for Delivering Severe Weather Early Warnings to Fishermen on Lake Victoria. *PLoS Currents*, *9*. https://doi.org/10.1371/currents.dis.d645f658cf20bc4a23499be913f1cbe1

United States Lifesaving Association. (n.d.). *Rip Currents*. United States Lifesaving Association. Accessed January 31, 2024. See www.usla.org/page/ripcurrents

Water Safety USA. (n.d.). *Water Safety USA Leading the US in Water Safety and Drowning Prevention*. Water Safety USA. Accessed January 31, 2024. See www.watersafetyusa.org/

World Health Organization. (2014). *Global Report on Drowning: Preventing a Leading Killer*. See www.who.int/publications/i/item/global-report-on-drowning-preventing-a-leading-killer

BIBLIOGRAPHY

American Academy of Pediatrics. (2021). *Drowning Prevention and Water Safety*. American Academy of Pediatrics. See www.aap.org/en/patient-care/drowning-prevention-and-water-safety/

Barcala-Furelos, R., Szpilman, D., Palacios-Aguilar, J., Costas-Veiga, J., Abelairas-Gomez, C., Bores-Cerezal, A., López-García, S., & Rodríguez-Nuñez, A. (2016). Assessing the Efficacy of Rescue Equipment in Lifeguard Resuscitation Efforts for Drowning. *The American Journal of Emergency Medicine, 34*(3), 480–485. https://doi.org/10.1016/j.ajem.2015.12.006

Doocy, S., Daniels, A., Murray, S., & Kirsch, T. D. (2013). The Human Impact of Floods: A Historical Review of Events 1980–2009 and Systematic Literature Review. *PLoS Currents, 5.* See www.researchgate.net/publication/249647995_The_Human_Impact_of_Floods_A_Historical_Review_of_Events_1980-2009_and_Systematic_Literature_Review

Dunne, C. L., Madill, J., Peden, A. E., Valesco, B., Lippmann, J., Szpilman, D., & Queiroga, A. C. (2021). An Underappreciated Cause of Ocean-Related Fatalities: A Systematic Review on the Epidemiology, Risk Factors, and Treatment of Snorkelling-Related Drowning. *Resuscitation Plus, 6*, 100103. https://doi.org/10.1016/j.resplu.2021.100103

Fernández-Méndez, F., Otero-Agra, M., Abelairas-Gómez, C., Sáez-Gallego, N. M., Rodríguez-Núñez, A., & Barcala-Furelos, R. (2019). ABCDE Approach to Victims by Lifeguards: How Do They Manage a Critical Patient? A Cross Sectional Simulation Study. *PLoS ONE, 14*(4), e0212080–e0212080. https://doi.org/10.1371/journal.pone.0212080

Fien, S., Lawes, J. C., Ledger, J., Drummond, M., Simon, P., Joseph, N., Daw, S., Best, T., Stanton, R., & de Terte, I. (2023). A Preliminary Study Investigating the Neglected Domain of Mental Health in Australian Lifesavers and Lifeguards. *BMC Public Health, 23*, 1–13. https://doi.org/10.1186/s12889-023-15741-5

Gentile, M. (2019). *Risk Management: Do I Need a Lifeguard*? Hospitalitylawyer.com. Accessed December 22, 2023. See https://hospitalitylawyer.com/wp-content/uploads/2019/01/lifeguard.pdf

Hawkins, S. C. (2018). *Dry Drowning, Delayed Drowning, Secondary Drowning Are Myths.* Water Safety USA. Accessed December 20, 2023. See www.watersafetyusa.org/mythical-drowning-terms.html

Houser, C., Barrett, G., & Labude, D. (2011). Alongshore Variation in the Rip Current Hazard at Pensacola Beach, Florida. *Natural Hazards (Dordrecht), 57*(2), 501–523. https://doi.org/10.1007/s11069-010-9636-0

Koopmans, K. (2022). *Mother Pours Grief from Her Child's Drowning into Creating Nonprofit, Film.* KOMO. Accessed December 22, 2023. See https://wpde.com/news/nation-world/mother-pours-grief-from-her-childs-drowning-into-creating-nonprofit-film-centers-for-disease-control-and-prevention-kid-deaths-drowning-in-silence-no-more-under-water-safety-education

Lawes, J. C., Koon, W., Berg, I., van de Schoot, D., & Peden, A. E. (2023). The Epidemiology, Risk Factors and Impact of Exposure on Unintentional Surfer and Bodyboarder Deaths. *PLoS ONE, 18*(5), e0285928. https://doi.org/10.1371/journal.pone.0285928

Laxton, V., & Crundall, D. (2018). The Effect of Lifeguard Experience Upon the Detection of Drowning Victims in a Realistic Dynamic Visual Search Task. *Applied Cognitive Psychology, 32*(1), 14–23. https://doi.org/10.1002/acp.3374

McCandless, C. (2013). *No Longer Just a Hole in the Ground: The Adaptive Re-use of Resource Depleted Quarries.* Massachusetts Institute of Technology. See https://web.mit.edu/people/spirn/Public/Ulises-11-308/Quarrying.pdf

Meddings, D. R., Scarr, J.-P., Larson, K., Vaughan, J., & Krug, E. G. (2021). Drowning Prevention: Turning the Tide on a Leading Killer. *The Lancet, 6*(9), e692–e695. https://doi.org/10.1016/S2468-2667(21)00165-1

Mell, H. K., Mumma, S. N., Hiestand, B., Carr, B. G., Holland, T., & Stopyra, J. (2017). Emergency Medical Services Response Times in Rural, Suburban, and Urban Areas. *JAMA Surgery, 152*(10), 983–984. https://doi.org/10.1001/jamasurg.2017.2230

Monteiro, E. H. (2010). *Factors Influencing Childhood Swimming Instruction: Evaluation of the Safe Water Initiative: Maryland Program for Drowning Prevention.* ProQuest Dissertations Publishing.

Peden, A. E., Franklin, R. C., & Leggat, P. A. (2020). Developing Drowning Prevention Strategies for Rivers Through the Use of a Modified Delphi Process. *Injury Prevention: Journal of the International Society for Child and Adolescent Injury Prevention, 26*(3), 240–247. https://doi.org/10.1136/injuryprev-2019-043156

Shakespeare, W. (2016). *Hamlet, Prince of Denmark.* B. Mowat & P. Werstine (Eds.). See www.folgerdigitaltexts.org/html/Ham.html#line-4.7.211 (Original work published 1599).

Stokes, R. (2007). *Drowning Prevention.* Maryland Department of Health. See https://health.maryland.gov/phpa/OEHFP/CHS/Shared%20Documents/Youth%20Camps/drwnng-prevtn%20ycoptrs.pdf

USA Swimming Foundation. (2017). *USA Swimming Foundation Surpasses 5 Million Children Served with Swim Lessons Through Make a Splash.* Accessed December 22, 2023. See www.usaswimming.org/news/2017/11/14/usa-swimming-foundation-surpasses-5-million-children-served-with-swim-lessons-through-make-a-splash

U.S. Consumer Product Safety Commission. (2023). *CPSC Issues New Drowning Report with Child Fatalities; Reminder for Extra Water Safety Vigilance.* U.S. Consumer Product Safety Commission. Accessed December 21, 2023. See www.cpsc.gov/Newsroom/News-Releases/2023/CPSC-Issues-New-Drowning-Report-with-Child-Fatalities-Reminder-for-Extra-Water-Safety-Vigilance

Water Safety Ireland. (2021). *Swimming Pool Safety Guidelines.* Government of Ireland. See https://watersafety.ie/wp-content/uploads/2021/05/Swimming-Pool-Safety-Guidelines-2021-Screen-Read.pdf

World Health Organization. (2014). *Global Report on Drowning: Preventing a Leading Killer.* See www.who.int/publications/i/item/global-report-on-drowning-preventing-a-leading-killer

12 Too Much Raw Water/Freshwater

Be still, sad heart! and cease repining;
Behind the clouds is the sun still shining;
Thy fate is the common fate of all,
Into each life some rain must fall,
Some days must be dark and dreary.

– Henry Wadsworth Longfellow[1]

12.1 OVERVIEW

This chapter is about too much raw water or freshwater – excess water in rivers, lakes, streams, etc. – even when pipes burst inside and outside of a building. This chapter will cover hazards from the threats of excess water from the air, including snow, sleet, hail, and ice. Earth's water cycle[2] has water constantly moving – through the atmosphere, watersheds, pools of water, and more. Hazards from the ocean and seawater – by itself – will be covered in Chapter 13.

The human-made movement of water from one point to another can also generate areas of space for future excess raw water hazards. Phantom lakes are one example of where it may appear as a drought condition now but become a flooding condition later. Decisions made in the past, about diverting raw water for urban drinking water or agricultural use elsewhere, can become threats today and exacerbated by Climate Change. In March of 2023, Tulare Lake "reappeared" in the San Joaquin Valley of California:

> Once the largest freshwater lake west of the Mississippi River, Tulare Lake was largely drained in the late 19th and early 20th centuries, as the rivers that fed it were dammed and diverted for agriculture.
>
> This month, after a historic series of powerful storms, the phantom lake has reemerged. Rivers that dwindled during the drought are swollen with runoff from heavy rains and snow, and are flowing full from the Sierra Nevada into the valley, spilling from canals and broken levees into fields that usually teem with lucrative plantings of tomatoes, cotton and hay.
>
> **(James and Rust, 2023, p. 1)**[3]

Almost every part of our world is subject to some hazards of flooding.[4] Even deserts flood.[5] And any amount of water on the ground has weight and force. Coupled with waves and wind, it can be harmful to people, places, and things. For example, the power of moving water can lift your vehicle off the ground and carry it away.[6]

 DOI: 10.1201/9781003474685-16

And, then when it rains – or heavily rains[7] – the flash flooding which will occur can generate hazards as well.

THE WEIGHT OF WATER

Water has weight – especially large amounts of water. While those on the metric system may recall that 1 m^3 of water weighs 1,000 kg, this translates to 2,205 pounds of water for about 3 ft^3 of water. A commercial building's roof is generally designed to hold 40–100 pounds/ft^2 (18–45 kg per 3 dm). A cubic foot of water weighs over 60 pounds (3 dm^3 = 3 kg), so there is not too much math involved in calculating how quickly structures will collapse under the weight of the *volume* of standing water. And these reference numbers are for raw or freshwater, under normal ambient temperatures. Seawater weighs more:

> The weight of sea water depends principally upon its salinity, which is defined as the number of grams of dissolved salts in 1,000 grams of sea water. Thus, S = 35.7 means that there are 35.7 grams of dissolved salts in each 1,000 grams of the sea water in question.
>
> The weight of sea water depends also upon its temperature, but to a far less extent than upon salinity. And whereas the weight varies directly with salinity for a given temperature it varies in a quite complicated manner with changes in temperature.
>
> **(Scarborough, 1930, p. 1)[8]**

And snow has weight, too – usually expressed as a percentage of the water weight based on the density of the snow. Large amounts of accumulated snow on that same roof (not to mention ice), can adversely impact its structural integrity – and removing the snow can be a hazardous activity, as well.[9]

As with other hazards, the question of causality for incidents involving too much raw water is immaterial to Emergency Management's Response and Recovery operations, except that there may be law enforcement, occupational health and safety, and/or environmental protection intelligence and investigation elements imbedded within the Incident Command System (ICS) which is established, specifically, for responding to and recovery from the adverse impacts of the incident. If this is the case, then actions of the other Command and General Staff branches/sections may need to be adjusted and/or expanded to support these intelligence and investigation elements. This could include investigators for criminal conduct laws, for example. There may be a parallel incident command system established by law enforcement officials and there may or may not be a liaison officer (LNO) linking the two – or more – ICSs. Each incident, emergency, disaster, etc. is different. Emergency managers should be involved in the full disaster cycle planning, organization, equipping, training, and exercising for too much raw water/freshwater incidents anywhere in their respective communities (including adjacent jurisdictions – such as a neighboring community's water supply system (for water main breaks) or shared rivers and waterways.

12.1.1 River/Stream/Lake Flooding

Rivers, streams, lakes – even runoffs from small watershed areas all flow – usually via gravity and toward larger bodies of water "down-stream". Note that where water flows locally, may not match global assumptions (i.e., that all water flows toward the equator). There are rivers which flow South to North, such as the Nile River in Egypt,[10] and others which flow West to East, such as the Olivia and Pipa rivers connecting the Beagle Channel in Tierra del Fuego, Argentina.[11] Human-made interventions, such as dams, canals, levees, aqueducts, etc. which may divert waterflow for human consumption, agricultural, energy, and/or recreational use, can also impact river and stream flow.

When these impacts become adverse – many times due to poor or insufficient floodplain management – the flow of excess raw water from rivers, streams, lakes, etc. can be overpowering. The hazards of flash flooding is noted below.

In the United States, several Federal[12] and state-level organizations monitor raw water bodies of all shapes and sizes for the potential of flooding. Some of these organizations have web tools,[13] maps, and other direct intelligence (see Chapter 2). Others have datasets available to bring into geospatial intelligence systems, which an emergency manager may have access to, in their own jurisdiction.

And as noted in Chapter 10, just because there is a drought, does not mean that flooding cannot occur. These two are mutually exclusive incidents, threats, hazards, etc. And River/Stream flooding can also occur as part of a Tropical Storm system (covered in Chapter 14).

12.1.2 Hazardous Materials Displacement Due to Flooding

A secondary disaster can occur when floodwaters extend into commercial and industrial areas and release pollutants and other toxins beyond their protective infrastructure:

> Increased pollutant discharge of industrial countries has the potential to cause long-term socio-economic and environmental impacts by itself but is exacerbated by the erosion and redistribution of historically contaminated sediments during such floods. Research has recently begun to consider the effects of climate change on specific aspects of toxic floods, such as the release of toxic elements in floodplain soils, and impacts of climate change on suspension of metal contaminated sediment in urban waterways, to name a few.
>
> **(Crawford et al., 2022)[14]**

12.1.3 Water Main Breaks

In some cases, water main breaks can be worse than flash floods. They can send millions of gallons/liters of water into streets, basements, etc. Most homeowner's insurance does not cover flooding from any source – and U.S. Federal flood insurance usually does cover water damage generated from broken water main and subsequent flood. Water main breaks have reached the level of federally declared emergencies.[15]

> Communities across the country face the challenge of aging and inadequate water infrastructure. Most of our underground water infrastructure was built 50 or more years ago, in the post-World War II era. In some older urban areas, many water mains are a century old. The implications of deteriorating infrastructure can be felt nationwide.
>
> - There are approximately 240,000 water main breaks per year in the U.S.
> - Approximately $2.6 billion[16] is lost as water mains leak trillions of gallons of treated drinking water.
>
> **(USEPA, 2023, p. 1)**[17]

These are hazards which cross local emergency management organization's missions with those of Water Management Districts. While the main response focus might be on the excess water from a water main break, it can also have low pressure and drought threats/hazards aspects as well. See Chapter 10 for more details. The IAP and unified command aspects should cover both operational areas (the water main break's cleanup effort as well as the distribution of emergency supplies and community backup resources needed for firefighting in the impacted area, for example. And if there are other contributing threats/hazards – such as the high heat of a summer day – this incident can become more complex and complicated very quickly.

There can be a water line break inside a building as well, which can become a disaster for the organizations in that building. When these lines break overnight[18] or on weekends[19] – and flood basements containing infrastructure elements such as electrical panels and boilers – their impacts can be long-standing. Cold weather can exacerbate the hazards, too.[20] Excess water can adversely impact buildings over time, as well. Improper roof drainage and interior leaks can weaken infrastructure such as corroding the metal support bars inside of concrete walls. As noted above, standing water has weight – many times greater than the load for which the structure was designed to hold – and if engineering elements have been compromised, disasters can result.

12.1.4 Avalanche Impacts

Avalanches are the rapid flow of snow and/or ice down a hill or mountainside:

> Although avalanches can occur on any steep slope given the right conditions, certain times of the year and types of locations are naturally more dangerous. While avalanches are sudden, there are typically a number of warning signs you can look for or feel before one occurs. In 90 percent of avalanche incidents, the snow slides are triggered by the victim or someone in the victim's party. Avalanches kill more than 150 people worldwide each year. The National Weather Service provides current weather conditions and forecast information to regional avalanche forecast centers that in-turn issue avalanche forecasts. Avalanche warnings and special advisories are included on NWS websites and broadcast over NOAA Weather Radio.[21]

Avalanches are significant threats in many other countries, as well. In Switzerland, avalanche warnings and other crisis communications is produced by the SLF, which also conducts research on natural hazards such as snow, permafrost, and mountain ecological systems.[22] As noted in Chapter 5, avalanches in the United States are monitored and reported on by the various states, and not a single Federal entity.

12.1.5 Boating Hazards: Freshwater, Open-water

As noted in the introduction to this book, incidents occurring on any oceans where the jurisdictional emergency management actions are **not** included in command and control or resourcing support, are not covered in this book. There are other inland incidents involving watercraft (houseboats, small sport watercraft, and other boat sinkings), which can have an impact on Emergency Management. Larger-scale watercraft incidents involving mass casualties will need extra transportation and treatment resources to area hospitals,[23] require family reunification, etc. Search and Rescue operations will be conducted, as will multiple agency investigations. Severe weather can have an impact on Emergency Management Response work.[24] The question of whether these incidents are underreported, as compared to other incidents, is being researched and reported.[25]

12.1.6 Dam Failures

Not only will a dam failure release large amounts of water into areas where it is not expected, there will be adverse impacts to other sectors when this happens, as shown in Figure 12.1:

Emergency Services

Law enforcement is among the first responders during Dams Sector asset failure or disruption. and their response capabilites can determine the extent of event consequences.

Food & Agriculture

About 60 percent of the country's farm exports travel through inland waterways for export overseas. Four percent of U.S. cropland is irrigated by dams.

Information Technology

Information technology systems control critical processes, manage day-to-day operations, and store sensitive information for the Dams Sector.

Transportation

The Nation's 12,000-mile inland marine network relies on navigation locks to move valuable products throughout the United States.

Communications

Uninterrupted Internet and telecommunication networks are essential for employee communications and remote monitoring and control.

Water

About 70 percent of all freshwater used in the United States comes from surface-water sources, including reservoirs created by dams.

Energy

More than 20 percent of coal used to produce U.S. electricity is shipped via inland waterways that rely on navigation locks.

Nuclear

Dams may store water for cooling operations near nuclear facilities.

Chemical

Chemicals and fertilizers are major commodities shipped via inland waterways.

FIGURE 12.1 Critical sector dependencies and interdependencies.

Source: CISA. (2021). *Dam Sector Profile*. CISA. See www.cisa.gov/sites/default/files/publications/dams-sector-profile-112221-508.pdf, p. 1. Public domain

The following Federal agencies, as shown in Table 12.1, are involved in the Protection phase associated with dams and dam security, usually aligned with those dams being monitored by state agencies and on the National Inventory of Dams (NID):

TABLE 12.1
U.S. Federal Agencies Involved with Dam Safety or Security

U.S. Department of Agriculture	Natural Resources Conservation Service • Provides technical and financial assistance for almost 27,000 NID dams and financial assistance for another 11,000 NID dams designed for agricultural water storage, sediment retention, and flood protection.
U.S. Department of Defense	U.S. Army Corps of Engineers • Oversees 716 dams, 239 locks, 75 hydropower projects, and 2,220 levee systems and provides technical assistance to flood-risk communities and the military.
U.S. Department of Energy	U.S. Department of Energy – Owns and operates 15 dams at three sites. Federal Energy Regulatory Commission.[26] – Regulates 2,600 non-federal hydropower dams.
U.S. Department of Homeland Security	Cybersecurity and Infrastructure Security Agency • Serves as the Dams Sector Risk Management Agency and collaboratively develops guidance, resources, and training for the Dams Sector. Federal Emergency Management Agency • Leads the National Dam Safety Review Board and the Interagency Committee on Dam Safety and is the head of the National Dam Safety Program.
U.S. Department of the Interior	Bureau of Reclamation • Maintains 475 dams and 348 reservoirs bringing water to more than 31 million people and operates 58 hydroelectric power plants producing enough electricity to serve 3.5 million homes. • Other U.S. Department of the Interior agencies involved with dam safety and security include Bureau of Indian Affairs, Bureau of Land Management, Fish & Wildlife Service, National Park Service, and Office of Surface Mining Reclamation and Enforcement.
U.S. Department of Labor	Mine Safety and Health Administration • Regulates the safety of the 1,640 mining industry dams in its inventory.
U.S. Department of State	International Boundary & Water Commission • Owns and operates dams and maintains more than 500 miles of levees and associated floodways along the lower portion of the Rio Grande River.

Source: CISA. (2021). *Dam Sector Profile*. CISA. See www.cisa.gov/sites/default/files/publications/dams-sector-profile-112221-508.pdf. Public domain.

12.1.7 Flash Flooding

Flash flooding is generally caused by slow-moving rainstorms, such as thunderstorms, and can be exacerbated by clogged roadway drains, prior rain events, and poor absorption of the rainwater into the soil (which can occur for a number of reasons and happen in any season).

> Flash floods generally occur within a short time period after a rain event – generally 6 hours or less. For this reason they are more life threatening. Areas most susceptible to flash flooding are mountainous streams and rivers, urban areas, low-lying area, storm drains, and culverts.
>
> **(NWS, n.d.a, p. 1)**[27]

As previously noted, water main breaks can sometimes be worse than flash floods. But flash flooding kills more people annually, globally. In the United States, more than half of the fatalities from flash flooding were vehicle-related.[28]

12.1.8 Ground Blizzards

Ground Blizzards (see Part 2/Chapter 5) are something to consider, when excessive snowfall occurs, and possibly warmer temperatures then happen. If extreme winds follow, the snow can get picked up and cause whiteout conditions, and frostbite and other impacts from the now much colder air. More than 200 people died in such a storm in 1888 in the U.S. Great Plains States.[29]

12.1.9 Iceberg Sections and Ice Dams

Iceberg sections breaking off into waterways, ice dams forming on bridge supports, etc. will be covered in Chapter 18 in Part 5, as they are cascading threats/hazards.

12.1.10 Sewer Line Breaks/Sanitary Sewer Overflows

Sanitary Sewer Overflows (SSOs),[30] which can be caused by sewer line breaks, have significant health hazards as those lines contain raw human waste. People and animals can be harmed through direct contact with water that is from an SSO (or mixed with floodwater) and also from consuming food, such as shellfish, which has been contaminated by the SSO.

> SSOs also damage property and the environment. When basements flood, the damaged area must be thoroughly cleaned and disinfected to reduce the risk of disease. Cleanup can be expensive for homeowners, and municipalities. Rugs, curtains, flooring, wallboard panels, and upholstered furniture usually must be replaced.
>
> A key concern with SSOs which enter rivers, lakes, streams, or brackish waters is their effect on water quality. When bodies of water cannot be used for drinking water, fishing, or recreation, society experiences an economic loss. Tourism and water front home values may fall. Fishing and shellfish harvesting may be restricted or halted. SSOs can also close beaches.
>
> **(USEPA, 1996, p. 4)**[31]

The public health impacts from SSOs are significant. Excessive stormwater flow can overwhelm a wastewater treatment plant and reduce its capability to prevent pathogen transmission[32] – as well as increase the number of patients at local hospitals for treatment from diseases in the SSO.[33]

12.1.11 Severe Winter Storms, Snow Fall, Snowpack

Severe Winter Storms can include hazards such as blizzards, heavy rainfall (also noted below), freezing rain, and other phenomena. In the United States, the weather forecasters at the National Weather Service's Storm Prediction Center[34] can issue a

> Mesoscale Discussion (MCD) statement anywhere from roughly half an hour to several hours before issuing a weather watch. SPC also puts out MCDs for hazardous winter weather events on the mesoscale, such as locally heavy snow, blizzards and freezing rain (see below). MCDs are also issued on occasion for heavy rainfall, convective trends, and other phenomena, when the forecaster feels he/she can provide useful information that is not readily available or apparent to field forecasters. MCDs are based on mesoscale analysis and interpretation of observations and of short term, high resolution numerical model output.
>
> The MCD basically describes what is currently happening, what is expected in the next few hours, the meteorological reasoning for the forecast, and when/where SPC plans to issue the watch (if dealing with severe thunderstorm potential). Severe thunderstorm MCDs can help you get a little extra lead time on the weather and allow you to begin gearing up operations before a watch is issued. The MCD begins with a numerical string that gives the LAT/LON coordinates of a polygon that loosely describes the area being discussed.
>
> **(NWS, n.d., p. 1)**[35]

Snowfall is an important element of these MCDs. And since it does not saturate into the ground as quickly as rain does, as well as being subject to drifts and heavy wind conditions, the accumulation of snow has weight, as noted previously. One inch of snow on a roof is about 5 pounds/ft^2, which means 10 inches of snow is about 50 pounds/ft^2.[36]

In mountainous geographies – including those with glaciers and snowpacks year-round – the reduction of the snowpack (in part due to Climate Change) can generate more rainwater into the ecosystem and at seasonal time frames not anticipated by the communities impacted. The State of Washington in the United States saw this in 2022 and 2023:

> Washington saw glimpses of its future last summer and fall, when a drought gripped the state and officials implored or required people to use less water. The same thing happened in 2015. These water restrictions, and other onerous effects, like rising utility rates, dying salmon, poor harvests and wildfires, are expected to become more common problems.
>
> As spring turns to summer, snowpack acts as a giant and frozen reservoir, melting gradually and flowing into the state's relatively small, dammed reservoirs, when there is room to spare, said Guillaume Mauger, a research scientist with UW's Climate Impacts Group.

The process involves a staggering amount of water.

Washington's annual peak mountain snowpack can hold as much as 25 million acre-feet, according to Mauger and his colleague, Matt Rogers. That would be enough to supply Seattle's drinking water for 181 years.

Shrinking snowpack could also melt four to six weeks earlier than normal. The rain that falls instead of snow fills the reservoirs fairly quickly and any excess precipitation must flow through the dams and into the sea.

(Swanson, 2024, p. 1)[37]

MANY FACTORS IMPACT RIVER LEVEL FORECASTING

In California, their Governor's Office of Emergency Services uses a customized tool built by the National Oceanic and Atmospheric Administration and the National Weather Service called the California Nevada River Forecast Center.[38]

The back-and-forth of water availability and consumption in and around California – going between the extremes of drought and flood and back again now regularly – the significant role of dams[39] and engineering, and the analysis that goes into estimating snowpack, thaw, and timing for the release of that snowmelt from the rivers to the Pacific Ocean are all quite complex.

12.1.12 Sandbags

Sandbagging is one protection mechanism against the adverse impacts of area flooding. This activity may be part of a Public Works (ESF#3) set of roles/missions, but assistance may be available from community groups (civic organizations, scout troops, etc.), as well as local/county/parish correctional facilities (i.e., prisons and jails). All of this requires pre-planning and coordination (the POETE process) before an incident occurs.

12.1.13 Torrential Rainfall Impacts

Research is now showing that torrential rainfall threats[40] are lasting longer, becoming more frequent, and getting heavier.[41] Heavy rain can be hazardous to communities,[42] roadways,[43] etc. including the dams[44] on lakes and other waterways.

Excess water in roadways and other areas not normally covered in water (streams, storm basins, etc.) can pose a danger to people – including public works staff – trying to clean out storm drains. When water flow is blocked, the excess water pressure once the blockage is cleared, can create a vacuum effect, and even kill people who get trapped in the drain works.[45]

The oversaturation of crisis communications regarding heavy rainfall predictions, watches, and warnings should be performed by Emergency Management officials. Research has shown that the more diverse the message delivery (i.e., same message coming from different trusted sources) can increase the numbers of people who take protective actions, such as evacuation.[46]

12.1.14 Seiches

While this threat is covered in Chapter 3, the hazards from seiches are like those from SLOSH, Storm Surge, and Tidal events. Similar aspects of a notice incident can exist; as there is generally a weather forecast indicating high winds, and the oscillation effect on the water can take hours to cross large lakes and other bodies of water.

12.1.15 Snow Squall Hazards: Transportation

Highway pileups are a significant hazard from snow squalls (see Part 1/Air), as the strong winds and sudden drop in temperatures, coupled with heavy and sudden bursts of snow in a very short time frame – and with no apparent warnings – can create icy road conditions, whiteouts, and other LIPER hazards. The U.S. National Weather service has an entire week of snow squall preparedness material, which can be found at www.weather.gov/otx/spokanesnowsquall.[47]

12.1.16 Swiftwater or Whitewater Rescues

Regardless of how excess water originates, when people are not prepared to evacuate themselves, then first responders may need to assist. Swiftwater rescues are very complex[48] and pose significant dangers to the rescuers themselves. FEMA has developed resource typing standards for Swiftwater/Flood Search and Rescue Teams to include these functionalities:

1. Searches for and rescues individuals who may be injured or otherwise in need of medical attention
2. Provides emergency medical care including Basic Life Support (BLS)[49]
3. Provides animal rescue
4. Transports humans and animals to the nearest location for secondary land or air transport
5. Provides shore-based and boat-based water rescue for humans and animals
6. Supports helicopter rescue operations and urban SAR in water environments for humans and animals
7. Operates in environments with or without infrastructure, including environments with compromised access to roadways, utilities, transportation, and medical facilities and with limited access to shelter, food, and water

(FEMA, 2020, p. 1)[50]

The USCG has jurisdiction for ocean rescues and some inland operations, but it is typically tactical rescue teams affiliated with local fire departments used for no-notice flooding rescues from streets, overflowing rivers, etc. Generally, this work falls under ESF#9 – Search and Rescue.[51]

12.1.17 Tropical Storms

Tropical Storms are covered in Chapter 14, in Part 5, as they are cascading threats/hazards.

12.2 CASE STUDIES

12.2.1 Buffalo, New York, Blizzard of 2022

In December 2022, a major snowstorm – a blizzard with lake effect snowfall[52] – hit Buffalo and the surrounding area of Erie County, New York. At least 44 people were killed. Surprisingly at the time, Buffalo did not have a formal emergency manager or staff (they have since hired one person – a former NYC fire official, as an Emergency Services Manager),[53] and relied then on their police, fire, and public works emergency services staff to coordinate, collaborate, cooperate, and communicate with county and state officials for resource requests, assistance, public messaging, etc. Those efforts were siloed and uncoordinated, uncooperative, and had significant failures in both communicating crisis messaging to the public and curating emergency management intelligence from whole-community partners.[54] There have been some after-action reports written[55] – including at least one independent one[56] – but with little positive actionable improvement planning successes, so far. Political in-fighting, lack of unified command, missing unity of effort, and an overall failure to prioritize the LIPER remain the status quo for this community.

12.2.2 Great Chicago Flood of 1992

While many may have heard of the Great Chicago Fire of 1871,[57] there was a Great Chicago Flood of 1992. Approximately 250 million gallons (946 million liters) of water from the Chicago River escaped down a crack into tunnels under their main downtown business district (the Loop).

> To protect workers as utility tunnels flooded, electrical service was immediately cut from the Chicago River south to Adams and from Michigan Avenue west to Dearborn Street. Hundreds of buildings were left without power for days. But an area far greater than that – south to Taylor Street – had to be evacuated.
>
> In the County Building, workers saved reams of record books going back more than 100 years, and computers were stacked under a tarp in the main hallway like boxes on a moving day. But the water swallowed up millions of dollars of office supplies and duplicate records as it turned entire floors into aquariums.
>
> The actual losses from the flood were estimated at $1.95 billion.
>
> **(CBS, 2022, p. 1)[58]**

"The flood marked a turning point for the city and downtown building owners. The unexpected crisis prompted officials to put emergency plans and other safety measures in place that have kept the tunnels in use and avoided other disasters" (Rumore, 2022, p. 1).[59]

12.2.3 Germany's Ahr Valley Flooding of 2021

Flash flooding over just two days in July of 2021 killed more than 220 people across Europe. A year later and towns in the German Ahr valley still have temporary schools, roads, and other infrastructure.

> In Bad Neuenahr-Ahrweiler, residents are still waiting for the return of normal life a year after the devastation of deadly flash flooding. About 18,000 inhabitants, or more than half the local population, were affected by the disaster in this once picturesque town in western Germany known for its thermal baths. The town's mayor, Guido Orthen, will be able to show Scholz roads cleared of the muck and debris strewn by the floodwaters that submerged Bad Neuenahr-Ahrweiler. None of the 18 bridges that used to cross the Ahr river is functional yet, with three temporary crossings installed in their place. The traces of the flood are everywhere, from the collapsed banks by the roadside to the high-water mark on many of the buildings.
>
> While officials may want to rebuild things as quickly as possible, they are also under pressure to make sure residents are protected from future floods. As it stands, "we are still living in the same dangerous situation as a year ago", Orthen says. This puts residents in a state of anxiety any time bad weather is forecast, he adds.
>
> **(Agence France Presse and Associated Press, 2022, p. 1)**[60]

A year later in 2023, some parts of Bad Neuenahr-Ahrweiler are still not restored. Insurers indicate this was the most expensive natural hazard incident in Germany, costing €8.75 billion ($9.6 billion). "Scientists at World Weather Attribution said the Ahr's extreme rain was up to nine times more probable because of global warming. Experts say such events will only become more probable as climate change intensifies" (Stickings, 2023, p. 1).[61]

12.2.4 Miami Building Collapse in 2021

The long-term water damage from rainwater and other elements on faulty building construction[62] led to a sudden condominium collapse in Miami, Florida which killed 98 people. Search and Rescue teams[63] spent two weeks recovering bodies. Multiple teams from around the world[64] came to assist – and the stress and trauma to those workers was significant.[65] There were multiple lawsuits filed, which were eventually settled out of court for more than $1 billion (€928 billion).[66]

C4 – CARBON COUNTY COORDINATION CENTER IN WYOMING

One example of using Mitigation work from one type of disaster to help Response work on another, occurred in Carbon County, Wyoming. In September of 2020, in that county's town of Hannah, there was a major wildfire (the Hannah – 316 – fire),[67] which forced the evacuation of the entire town.[68] At that time, the county had no Emergency Operations Center (EOC) to coordinate Response efforts, resource requests, communications and situational awareness/emergency management intelligence among the various partners and stakeholders.

Over the course of the next 12 months, Carbon County Emergency Management officials built an EOC in Rawlins, which they call C4,[69] from scratch using donations and matching grants. They also were successful in having each county department provide one staff member to support their EOC, which supports their overall ICS and ESFs.[70]

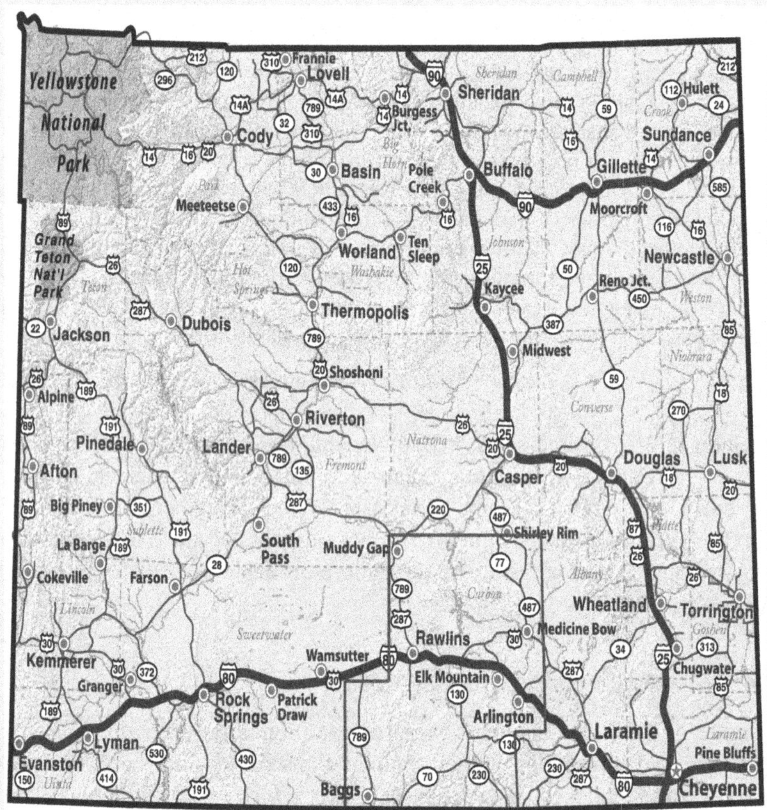

FIGURE 12.2 Wyoming Map with Carbon County highlighted.

Source: Wyoming Department of Transportation. Wyoming Department of Transportation Maps. (n.d.). State of Wyoming. Accessed April 26, 2024. https://www.dot.state.wy.us/home/travel/travel_maps.html

In March 2022, a series of cascading water-related hazards – including multiple water main breaks – occurred in Rawlins, as well as the next-door Town of Sinclair.

> During this event, the C4 had over 30 people who worked to meet the needs of our community so that the City of Rawlins Public Works staff could focus on the repairs. The C4 Staff reached out to multiple Private, City of Rawlins, Town of Sinclair, Carbon County, and State partners requesting resources and sharing a common operating picture so that all agencies were working with the same understanding, aligned with the same priority to accomplish the same set of goals. Some of these agencies included local churches, city and county departments, Wyoming Highway Patrol,[71]

Wyoming Department of Transportation,[72] Wyoming Office of Homeland Security,[73] Wyoming EPA,[74] Wyoming DEQ,[75] and other local and state partners.

(City of Rawlings, 2022, p. 4)[76]

Memorial Hospital of Carbon County[77] in Rawlins initially thought they needed to evacuate. It turns out there were a few critical care patients who only needed wastewater management support (i.e., filling toilets with water manually), and they were able to continue to operate in the community. Multiple boil alerts, points of distribution for water donations, and other community support elements were implemented.[78]

12.2.5 FUTURES 3.0: Urban Flood Risk Research

In 2023, researchers at the North Carolina State University's Center for Geospatial Analytics have created an open-source model for flood risk in urban areas utilizing exposure, hazard, and vulnerability: The greater Charlestown, South Carolina, as their prototype city. Their open-source FUTURES software is accessible through the GitHub repository[79] with the identifier doi:10.5281/zenodo.6607097.

12.2.6 Japanese River Management

In Japan, they have elevated the importance of river water quality, by implementing a River Bureau, within their Ministry of Land, Infrastructure, Transport, and Tourism.

In accordance with Japan's *River Law*, river administration is done by classifying rivers, breaking them into sections, and delegating responsibility for the administration of their various subdivisions. River systems deemed important for the national economy and people's lives are designated as "Class A river systems" and administrated by the Minister of Construction. The others are designated as "Class B river systems" and administrated by the prefectural governors.[80] See the Additional Reading section at the end of this chapter for a reference on research into evacuation notifications in Japan, from torrential rain hazards.

12.2.7 New York State's Resiliency and Economic Development Initiative

In 2023, New York State announced that – as part of their overall $300 million (€278 million) Lake Ontario Resiliency and Economic Development Initiative (REDI) – they would earmark funds for a flooding and high-water protection project for Camp Hollis, in Oswego, New York. This summer camp supports more than 2,000 at-risk children per year.

In response to the extended pattern of flooding along the shores of Lake Ontario and the St. Lawrence River, New York State established REDI to increase the resilience

> of shoreline communities and bolster economic development in the region. Five REDI Regional Planning Committees, comprised of representatives from eight counties (Niagara and Orleans, Monroe, Wayne, Cayuga and Oswego, and Jefferson and St. Lawrence) were established to identify local priorities, at-risk infrastructure and other assets, and public safety concerns. Through REDI, the State has committed up to $300 million, to benefit communities and improve resiliency in regions along Lake Ontario and the St. Lawrence River.
>
> **(New York State Department of Environmental Conservation, 2023, p. 1)**[81]

The question of where flooding has occurred is important to Water Management Districts – they too have a responsibility for flood plain management. The cost of flood insurance in Europe is moving populations away from floodplain areas.[82]

12.2.8 Red River Flooding: North Dakota, 1997

In 1997, wetter than normal conditions plus early ground freezing and heavy Spring rains created a massive flood in the Red River Valley area of North Dakota, devastating the city of Grand Forks and other places along the Minnesota–North Dakota state line. This was an incident where the weather predictions about how high the river would rise were dramatically off.

> First, hydrologists calculate the volume of water flowing past points along the Red River, using a system of river gauges. Then they used a mathematical technique based on past flood measurements to predict just how high the river would get.
>
> The river's level at East Grand Forks surpassed the previous record by a whopping four feet.
>
> Today, better river gauges and satellite data, plus new soil temperature data, will help the region know what kind of Red River to expect each spring.
>
> **(Dunbar, 2017, p. 1)**[83]

USACE – TRIBAL LANDS, FLOODING MITIGATION

In 2023, the U.S. Army Corps of Engineers (USACE) approved a flood mitigation plan for the sovereign tribal nation land of the Hopi Tribe. This was a collaborative effort across multiple state, federal, and local jurisdictions – and for this geography, also included tribal elders.

> Activities range throughout the flood risk management lifecycle and promote shared responsibility. Team activities have included state hazard mitigation plan updates, inundation mapping, tabletop exercises, and development of emergency action plans, risk communication workshops, coordination of perishable data collection, and many others.
>
> **(USACE, 2023, p. 1)**[84]

TABLE 12.2
Impacts to the Disaster Phase Cycles from Too Much Raw Water/Freshwater

Preparedness/Protection/Prevention
Work toward a continuous cycle of messaging about increasing the public's capacity to be better prepared to Respond themselves to flooding hazards, as well as protecting their homes and places of business. Government will need to support flooding protection work, such as dams and levees.
Response
Flooding incidents – whether from raw water or freshwater; notice (weather-related) or no-notice (pipe bursts, water main breaks) have massive Response phase work. There are workforce safety impacts – and many of the emergency responders who live in your community may be directly adversely impacted themselves. Communities may be overstressed from repeated flooding incidents and mitigation work takes years to complete. Resource requests, including staffing for the "disasters within the disaster" such as home fires which will occur while flooding is still in the community – can overtax first responder groups.
Recovery
Recovery work is also on a large scale – and may not reach the level of assistance from the U.S. Federal Government (i.e., Presidentially declared emergency or disaster). The thresholds to reach this vary by state/tribal/territory and adjust over time. Blizzards and other heavy snow incidents are one example where the storm must exceed the maximum of the prior storms to even generate an "ask" from the governor for Federal assistance. Without this assistance, states, territories, and tribal nations are left to finance any recovery work themselves – and private companies, individuals, etc. must have insurance coverage – especially flood insurance – in order to properly recover.
Mitigation
There are many opportunities and options for Mitigation work, from flooding incidents of any kind. Infrastructure – including personal homes – elevation[85] is one aspect. Purchasing homes in recurrent flood areas, by government, to demolish the home and restrict the land's use against further building, is another option. Entire towns are being relocated, due to recurrent flooding.[86] Academic research is being aligned to the needs of rural and other socially vulnerable populations to address flooding risks.[87] The Part 4 section text has a series of "Mitigation Paradoxes" associated with flooding.

Source: Author.

12.3 IMPACTS TO THE DISASTER PHASE CYCLES

As shown in Table 12.2, there are specific impacts to all four of the disaster cycle phases, from **Too Much Raw Water/Freshwater**.

12.4 ADVERSE IMPACTS TO THE INCIDENT COMMAND SYSTEM

There may be unique hazards directly impacting Responses and Responders, which should be considered adverse impacts to the Incident Command System (ICS). For too much raw water/freshwater, Table 12.3 details what may be occurring:

TABLE 12.3
Adverse Impacts to the Incident Command System

Command (including SO, PIO, and LNOs)

Safety Officers (SOs) need emergency management intelligence about this threat, as soon as possible – especially for hazardous open-water situations (swift water, river/street overruns, culverts, near utility drainage, etc.). These workforce safety elements are paramount. While there are always risks to responders (including pathogens in SSOs), there may be additional hazards such as oil and hazardous materials in the water, and uncharted risks for search and rescue, as well as recovery work. Dive teams, for example, will need counseling after recovering any drowning victims – so their own life safety from mental health/wellness concerns is sustained.

Public Information Officers (PIOs) need to urgently activate accurate templated messages to the public which provide critical crisis communications in the appropriate languages and modalities for oversaturation. If an "avoid the area" or "Turn Around, Don't Drown®"[88] warning is needed, for example, it needs to be communicated far and wide, and quickly.

Liaison Officers (LNOs) may be in place on this type of hazard, crossing all of the various groups which operate their own incident command structures/systems. For example, on major flooding incidents, the American Red Cross established their own NIMS-like structure and provides an LNO as the conduit between the two systems.

Intelligence

Sources of Emergency Management Intelligence for incidents involving too much raw water/freshwater will need to include the topography of the water source (river, lake, broken water main, etc.), especially if there are **now** *new* underwater hazards (oil/chemical discharges, building collapses, debris, tree branches, etc.) as well as cascading human-made hazards (power outages shutting down traffic lights near flooded areas where people are evacuating, fires, etc.). Current flow, temperature, and fluctuating water depth are but some of the factors which help with a Common Operating Picture. Knowledge of mutual-aid support capabilities for both search and rescue, as well as recovery is needed.

In 2023, researchers combined Artificial Intelligence (AI) associated with evacuation planning, along with existing flood maps in South Carolina, to produce http://floodevacuationtool.clemson.edu/.[89]

Finance/Administration

Generally, there are no adverse impacts to the Finance/Administration branch, specifically associated with a flooding hazard, if the community is properly prepared for them. Overtime and other staffing costs may exceed what is covered in normal fiscal budgets (Search and Rescue efforts from flash flooding may not reach the level of a state-declared emergency and Federal support) and/or emergency assistance reimbursements from higher governmental entities on Declared Emergencies or Disasters (Search and Rescue efforts during a Tropical Storm, for example).

Logistics

Generally, there are no adverse impacts to the Logistics branch from this hazard; again if the jurisdiction is prepared with rescue equipment, sandbags, trained staff and/or mutual aid agreements to support all of this.

Operations

Generally, there are no adverse impacts to the Operations branch – if it is properly staffed and equipped for this Response and Recovery work.

Planning

There are no specific adverse impacts to the Planning section, from this hazard. There is a plethora of planning resources available globally,[90] for flooding scenarios for water utilities,[91] local Emergency Management entities,[92] corporations, non-governmental organizations, and more.

Source: Author.

12.4.1 POETE Process Elements for This Hazard

Table 12.4 shows the specifics for too much raw water/freshwater.

TABLE 12.4
POETE Process Elements for This Hazard

Planning	As noted in the overview section and case examples in this chapter, there are many plans which need to protect against, prevent, respond to, recover, and mitigate against the adverse impacts from hazards involving too much raw water or freshwater. Plan for any area, site, building, facility, home, etc., to become flooded. Integrate and coordinate plans both across timespans (Preparedness through Response through Mitigation), nearby jurisdictions (flooding never stops at the town's border), and as a whole-community effort. There very well may be a separate annex or appendix to any jurisdiction's emergency operations plan for this specific hazard, since flooding is universal. There should also be elements of other main plan portions and annexes/appendices which cover the planning needed for a raw water/freshwater flooding incident, which can include extraordinary measures, such as search and rescue. Flooding can occur in all types of weather, indoors or outdoors, etc. From this chapter, there are adverse impacts of death and near-death for both the public and responders, historically unique references of past incidents to learn from, and elements of exercises which can be performed, in order to facilitate positive changes and modifications to preparedness, response, recovery, mitigation, and continuity of operations plans.
Organization	The **trained** staff needed to effect plans, using proper equipment – especially first responder life safety equipment[93] – should be exercised and evaluated on a regular basis each year. Actual incident response and recovery can help determine elements of POETE, including how these types of incidents are organized for staffing. Also, as part of both deployments and exercises, staff involved should be evaluated for both further training needed and also the possibility of advancement in their disaster state roles to leadership positions. A floodplain manager should be part of any jurisdiction's Emergency Management team. Knowledge of the jurisdiction's specific flood-prone areas, familiarity with the National Flood Insurance Program (NFIP), local Flood Insurance Rate Maps, Substantial Improvement/Substantial Damage concepts, and Regulatory Floodways are just some of the skills of a floodplain manager.[94]
Equipment	Equipment needed for this hazard specifically includes anything needed to rescue someone from any type of water source (ocean, floodwater, quarry, frozen lake, etc.). This should also include decontamination equipment. Also included for this hazard are flood protection devices such as floodwalls and sandbags. See the tools in this chapter and others, to monitor the levels of streams, rivers, etc. Also, incorporate this data into the jurisdiction's GIS for steady-state and disaster state planning capability. There are many types of temporary and permanent floodwater barriers, and in the long-term, elevating infrastructure can be an effective method to mitigate flooding. Consider higher levels than required by current floodplain management standards, to account for climate change impacts in the future. Also, consider the impacts to people with DAFN, with any elevations (ramps needed, in place of stairs, for example). Look to pre-stage equipment as part of the ramp-up process, to reduce transportation risks and threats which come with flooding.

(Continued)

TABLE 12.4 *(Continued)*
POETE Process Elements for This Hazard

Training	In addition to standardized Incident Command System training, specific awareness training on building capacity for swiftwater rescue should be taken by emergency management officials, from an all-hazards perspective. FEMA has both in-person and online self-study courseware in support of flooding, across the disaster phase cycle. This includes training on the NFIP,[95] FIRM interpretation,[96] and other Mitigation[97] phase work.
Exercises and evaluation	A full cycle of discussion and performance exercises should be conducted each year for flooding incidents: from human-made hazards inside buildings to massive river flooding/dam breeches, etc. These exercises should stand on their own and be part of various cascading incident scenarios – and cover continuity of operations/continuity of government, as well. Exercises should involve the whole community in an exercise planner's invitations for both additional exercise planners/evaluators and actual players, especially since many private organizations (water and power utilities, etc.) are separate legal entities from the local governments they support. In the same way, there are adversely impacted ESFs, RSFs, and CLs, as noted in the Part 4 overview, those elements should be exercised and evaluated so that the elements of the POETE cycle are reviewed, and revised as needed (i.e., agreed upon for change, as part of the Improvement Plan components). The AAR/IP is a critical part of this process step, and this chapter's References/Additional Reading section contains a few examples of open-source raw water/freshwater flooding incident AAR/IPs, and one flood exercise AAR/IP. Work with federal partners[98] on exercises. For example, the U.S. Geological Survey has a Multi-Hazards Demonstration Project, which has included severe weather impacts. Their Arkstorm Scenario is noted in the References/Additional Reading section.

Source: Author.

12.5 CHAPTER SUMMARY/KEY TAKEAWAYS

Adverse impacts, especially flooding, from too much raw water/freshwater is a recurring high probability, high impact threat. And with Climate Change, these incidents will be more intense and occur more frequently. This chapter was designed not to be a "full list" of all aspects of Emergency Management related to flooding, but rather organize some of the little known or critical elements needed before, during and after these hazards strike a community.

Please review the other chapters in this Water Quantity Part, as well as Chapter 14. In many cases, it is not the type of water (fresh vs. salt vs. other), but the fact that there is water where there should not be water – which causes the adverse impacts.

Further Emergency Management Intelligence research conducted on a continuous basis – including participating in the annual THIRA/SPR and Natural Hazard Mitigation Plan efforts (see Chapter 2 for specifics) – should be performed.

NOTES

1 Longfellow, H. (1842). The Rainy Day. *Henry Wadsworth Longfellow: Selected Works* (Lit2Go Edition). Accessed February 9, 2024. See https://etc.usf.edu/lit2go/71/henry-wadsworth-longfellow-selected-works/5030/the-rainy-day/
2 See www.usgs.gov/special-topics/water-science-school/science/water-cycle
3 James, I., & Rust, S. (2023). Tensions High as Tulare Lake Reappears on Prime Farmland; Amid Unprecedented Storms, Agricultural Communities Grapple with How to Deal with Swiftly Rising Floodwaters. *Los Angeles Times.* See www.latimes.com/environment/story/2023-03-24/as-tulare-lake-reappears-floodwaters-raise-tensions-in-san-joaquin-valley
4 See www.preventionweb.net/files/52828_04floodhazardandriskassessment.pdf
5 See www.preventionweb.net/news/mali-when-worlds-deserts-flood
6 See https://community.fema.gov/ProtectiveActions/s/article/Flood-Vehicle-Do-Not-Drive-in-Floodwaters-Turn-Around-Don-t-Drown
7 See www.latimes.com/california/story/2023-09-03/death-valley-national-park-faces-months-of-storm-repairs
8 Scarborough, J. B. (1930). The Weight of Sea Water and Its Variation with Salinity and Temperature. *United States Naval Institute Proceedings*, *56*(4/326). US Naval Institute. See www.usni.org/magazines/proceedings/1930/april/weight-sea-water-and-its-variation-salinity-and-temperature
9 See www.weather.gov/media/ajk/articles/snowloads.pdf
10 See https://education.nationalgeographic.org/resource/nile-river/
11 See https://museomaritimo.com/en/geography
12 See www.weather.gov/
13 See www.usgs.gov/products/web-tools
14 Crawford, S. E., Brinkmann, M., Ouellet, J. D., Lehmkuhl, F., Reicherter, K., Schwarzbauer, J., . . . & Hollert, H. (2022). Remobilization of Pollutants During Extreme Flood Events Poses Severe Risks to Human and Environmental Health. *Journal of Hazardous Materials*, *421*, 126691. https://doi.org/10.1016/j.jhazmat.2021.126691
15 See www.fema.gov/es/disaster/3312
16 2.4 Billion Euros (as of 2/10/24).
17 USEPA. (2023). *About the Water Infrastructure and Resiliency Finance Center.* Accessed February 3, 2024. See www.epa.gov/waterfinancecenter/about-water-infrastructure-and-resiliency-finance-center
18 See www.channel3000.com/news/water-pipe-burst-flooded-uw-madison-engineering-building-with-55k-gallons-of-water/article_c3dac440–587d-11ee-b990-f396a2f1f746.html
19 See www.wkow.com/archive/burst-pipe-closes-second-uw-madison-building/article_4eab08af-d6c6-5056-9cdc-858475d7bf15.html
20 See www.tmj4.com/news/local-news/winter-woes-schools-shut-down-over-broken-pipes-no-heat
21 See www.weather.gov/safety/winter-avalanche
22 See www.slf.ch/en/about-the-slf/
23 See www.ncbi.nlm.nih.gov/pmc/articles/PMC10431368/
24 Hoss, F., & Fischbeck, P. (2016). Increasing the Value of Uncertain Weather and River Forecasts for Emergency Managers. *Bulletin of the American Meteorological Society*, *97*(1), 85–97. https://doi.org/10.1175/BAMS-D-13-00275.1
25 See www.researchgate.net/publication/237384350_Recent_Research_on_Recreational_Boating_Accidents_and_the_Contribution_of_Boating_Under_the_Influence

26 See www.ferc.gov/industries-data/hydropower
27 NWS. (n.d.). *Floods and Flash Floods.* NOAA. Accessed February 7, 2024. See www.weather.gov/ffc/floods
28 See www.weather.gov/shv/awarenessweek_severe_flashflood
29 See www.weather.gov/unr/1888-01-12
30 See https://www3.epa.gov/npdes/pubs/ssodesc.pdf
31 USEPA Office of Wastewater Management. (1996). *Sanitary Sewer Overflows.* USEPA. See https://www3.epa.gov/npdes/pubs/ssodesc.pdf
32 Sojobi, A. O., & Zayed, T. (2022). Impact of Sewer Overflow on Public Health: A Comprehensive Scientometric Analysis and Systematic Review. *Environmental Research, 203*, 111609. https://doi.org/10.1016/j.envres.2021.111609
33 Jagai, J. S., DeFlorio-Barker, S., Lin, C. J., Hilborn, E. D., & Wade, T. J. (2017). Sanitary Sewer Overflows and Emergency Room Visits for Gastrointestinal Illness: Analysis of Massachusetts Data, 2006–2007. *Environmental Health Perspectives, 125*(11), 117007. https://doi.org/10.1289/EHP2048
34 See www.spc.noaa.gov/
35 NWS. (n.d.). *Mesoscale Discussion.* NOAA. Accessed February 7, 2024. See https://forecast.weather.gov/glossary.php?word=mesoscale%20discussion
36 See www.weather.gov/media/ajk/articles/snowloads.pdf
37 Swanson, C. (2024). Washington Drinking Water, Hydropower at Risk as Pacific Northwest Snowpack Shrinks. *The Chronicle.* See www.chronline.com/stories/washington-drinking-water-hydropower-at-risk-as-pacific-northwest-snowpack-shrinks,333895
38 See www.cnrfc.noaa.gov/
39 See https://arstechnica.com/science/2024/01/the-largest-us-dam-removal-effort-to-date-has-begun/
40 See https://skydayproject.com/torrential-rain
41 See www.sciline.org/climate/climate-change/torrential-rain/
42 See www.scientificamerican.com/article/new-york-citys-floods-and-torrential-rainfall-explained/
43 Pregnolato, M., Ford, A., Wilkinson, S. M., & Dawson, R. J. (2017). The Impact of Flooding on Road Transport: A Depth-Disruption Function. *Transportation Research Part D: Transport and Environment, 55*, 67–81. https://doi.org/10.1016/j.trd.2017.06.020
44 See www.nytimes.com/2023/09/12/world/middleeast/libya-floods-dams-collapse.html
45 See www.nbcphiladelphia.com/news/local/family-friends-mourn-man-who-died-in-storm-drain/1910859/
46 Kakimoto, R., & Yoshida, M. (2022). Evacuation Action During Torrential Rain Considering Situation Awareness Error Using Protection Motivation Theory. *International Journal of Disaster Risk Reduction, 82*, 103343. https://doi.org/10.1016/j.ijdrr.2022.103343
47 See www.weather.gov/otx/spokanesnowsquall
48 See www.domesticpreparedness.com/articles/the-importance-of-swift-water-rescue-teams
49 See www.acep.org/wilderness/newsroom/newsroom-articles/oct2019/swiftwater-rescue-a-basic-introduction
50 FEMA. (2020). *Resource Typing Definition for Response Mass Search and Rescue Operations – Swiftwater/Flood Search and Rescue Team.* FEMA. See www.fema.gov/sites/default/files/2020-08/fema_swiftwater-flood-sar-team_definition_08-03-2020.pdf
51 See www.fema.gov/sites/default/files/2020-07/fema_ESF_9_Search-Rescue.pdf
52 See www.weather.gov/buf/lesEventArchive?season=2022-2023&event=D
53 See www.wkbw.com/news/local-news/buffalo/city-of-buffalo-announces-the-hire-of-new-emergency-services-manager-and-fleet-manager
54 See https://cemir.org/store/ols/products/2022-buffalo-blizzard-aar-ip

55 See https://wagner.nyu.edu/impact/research/publications/lessons-learned-buffalo-blizzard-recommendations-for-strengthening
56 CEMIR. (2023). *2022 Buffalo Blizzard After-Action Report/Improvement Plan – An Independent Report.* York Drive, LLC. See https://cemir.org/store/ols/products/2022-buffalo-blizzard-aar-ip
57 See https://education.nationalgeographic.org/resource/chicago-fire-1871-and-great-rebuilding/
58 CBS Chicago. (2022). 30 Years Ago Today: Great Chicago Flood Paralyzes Loop Businesses. *CBS News.* Accessed February 8, 2024. See www.cbsnews.com/chicago/news/great-chicago-flood-30th-anniversary-loop-chicago-river/
59 Rumore, K. (2022). 30 Years Ago Today, the Great Chicago Flood Stunned the Loop – Pouring 124 Million Gallons of Water into City Basements. *Chicago Tribune.* See www.chicagotribune.com/2022/04/13/30-years-ago-today-the-great-chicago-flood-stunned-the-loop-pouring-124-million-gallons-of-water-into-city-basements/
60 Agence France Presse and Associated Press. (2022). After the Floods: Germany's Ahr Valley Then and Now – in Pictures. *The Guardian.* See www.theguardian.com/world/2022/jul/13/floods-then-and-now-photographs-germany-ahr-valley-flooding-disaster-july-2021
61 Stickings, T. (2023). Germany's Valley of Floods Is Braced to Withstand Future Disasters. *The National.* See www.thenationalnews.com/world/2023/08/03/germanys-valley-of-floods-is-braced-to-withstand-future-disasters/
62 Blanchard, R. L. (2022). *An Investigation and Analysis of the 2021 Surfside Condo Collapse.* University of New Orleans Theses and Dissertations, 2957. See https://scholarworks.uno.edu/td/2957
63 See www.nfpa.org/news-blogs-and-articles/nfpa-journal/2021/06/21/group-effort/miami-1
64 See www.reuters.com/world/us/israeli-mexican-rescuers-bring-distinct-experience-miami-building-collapse-2021-06-27/
65 Beidel, D. C., Rozek, D. C., Bowers, C. A., Newins, A. R., & Steigerwald, V. L. (2023). After the Fall: Responding to the Champlain Towers Building Collapse. *Frontiers in Public Health, 10,* 1104534. https://doi.org/10.3389/fpubh.2022.1104534
66 Logan, L., & Singh, A. (2023). High-Rise Condominium Collapse. *Journal of Legal Affairs and Dispute Resolution in Engineering and Construction, 15*(1), 05022008. https://doi.org/10.1061/(ASCE)LA.1943-4170.0000583
67 See www.fema.gov/press-release/20210318/fema-authorizes-funds-fight-wyomings-316-fire
68 See https://oilcity.news/wyoming/wildfire/2020/09/07/photos-fire-that-forced-hanna-residents-to-evacuate-grows-to-14201-acres-20-contained/
69 See www.carboncountywy.gov/1168/CC-OEM-Incident-Reporting
70 See https://bigfoot99.com/bigfoot99-news/governor-gordon-visits-emergency-management/
71 See https://whp.wyo.gov/
72 See www.dot.state.wy.us/home.html
73 See https://hls.wyo.gov/
74 See www.epa.gov/wy/environmental-information-wyoming
75 See https://deq.wyoming.gov/
76 City of Rawlings. (2022). *Water Infrastructure and 2022 Critical Water Event Report.* Rawlings, Wyoming. See www.rawlinswy.gov/DocumentCenter/View/16622/Water-Infrastructure-and-2022-Critical-Water-Event-Report
77 See www.imhcc.com/
78 See www.rawlinswy.gov/374/2022-Critical-Water#:~:text=The%20City%20of%20Rawlins%20water,lower%20than%20normal%20water%20intake
79 See https://github.com/ncsu-landscape-dynamics/GRASS_FUTURES
80 See www.mlit.go.jp/river/basic_info/english/admin.html

81 New York State Department of Environmental Conservation. (2023). *DEC Announces Start of Construction of Project to Protect Camp Hollis from Future Flooding and High Water.* New York State. See https://dec.ny.gov/news/press-releases/2023/11/dec-announces-start-of-construction-of-project-to-protect-camp-hollis-from-future-flooding-and-high-water
82 Tesselaar, M., Wouter Botzen, W. J., Tiggeloven, T., & Aerts, J. C. J. H. (2023). Flood Insurance Is a Driver of Population Growth in European Floodplains. *Nature Communications, 14*(1), 7483. https://doi.org/10.1038/s41467-023-43229-8.
83 Dunbar, E. (2017). *The 1997 Red River Flood: What Happened*? Minnesota Public Radio. See www.mprnews.org/story/2017/04/17/1997-red-river-flood-what-happened
84 USACE. (2023). Army Corps of Engineers Approves Flood Mitigation Plan for First Mesa and Moenkopi Villages. *Navajo-Hopi Observer.* See www.nhonews.com/news/2023/dec/05/army-corps-engineers-approves-flood-mitigation-pla/
85 See www.domesticpreparedness.com/articles/rising-above-the-flood-a-decision-tool-for-structural-safety
86 See www.bbc.com/future/article/20240130-this-louisiana-town-moved-to-escape-climate-disaster
87 See www.esf.edu/news/2023/flood_recovery_research_earns_nsf_award.php
88 Turn Around, Don't Drown® is a registered trademark of the National Weather Service. Used with permission, per see www.weather.gov/safety/flood-turn-around-dont-drown
89 See http://floodevacuationtool.clemson.edu/
90 See https://assets.gov.ie/117497/503a0c66-7e86-4ebc-8b1d-02bbcbd02195.pdf
91 See www.epa.gov/system/files/documents/2021-10/incident-action-checklist-flooding_508c-final.pdf
92 See https://community.fema.gov/ProtectiveActions/s/article/Flood-Response-Planning-Evacuation
93 See www.fema.gov/sites/default/files/2020-08/fema_swiftwater-flood-sar-team_definition_08-03-2020.pdf
94 See www.fema.gov/about/organization/region-2/floodplain-management-training
95 See https://nfipservices.floodsmart.gov/training
96 See https://training.fema.gov/is/courseoverview.aspx?code=IS-273&lang=en
97 See https://training.fema.gov/is/courseoverview.aspx?code=is-322&lang=en
98 See https://preptoolkit.fema.gov/web/em-toolkits/archive1

REFERENCES/ADDITIONAL READING

Association of State Dam Safety Officials. (n.d.). *National Dam Safety Program Research Needs: Reports and Research Papers.* Association of State Dam Safety Officials. Accessed February 2, 2024. See https://damsafety.org/national-dam-safety-program-research-needs

Austin, J. T. (2015). *Floods and Droughts in the Tulare Lake Basin* (2nd ed.). Sequoia Parks Conservancy.

Barua, P., Mitra, A., & Eslamian, S. (2022). Disaster Risk Reduction: Detecting Himalayan Glacial Lake Outburst Floods. In S. Eslamian & F. Eslamian (Eds.), *Disaster Risk Reduction for Resilience: Disaster and Social Aspects* (pp. 393–403). Springer International Publishing. https://doi.org/10.1007/978-3-030-99063-3_16

CISA. (2021). *Dam Sector Profile.* USDHS. See www.cisa.gov/sites/default/files/publications/dams-sector-profile-112221-508.pdf

City of Los Angeles Emergency Management Department. (2023). *Adverse Weather: January 2023 Storms Emergency Operations Center Activation.* City of Los Angeles. See https://ens.lacity.org/epd/eobagendas/epdeobagendas210171812_05222023.pdf

Denning, E. J. (1992). *Hazardous Materials as Secondary Results of Flooding: A Case Study of Planning and Response.* FMHI Publications, 3. See https://digitalcommons.usf.edu/fmhi_pub/3

Evans, H., Thies, C., Jantzen, L., & Maidment, D. (2022). *Report P6A1 Beaumont Flood Emergency Response Exercise.* TxDOT and University of Texas. See www.caee.utexas.edu/prof/maidment/StreamflowII/Documents/ReportP6A1Project07095.pdf

Federal Energy Regulatory Commission. (2005). *Dam Safety Performance Monitoring Program (DSPMP) and Potential Failure Modes Analysis (PFMA).* Department of Energy. See www.ferc.gov/dam-safety-and-inspections/dam-safety-performance-monitoring-program-dspmp-and-potential-failure

French, J., Ing, R., Von Allmen, S., & Wood, R. (1983). Mortality from Flash Floods: A Review of National Weather Service Reports, 1969–81. *Public Health Reports, 98*(6), 584–588. See https://pubmed.ncbi.nlm.nih.gov/6419273/

Garone, P. (2019). *The Fall and Rise of the Wetlands of California's Great Central Valley.* University of California Press. https://doi.org/10.1525/9780520948495

Geoscience Australia. (n.d.). *Australian Flood Risk Information Portal (AFRIP).* Australian Government. Accessed January 18, 2024. See www.community-safety.ga.gov.au/data-and-products/afrip

Iowa Department of Homeland Security and Emergency Management (HSEMD). (2011). *Iowa 2011 Missouri River Floods After Action Report (AAR).* Iowa HSEMD. See https://homelandsecurity.iowa.gov/wp-content/uploads/2020/11/HSEMD_AAR_2011_MoRiverFlood.pdf

Jaslow, D., Zecher, D., Synnestvedt, R., Melly, K., & Overberger, R. (2019). A New Strategy for Swiftwater Rescue from Roadways during Urban and Small Stream Flash Flooding. *Prehospital and Disaster Medicine, 34*(s1), s65–s66. https://doi.org/10.1017/S1049023X19001468

Kakimoto, R., & Yoshida, M. (2022). Evacuation Action During Torrential Rain Considering Situation Awareness Error Using Protection Motivation Theory. *International Journal of Disaster Risk Reduction, 82*, 103343. https://doi.org/10.1016/j.ijdrr.2022.103343

Kerrigan, R. W. (2023). *Megastorms, California, and You: Navigating Extreme West Coast Weather.* Dryas Press.

Maltese, A., Pipitone, C., Dardanelli, G., Capodici, F., & Muller, J.-P. (2021). Toward a Comprehensive Dam Monitoring: On-Site and Remote-Retrieved Forcing Factors and Resulting Displacements (GNSS and PS – InSAR). *Remote Sensing, 13*(8), 1543. https://doi.org/10.3390/rs13081543

Metropolitan Government of Nashville and Davidson County. (2011). *Severe Flooding May 2010 After Action Report and Improvement Plan for the Metropolitan Government of Nashville and Davidson County.* Metropolitan Government of Nashville and Davidson County. See https://filetransfer.nashville.gov/portals/0/sitecontent/OEM/docs/AARIP.pdf

Miguez, M. G., Veról, A. P., Rezende, O. M., de Sousa, M. M., & Guimarães, L. F. (2022). Flood Resilient Cities. In S. Eslamian & F. Eslamian (Eds.), *Disaster Risk Reduction for Resilience: Disaster and Social Aspects* (pp. 329–356). Springer International Publishing. https://doi.org/10.1007/978-3-030-99063-3_14

National Academies of Sciences, Engineering, and Medicine. (2024). *Modernizing Probable Maximum Precipitation Estimation.* The National Academies Press. https://doi.org/10.17226/27460

Natural Hazards Research Australia. (2023). *Hazard Note 2: Learning from Resident's Experiences of the January-July 2022 Floods in NSW and Qld.* Australian Government. See www.naturalhazards.com.au/hazardnotes/002

Porter, K., et al. (2011). *Overview of the ARkStorm Scenario: U.S. Geological Survey Open-File Report 2010–1312*. USGS. See https://pubs.usgs.gov/of/2010/1312/

Siders, A. R., & Gerber-Chavez, L. (2021). *Floodplain Buyouts: Challenges, Practices, and Lessons Learned*. The Nature Conservancy/University of Delaware. See www.nature.org/content/dam/tnc/nature/en/documents/Buyouts_Lessons_Learned_Siders_Gerber_Chavez_TNC_Full_Report_2021.pdf

Texas Commission on Environmental Quality. (2006). *Guidelines for Operation and Maintenance of Dams in Texas*. Texas Commission on Environmental Quality. See www.tceq.texas.gov/downloads/publications/gi/dam-guidelines-gi-357.pdf

Underwriters Laboratories. (n.d.). *Storm Safety*. UL. Accessed February 1, 2024. See https://code-authorities.ul.com/storm-safety/

Zhu, Y., Ma, W., Feng, H., Liu, G., & Zheng, P. (2022). Effects of Preparedness on Successful Emergency Response to Ship Accident Pollution Using a Bayesian Network. *Journal of Marine Science and Engineering*, *10*(2), 179. https://doi.org/10.3390/jmse10020179

BIBLIOGRAPHY

Agence France Presse and Associated Press. (2022). After the Floods: Germany's Ahr Valley Then and Now – In Pictures. *The Guardian*. See www.theguardian.com/world/2022/jul/13/floods-then-and-now-photographs-germany-ahr-valley-flooding-disaster-july-2021

Beidel, D. C., Rozek, D. C., Bowers, C. A., Newins, A. R., & Steigerwald, V. L. (2023). After the Fall: Responding to the Champlain Towers Building Collapse. *Frontiers in Public Health*, *10*, 1104534. https://doi.org/10.3389/fpubh.2022.1104534

Blanchard, R. L. (2022). *An Investigation and Analysis of the 2021 Surfside Condo Collapse*. University of New Orleans Theses and Dissertations, 2957. See https://scholarworks.uno.edu/td/2957

Carbon County Visitors Council. (n.d.). *Carbon County Maps – Carbon County Maps Project*. Carbon County Visitors Council. Accessed February 10, 2024. See www.wyomingcarboncounty.com/resources/maps

CBS Chicago. (2022). 30 Years Ago Today: Great Chicago Flood Paralyzes Loop Businesses. *CBS News*. Accessed February 8, 2024. See www.cbsnews.com/chicago/news/great-chicago-flood-30th-anniversary-loop-chicago-river/

CEMIR. (2023). *2022 Buffalo Blizzard After-Action Report/Improvement Plan – An Independent Report*. York Drive, LLC. See https://cemir.org/store/ols/products/2022-buffalo-blizzard-aar-ip

CISA. (2021). *Dam Sector Profile*. CISA. See www.cisa.gov/sites/default/files/publications/dams-sector-profile-112221-508.pdf

City of Rawlings. (2022). *Water Infrastructure and 2022 Critical Water Event Report*. Rawlings, Wyoming. See www.rawlinswy.gov/DocumentCenter/View/16622/Water-Infrastructure-and-2022-Critical-Water-Event-Report

Crawford, S. E., Brinkmann, M., Ouellet, J. D., Lehmkuhl, F., Reicherter, K., Schwarzbauer, J., . . . & Hollert, H. (2022). Remobilization of Pollutants During Extreme Flood Events Poses Severe Risks to Human and Environmental Health. *Journal of Hazardous Materials*, *421*, 126691. https://doi.org/10.1016/j.jhazmat.2021.126691

Dunbar, E. (2017). *The 1997 Red River Flood: What Happened*? Minnesota Public Radio. See www.mprnews.org/story/2017/04/17/1997-red-river-flood-what-happened

FEMA. (2020). *Resource Typing Definition for Response Mass Search and Rescue Operations – Swiftwater/Flood Search and Rescue Team*. FEMA. See www.fema.gov/sites/default/files/2020-08/fema_swiftwater-flood-sar-team_definition_08-03-2020.pdf

Hoss, F., & Fischbeck, P. (2016). Increasing the Value of Uncertain Weather and River Forecasts for Emergency Managers. *Bulletin of the American Meteorological Society*, *97*(1), 85–97. https://doi.org/10.1175/BAMS-D-13-00275.1

Jagai, J. S., DeFlorio-Barker, S., Lin, C. J., Hilborn, E. D., & Wade, T. J. (2017). Sanitary Sewer Overflows and Emergency Room Visits for Gastrointestinal Illness: Analysis of Massachusetts Data, 2006–2007. *Environmental Health Perspectives*, *125*(11), 117007. https://doi.org/10.1289/EHP2048

James, I., & Rust, S. (2023). *Tensions High as Tulare Lake Reappears on Prime Farmland; Amid Unprecedented Storms, Agricultural Communities Grapple with How to Deal with Swiftly Rising Floodwaters.* Los Angeles Times. See https://pageturner.us/storage/classic-marketing-services/1708015311.e126f2c7.pdf

Kakimoto, R., & Yoshida, M. (2022). Evacuation Action During Torrential Rain Considering Situation Awareness Error Using Protection Motivation Theory. *International Journal of Disaster Risk Reduction*, *82*, 103343. https://doi.org/10.1016/j.ijdrr.2022.103343

Logan, L., & Singh, A. (2023). High-Rise Condominium Collapse. *Journal of Legal Affairs and Dispute Resolution in Engineering and Construction*, *15*(1), 05022008. https://doi.org/10.1061/(ASCE)LA.1943-4170.0000583

Longfellow, H. (1842). The Rainy Day. *Henry Wadsworth Longfellow: Selected Works* (Lit2Go Edition). Accessed February 9, 2024. See https://etc.usf.edu/lit2go/71/henry-wadsworth-longfellow-selected-works/5030/the-rainy-day/

New York State Department of Environmental Conservation. (2023). *DEC Announces Start of Construction of Project to Protect Camp Hollis from Future Flooding and High Water.* New York State. See https://dec.ny.gov/news/press-releases/2023/11/dec-announces-start-of-construction-of-project-to-protect-camp-hollis-from-future-flooding-and-high-water

NWS. (n.d.a). *Floods and Flash Floods.* NOAA. Accessed February 7, 2024. See www.weather.gov/ffc/floods

NWS. (n.d.b). *Mesoscale Discussion.* NOAA. Accessed February 7, 2024. See https://forecast.weather.gov/glossary.php?word=mesoscale%20discussion

Pregnolato, M., Ford, A., Wilkinson, S. M., & Dawson, R. J. (2017). The Impact of Flooding on Road Transport: A Depth-Disruption Function. *Transportation Research Part D: Transport and Environment*, *55*, 67–81. https://doi.org/10.1016/j.trd.2017.06.020

Rumore, K. (2022). 30 Years Ago Today, the Great Chicago Flood Stunned the Loop – Pouring 124 Million Gallons of Water into City Basements. *Chicago Tribune.* See www.chicagotribune.com/2022/04/13/30-years-ago-today-the-great-chicago-flood-stunned-the-loop-pouring-124-million-gallons-of-water-into-city-basements/

Scarborough, J. B. (1930). The Weight of Sea Water and Its Variation with Salinity and Temperature. *United States Naval Institute Proceedings*, *56*(4/326). US Naval Institute.

Sojobi, A. O., & Zayed, T. (2022). Impact of Sewer Overflow on Public Health: A Comprehensive Scientometric Analysis and Systematic Review. *Environmental Research*, *203*, 111609. https://doi.org/10.1016/j.envres.2021.111609

Stickings, T. (2023). Germany's Valley of Floods Is Braced to Withstand Future Disasters. *The National.* See www.thenationalnews.com/world/2023/08/03/germanys-valley-of-floods-is-braced-to-withstand-future-disasters/

Swanson, C. (2024). Washington Drinking Water, Hydropower at Risk as Pacific Northwest Snowpack Shrinks. *The Chronicle.* See www.chronline.com/stories/washington-drinking-water-hydropower-at-risk-as-pacific-northwest-snowpack-shrinks,333895

Tesselaar, M., Wouter Botzen, W. J., Tiggeloven, T., & Aerts, J. C. J. H. (2023). Flood Insurance Is a Driver of Population Growth in European Floodplains. *Nature Communications*, *14*(1), 7483. https://doi.org/10.1038/s41467-023-43229-8

USACE. (2023). Army Corps of Engineers Approves Flood Mitigation Plan for First Mesa and Moenkopi Villages. *Navajo-Hopi Observer.* See www.nhonews.com/news/2023/dec/05/army-corps-engineers-approves-flood-mitigation-pla/

USEPA. (2023). *About the Water Infrastructure and Resiliency Finance Center.* Accessed February 3, 2024. See www.epa.gov/waterfinancecenter/about-water-infrastructure-and-resiliency-finance-center

13 Too Much Seawater

"Do you know a cure for me?"
"Why yes", he said, "I know a cure for everything. Salt water".
"Salt water?" I asked him.
"Yes", he said, "in one way or the other. Sweat, or tears, or the salt sea".

– Poet Isak Dinesen[1]

13.1 OVERVIEW

This chapter covers how there can be water quantity hazards from incidents which generate excess seawater in any community or communities. This chapter will also cover some hazards on the open seas – those **with** Emergency Management impacts – and when seawater encroaches into land areas, it normally is not found, including from sea level rise due to Climate Change. This chapter will not cover the complex threat/hazard of storms impacting oceanwater, which is covered in Chapter 14. There are many ways that this excess seawater *by itself* can be a hazard. There are similar hazards from raw water/freshwater, but those will be covered in Chapter 12. Adverse impacts of too much seawater can apply to all parts of the community affected, including commercial, industrial, and agricultural uses, as well as navigable sea waterways. These hazards will most likely require Emergency Management actions, across the full disaster phase cycle.

This chapter may seem to be applicable only to emergency managers who have coastal jurisdictions, but that is not the reality of incidents involving too much seawater. A rise in seawater can move inland very quickly, through rivers, streams, estuaries, etc. Water has a way of finding itself everywhere. U.S. President John F. Kennedy used the phrase many times of "A rising tide lifts all the boats"[2] – when talking about the economic benefits of actions in one part of the country impacting the others. This idea can also be applied more directly to disasters: A rising tide impacts all the boats – and all the houses, schools, hospitals, and everything else in its way. Whether too much seawater is causality-linked to Climate Change[3] or not – it can become a major disaster quickly. Many emergency managers and community leaders will see it happening bit-by-bit over time, before they sound the alarm to do something about it, through effective Prevention, Protection, and Preparedness work.

DOI: 10.1201/9781003474685-17

SEAWATER HAZARDS ALONE

This chapter is not going to delve into the threat/hazard specifics of Tropical Storms (Tropical Depressions, Hurricanes, Typhoons, etc.) – since their impacts for Emergency Management are compounded by wind, and they can be a combination of **both** too much freshwater **and** too much seawater. These types of complex threats/hazards will be covered in Chapter 14.

And while there is a short item in this chapter about boating hazards on the ocean, it is included because there can be a land-based community impact involving Emergency Management. There are other books on ocean and marine safety,[4] of which some are referenced at the end of this chapter.

This chapter will focus solely on terms and topics associated with the hazards of too much seawater by itself, adversely impacting a community. This will be much like the flooding impacts from raw water/freshwater, except there are the *additional* corrosive and ecologically harmful effects from saltwater, which add to the hazard.

As with other hazards, the question of causality for incidents involving too much seawater is immaterial to Emergency Management's Response and Recovery operations, except that there may be law enforcement (including the U.S. Coast Guard), natural resources, and/or environmental protection intelligence and investigation elements imbedded within the Incident Command System (ICS) which is established, specifically, for responding to and recovery from the adverse impacts of the incident. If this is the case, then actions of the other Command and General Staff branches/sections may need to be adjusted and/or expanded to support these intelligence and investigation elements. This could include investigators for criminal conduct laws including acts of terrorism, for example. There may be a parallel incident command system established by law enforcement officials and there may or may not be a liaison officer (LNO) linking the two – or more – ICSs. Each incident, emergency, disaster, etc. is different. Emergency managers should be involved in the full disaster cycle planning, organization, equipping, training, and exercising for too much seawater incidents anywhere in their respective communities (including those from adjacent jurisdictions – such as impacting a neighboring community's seawalls or shared oceanfront inlets or beaches).

13.1.1 Boating Hazards: Ocean Open-Water

There are boating and shipping hazards on the open-waters of the ocean, which impact land-based/community Emergency Management. For example, an oil spill from oil tanker or drilling platform can have significant ecological hazards.[5] Rogue waves have damaged navigation systems on ships.[6] Sometimes, the incidents occur close enough to shore[7] to involve community-based Emergency Management officials in the Response efforts for rescue and/or recovery.

13.1.2 Bridge Collapses over Ocean Waterways

This hazard is included in this chapter because while human-made accidents/incidents[8] can certainly damage or destroy a bridge, normal weather conditions – including wave actions – in the ocean can generate decay and erosion to bridge structural elements and become a hazard. These hydrodynamic threats can become hazards to elevated pile cap foundations[9] and other bridge substructure.[10]

This is *in addition to* any storm-related damage,[11] which may compromise the integrity of the bridge later, even years after the storm occurred. Bridge collapses have significant Response and Recovery work associated with them, and can harm the economy of the impacted community as well.

13.1.3 Desalination Processes

Saltwater has been converted into freshwater around the world – most of the Middle East or Gulf countries use desalination to reduce potable water supply/demand stress.

> By 2030, desalination capacity in Middle Eastern countries is expected to almost double, as part of plans announced in the region to prepare these economies for their transition to "post-oil" and to foster resilience. Saudi Arabia's desalination capacity is set to increase from 5.6 million cubic meter (m^3) per day in 2022 to 8.5 million m^3 per day in 2025, and it will have to cover more than 90% of the country's water consumption. The same holds for the UAE, Kuwait, Bahrain and Israel, where the production of desalinated water will more than double by 2030.
>
> **(Eyl-Mazzega and Cassignol, 2023, p. 1)**[12]

The discharged water[13] from these desalination plants can have an adverse effect on the marine environment,[14] and how these plants are located – including their ability to responsibly manage their wastewater processes – as it relates to sea level rise, both are concerns for global sustainability.[15]

13.1.4 Dunes

Sand Dunes and Sand Bars can be both natural and human-made, and they help prevent seawater from entering communities. Both can become degraded and even destroyed by both tidal and wave action and severe storms. And those adverse impacts may not be visible from the surface, eroding from below the waterline.[16]

The U.S. Army Corps of Engineers (USACE) has the missions for beach replenishment after disasters, which can include replacing or supplementing both Sand Dunes and Sand Bars.[17]

> A wide, nourished beach system absorbs wave energy, protects upland areas from flooding, and mitigates erosion. The beach provides a buffer between storm waves and landward areas, and it can prevent destructive waves from reaching the dunes and upland developments. When sediment is naturally moved offshore from a nourished beach, it causes waves to break farther from the shoreline, which weakens their energy before reaching the shore.

Before a project can be implemented, however, project designers must determine the necessary amount of sand to nourish the beach. Engineers develop sediment budgets, which do not involve monetary figures but rather inflows and outflows of sediment in a given coastal system.

Once a sediment budget has been calculated, sources must be found to provide the needed sediment. Beach fill material must closely match the sand on the native beach so that when waves and currents naturally distribute the fill, most of it remains on the beach and is not swept offshore. Concurrently, fill material must usually be of a texture acceptable for beach-goers.

Beach nourishment projects must be supplemented with additional quantities of sand to counteract the natural removal of sediment by waves and currents. This periodic renourishment is calculated in sediment budgets, and it results in the placement of sand at a project location usually every few years.

(USACE, n.d., p. 1)[18]

13.1.5 Intracoastal Waterway

In the United States, there is an Atlantic Intracoastal Waterway, which runs from Massachusetts to Texas, using a number of rivers, canals, bays, lagoons, and sounds. The Waterway supports all of the ports along the Atlantic Ocean and is deep enough for many commercial vessels to travel the bulk of it – with the shallowest point in the waterway at just over 6 ft (less than 2 m).

Today, the Intracoastal Waterway is used by both commercial and recreational traffic. North of Norfolk, Virginia, the waterway functions primarily as a channel for barges and other commercial vessels. South of Norfolk, you'll mostly find crafts using the waterway to travel south to resorts or enjoy the passage for a pleasurable sail.

The Gulf portion of the Intracoastal Waterway stretches for 1,100 miles between Apalachee Bay, Florida to Brownsville. The channel is 12 feet deep and 150 feet wide. Most of the waterway and the ports around Florida are used by recreational sailors. There is greater commercial activity around New Orleans and the routes east to Mobile Bay, Alabama and west to Texas. The Intracoastal Waterway connects to the Mississippi River from New Orleans and is a major shipping route for petroleum and oil field supplies.

(Hays, 2023, p. 1)[19]

And like a highway for vehicles, the Intracoastal Waterway has map markings on NOAA nautical charts, for where it traverses the water. The main path is called the "Magenta Line" – see Figure 13.1 – and is the general path (other nautical navigational tools should be used to accurately navigate the Waterway) to stay on course.

13.1.6 King Tides

King Tides are a non-scientific or colloquial term for a very high tide:

Tides are long-period waves that roll around the planet as the ocean is "pulled" back and forth by the gravitational pull of the moon and the sun as these bodies interact with the Earth in their monthly and yearly orbits. Higher than normal tides typically occur

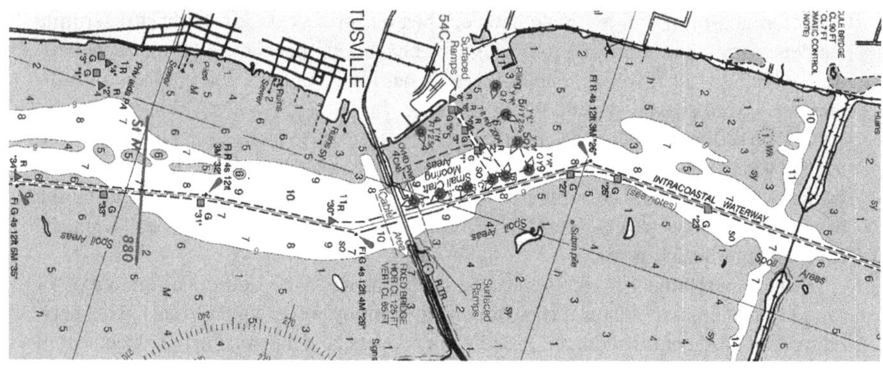

FIGURE 13.1 The Magenta Line.

Source: NOAA. Public domain.

> during a new or full moon[20] and when the Moon is at its perigee,[21] or during specific seasons around the country.
>
> **(National Ocean Service-NOAA, 2023, p. 1)[22]**

Use of the term "king tide" originated in Australia, New Zealand, and other Pacific nations to refer to an especially high tide, one which occurs only a few times per year. The term is now used in the United States as well. Climate Change and Sea Level Rise both are bringing about more King Tidal events.[23]

> The moon moves around Earth in an elliptic orbit that takes about 29 days to complete. The gravitational force is greatest when the moon is closest to Earth. The sun's gravitational force is greatest when the Earth is closest to the sun, in early January. King tides occur when the earth, moon and sun are aligned at perigee and perihelion, resulting in the largest tidal range seen over the course of a year.
>
> **(Shore Steward News, 2017, p. 6)[24]**

13.1.7 Marine Weather Forecasts

The National Weather Service in the United States provides a number of marine forecasts and warnings[25] for the coastal[26] areas of the United States, the Great Lakes,[27] and offshore[28] and high seas[29] in the North Atlantic and North Pacific. As a division of the U.S. Department of Commerce, the commercial shipping, fishing, and other industries rely on this data to support the U.S. economy. When incidents – such as oil spills, or even large-scale marine rescues – occur, knowing how the weather can impact the Response and Recovery work is critical Emergency Management Intelligence.

13.1.8 Rip Currents

Rip Currents – also known as Rip Tides – are described in this chapter, but they could have been included in Chapter 11, as they are dangerous – even deadly – for non-expert swimmers, and the leading hazard for surf zone rescues.

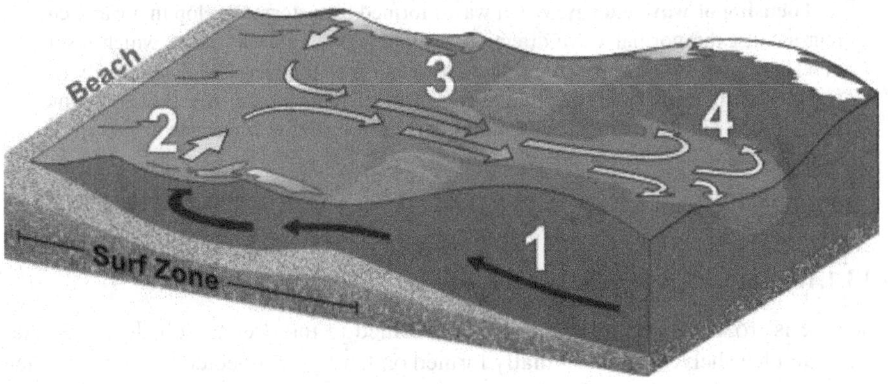

FIGURE 13.2 Anatomy of the rip current.

Source: NOAA. (2023). *Rip Currents.* U.S. Department of Commerce. See www.noaa.gov/media/image_download/fa0293ef-3e1b-4a5a-92b4-c92f17dfaf31. Public domain.

> Rip currents are powerful, channeled currents of water flowing away from shore. They typically extend from the shoreline, through the surf zone, and past the line of breaking waves. Rip currents can occur at any beach with breaking waves, including the Great Lakes.
>
> Rip currents most commonly form at low spots or breaks in sandbars and near structures such as groins, jetties, and piers. Rip currents can be very narrow or be hundreds of yards wide. The seaward pull of rip currents also varies: sometimes the rip current ends just beyond the line of breaking waves, but other times, rip currents continue to push hundreds of yards [meters] offshore.
>
> **(NOAA, 2023c, p. 1)[30]**

13.1.9 Rogue Waves

Rogue waves – once called "killer" or "freak" waves – were marine folklore for many years, but in the past few decades have had scientific evidence produced of their existence.[31] Generally described as a wave which is twice the size (height) of the nearby waves, they are very unpredictable and can even come from a different direction than the other waves in the area.

> Most reports of extreme storm waves say they look like "walls of water". They are often steep-sided with unusually deep troughs.
>
> Since these waves are uncommon, measurements and analysis of this phenomenon is extremely rare. Exactly how and when rogue waves form is still under investigation, but there are several known causes:
>
> **Constructive interference**. Extreme waves often form because swells, while traveling across the ocean, do so at different speeds and directions. As these swells pass through one another, their crests, troughs, and lengths sometimes coincide and reinforce each other. This process can form unusually large, towering waves that quickly disappear. If the swells are travelling in the same direction, these mountainous waves may last for several minutes before subsiding.

> **Focusing of wave energy**. When waves formed by a storm develop in a water current against the normal wave direction, an interaction can take place which results in a shortening of the wave frequency. This can cause the waves to dynamically join together, forming very big "rogue" waves. The currents where these are sometimes seen are the Gulf Stream[32] and Agulhas current.[33] Extreme waves developed in this fashion tend to be longer lived.
>
> **(NOAA, 2023, p. 1)[34]**

13.1.10 Sea Ice and Icebergs

Sea Ice is frozen ocean water, which was formed in the ocean itself. Icebergs, glaciers, and ice shelves were originally formed on land (or connected to land) and then broke off into the ocean.

> Sea ice keeps the polar regions cool and helps moderate global climate. Sea ice has a much brighter surface compared to many other Earth surfaces, particularly the surrounding ocean. The darker ocean reflects only 6 percent of the sun's energy and absorbs the rest, while sea ice reflects 50 to 70 percent of the incoming energy.
>
> Snow has an even higher ability to reflect solar energy than sea ice. Snow-covered sea ice reflects as much as 90 percent of the incoming solar radiation. This serves to insulate the sea ice, maintaining cold temperatures and delaying ice melt in the summer.
>
> As sea ice melts in the summer, it exposes the dark ocean surface. Instead of reflecting 80 percent of the sunlight, the ocean reflects only about 10 percent. So, oceans absorb about 90 percent of the sun's energy, causing them to heat while surrounding temperatures rise further.
>
> Thus, a small temperature increase at the poles leads to still greater warming over time, making the poles the most sensitive regions to climate change on Earth. According to scientific measurements, both the thickness and extent of summer sea ice in the Arctic have shown a dramatic decline since the 1980s. This is consistent with observations of a warming Arctic. The loss of sea ice also has the potential to accelerate global warming trends and to change climate patterns.
>
> **(National Snow and Ice Data Center, n.d., p. 1)[35]**

Icebergs can damage ships,[36] as well as infrastructure such as bridges. Icebergs can create ice dams or jams, too. Icebergs in North America are monitored by the USCG's International Ice Patrol.[37] As ocean temperatures rise, more sea ice is shrinking,[38] and more icebergs are forming.[39]

13.1.11 Sea Level Rise

Hazards from Sea Level Rise include increased sporadic flooding (tidal, storm-related), as well as systemic permanent seawater level rise. Sea Level Rise – which can be a resulting hazard from the threat of Climate Change – is when the ocean water level – commonly referred to as "sea level" permanently elevates. Infrastructure such as roads, bridges, homes, schools, etc. which were built so many feet (decimeters) above the sea level at the time, may flood today, through coastal inundation. It is critical for emergency managers to be involved in the urban planning efforts which are being adversely impacted by Sea Level Rise. NOAA has collected[40] several science-based online tools to help communities make better informed decisions about urban planning and emergency management. One of these is an interactive tool for selected coastal communities in the continental United States. It is their Sea Level Rise Viewer, and can be found at https://coast.noaa.gov/slr/[41]

13.1.12 Seawalls

Unlike dunes, which are predominately **temporary** constructs made of sand – they usually do not survive major storms and have to be continuously monitored and replenished – seawalls (also known as revetments) are constructed of building materials[42] and rocks. And they, too, may be temporary and not last lifetimes.[43] One of the factors involved in the different levels of coastal flood categories,[44] issued by the National Weather Service, is what adverse impacts, if any, are to seawalls.

Proposals for seawalls in urban cities can be a challenge, as they block views and access to the water: a "Catch-22"[45] in that the threat and hazard are what people seek out.[46] Their primary purpose was economic and used commercial building engineering (i.e., taller, harder), as their architecture. Today, an emphasis[47] on biodiversity and the ecology[48] (wider, softer) is gaining momentum as improved Mitigation work.

> Along many developed coastlines, the majority of nearshore ecosystems have been drastically altered as a result of shoreline hardening. In order to help restore these damaged ecosystems, scientists and engineers have begun to apply living shoreline principles on developed shorelines. Known as "green shores" or "living shores" or "nature-based shorelines", these ecoshorelines attempt to balance the needs of the natural and built environment. When successful, these ecoshorelines often create not only ecological benefits, but also recreational, aesthetic, educational, monetary, and even structural benefits.
>
> **(Miller et al., 2022, p. 1)**[49]

13.1.13 SLOSH

Since SLOSH – Sea, Lake, and Overland Surges from Hurricanes – is a *complex* model, calculated using atmospheric pressure, tides, and multiple wind factors,[50] it will be covered in Chapter 14.

13.1.14 Standard Seawater

The salinity – or salt – concentration levels in seawater is important to the ecology of the water, and for marine research purposes. That salinity can vary across the globe. The concept of a world "standard" to measure against, dates to 1899.

> But what is standard seawater? It's natural open-ocean seawater, collected from the ocean surface at the edge of the Sargasso Sea in the North Atlantic where there are no

land boundaries. Its boundaries are ocean currents. The Gulf Stream lies to the west, the North Atlantic Current to the north, the Canary Current to the east, and the North Atlantic Equatorial Current to the south.

Seawater from this patch of ocean is transported in containers to the Standard Seawater Service facility in England. There, it is filtered and circulated for several weeks and gradually diluted with distilled water to reach a final salinity near 35. That's Standard Atlantic seawater.

(NOAA Fisheries, 2021, p. 1)[51]

13.1.15 Tides and Ocean Currents

As noted in the SLOSH section, Tides have a significant impact on seawater levels – and therefore can be a major factor in coastal flooding and storm surges from severe weather incidents. Tides are caused by the navigational pull of the moon and the sun, which are

very long-period waves that move through the ocean and progress toward the coastlines where they appear as the regular rise and fall of the sea surface. The same happens in the Great Lakes, although the largest tides in the Great Lakes are only about 5 cm and are mostly impacted by precipitation, evaporation and runoff.

(NOAA, n.d.b, p. 1)[52]

NOAA maintains a 200+ permanent water-level sensors on the U.S. coastline and in the Great Lakes. From these data sources, they can provide the official national tidal predictions.

Ocean currents are the movement of water within and from one ocean to another. The water is moved[53] by wind, water density and temperature differences, and the tides. The Gulf Stream is an example of a large current:

The Gulf Stream is an intense, warm ocean current in the western North Atlantic Ocean. It moves north along the coast of Florida and then turns eastward off of North Carolina, flowing northeast across the Atlantic.

The velocity of the Gulf Stream current is fastest near the surface, with the maximum speed typically about nine kilometers per hour (5.6 miles per hour). The average speed of the Gulf Stream, however, is four miles per hour (6.4 kilometers per hour). The current slows to a speed of about 1.6 kilometers per hour (one mile per hour) as it widens to the north. The Gulf Stream transports an amount of water greater than that carried by all of the world's rivers combined.

(NOAA, 2023, p. 1)[54]

NOAA provides U.S. commercial vessel operators with tide and current predictions, as well as operational forecasts. "More than 95% of all U.S. trade involves some type of maritime transport, and ships move $11.4 trillion [€10.6 trillion] worth of products in and out of U.S. ports every year" (NOAA – Center for Operational Oceanographic Products and Services, n.d.b, p. 1)[55]

NOAA'S PHYSICAL OCEANOGRAPHIC REAL-TIME SYSTEM – PORTS®

NOAA has a real-time program called PORTS®, which provides navigational safety, and helps ensure coastal marine resources are protected.

> The National Ocean Service (NOS) is responsible for providing real-time oceanographic data and other navigation products to promote safe and efficient navigation within U.S. waters. The need for these products is great and rapidly increasing; maritime commerce has tripled in the last 50 years and continues to grow. Ships are getting larger, drawing more water and pushing channel depth limits to derive benefits from every last inch of draft. By volume, more than 95 percent of U.S. international trade moves through the nation's ports and harbors, with about 50 percent of these goods being hazardous materials. A major challenge facing the nation is to improve the economic efficiency and competitiveness of U.S. maritime commerce, while reducing risks to life, property, and the coastal environment. With increased marine commerce comes increased risks to the coastal environment, making marine navigation safety a serious national concern. From 1996 through 2000, for example, commercial vessels in the United States were involved in nearly 12,000 collisions, allisions, and groundings.
>
> PORTS® has the potential to save the maritime insurance industry from multi-million dollar claims resulting from shipping accidents. PORTS® is accessible to maritime users in a variety of user-friendly formats, including telephone voice response and Internet. PORTS® also provides forecasts via numerical circulation models. Telephone voice access to accurate real-time water-level information allows U.S. port authorities and maritime shippers to make sound decisions regarding loading of tonnage (based on available bottom clearance), maximizing loads, and limiting passage times without compromising safety.
>
> **(NOAA, n.d., p. 1)[56]**

13.1.16 Tropical Storms

Tropical Storms are covered in Chapter 14, in Part 5, as they are cascading threats and/or hazards.

13.1.17 Tsunamis and Tidal Waves

Generally, it is an earthquake threat which generates tsunami[57] hazards, they are covered in more detail in Chapter 14. Although both are sea waves, a tsunami and a tidal wave are two different and unrelated phenomena. A tidal wave is a shallow water wave caused by the gravitational interactions between the Sun, Moon, and the Earth. The wording of "Tidal wave" was used in the past to describe what is now called a tsunami.

> Tsunamis are ocean waves triggered by earthquakes that occur near or under the ocean, volcanic eruptions, submarine landslides, or onshore landslides in which large

volumes of debris fall into the water. Compared to other hazards such as hurricanes or forest fires that occur annually, large tsunamis are infrequent. However, when a large tsunami impacts a populated coastal region, the effects can be devastating. Waves of over 100 feet tall can bring massive volumes of water miles inland, destroying everything in their path.

(Applied Sciences Program – NASA, n.d., p. 1)[58]

There are generally no *long-term* additional **water-quality** hazards associated with either earthquakes or tsunamis unless the tsunami or tidal wave damaged a water supply system. Focusing on the direct adverse **water quantity** impact from these two hazards, it should be noted that tsunamis and tidal waves are not synonymous.

13.2 CASE STUDIES

13.2.1 Alaska – Bursting Glacial Ice Dam

When any dam breaks, excess water will flow out, causing adverse impacts to the communities through flooding. In Alaska's capital of Juneau in 2023,[59] a glacial ice dam burst in the Mendenhall Glacier,[60] which then sent water into the Mendenhall lake, which overflowed into the Mendenhall river.

Glacial ice dam breeches, also known as jökuhlaup, target over 15 million people globally, and more than half of that group live in only four countries (India, Pakistan, Peru, and China).[61]

13.2.2 Coastal U.S. Communities Will Experience Higher-End Sea Level Rise

Researchers in 2022, identified that the trajectory for sea level rise in coastal U.S. communities will rise toward the higher end of the predictive models.[62]

By 2050, sea level along contiguous U.S. coastlines could rise as much as 12 inches (30 centimeters) above today's waterline, according to researchers who analyzed nearly three decades of satellite observations. The results from the NASA Sea Level Change Team could help refine near-term projections for coastal communities that are bracing for increases in both catastrophic and nuisance flooding[63] in coming years.

Global sea level has been rising for decades in response to a warming climate, and multiple lines of evidence indicate the rise is accelerating. The new findings support the higher-range scenarios outlined in an interagency report[64] released in February 2022. That report, developed by several federal agencies – including NASA, the National Oceanic and Atmospheric Administration (NOAA), and the U.S. Geological Survey – expect significant sea level rise over the next 30 years by region. They projected 10 to 14 inches (25 to 35 centimeters) of rise on average for the East Coast, 14 to 18 inches (35 to 45 centimeters) for the Gulf Coast, and 4 to 8 inches (10 to 20 centimeters) for the West Coast.

(Younger, 2022, p. 1)[65]

13.2.3 Sierra Leone Island Is Disappearing

Sierra Leone's Nyangai Island was known for its fishing villages and lush forests as part of its almost half-mile (700 m) landscape. The island is quickly disappearing. In 2023, the land is now only 300 ft × 250 ft (91 m × 76 m) and many parts of the island are submerged when the tide rises.

> With nearly a third of its population living in coastal areas, and its heavy reliance on subsistence agriculture and fishing, Sierra Leone has been identified as one of the world's most vulnerable countries to the impacts of climate change, despite having contributed just a tiny fraction of global CO2 emissions. With a GDP per capita of barely $2,000 [€1,875], it is also one of the least prepared to deal with those impacts.
>
> With global sea levels projected to rise by anywhere between 1 and 3 feet by the end of the century, along with an increase in extreme weather events, the experience of this West African island offers a glimpse into the possible fate of countless other low-lying areas around the world.
>
> **(Trenchard, 2023, p. 1)[66]**

13.2.4 Kiribati and Tuvalu Are Fighting against Rising Seas in the Pacific

In many of the same ways as the islands off Siera Leone are adversely impacted by sea level rise, Kiribati and Tuvalu are facing challenges from King Tides, in the Pacific.

> Professor James Renwick from Victoria University of Wellington says sea levels in the western tropical Pacific have risen faster than just about anywhere on Earth.
>
> "As the sea-level rises even by a few centimetres, a king tide or a storm surge can come much further inland".
>
> It is very hard to predict at what point the low-lying atolls will become uninhabitable, Prof Renwick says.
>
> One of the things about the way some of these physical systems work in the tropics, the strengthening of trade winds and so on, can actually lead to the accretion of sand on some of the atolls, and some of the islands in the Pacific have actually been observed to be getting bigger.
>
> "So it's a complicated story, but as the sea-level continues to rise and if we don't limit greenhouse gas emissions soon and the rising sea-level accelerates . . . then the low-lying atolls could become uninhabitable by the end of the century at the very latest".
>
> **(Bramwell, 2016, p. 1)[67]**

NASA's Sea Level Change Team has partnered with the United Nations, to help research and provide scientific background to help project how fast the sea levels will rise, and how an increased amount of scientific intelligence can improve planning for future incidents.

> In a first step, the NASA team has released a new technical assessment[68] of the island nation's future environmental prospects. The report includes projections for how fast

sea levels are likely to rise and how frequently flooding might occur, and concludes that assessments and increased scientific monitoring could improve planning for potentially catastrophic events.

The NASA science report is designed to advance the goals of the Rising Nations Initiative (RNI), a project set in motion by the U.N. Global Centre for Climate Mobility.

"The RNI was launched by heads of state in September 2022, and works to protect the statehood of the Pacific atoll countries, preserve their sovereignty and safeguard their rights and heritage", said Professor Kamal Amakrane, managing director of the Global Centre for Climate Mobility. "It marks remarkable leadership by the atoll countries to determine their own futures, and we at the RNI are honored to support this, working closely with renowned global institutions like NASA".

Leaders in Tuvalu say increased scientific support is vital.

"I welcome the partnership between NASA and the Rising Nations Initiative to help provide the much-needed data, science and monitoring to support Tuvalu and the Pacific region", said Seve Paeniu, Tuvalu's minister of finance and climate change. "This effort is key to ensuring proper early warning and data-based and science-led policy design".

(Brennan, 2023, p. 1)[69]

13.2.5 The Impacts of Saltwater Anglers on Ocean Ecosystems and Coastal Economies in New Jersey

As disaster Recovery efforts include both environmental and economic impacts to communities, even the smallest human intervention can be a factor. New Jersey is registering saltwater anglers – those who fish on land and at sea, generally for recreational purposes, with a rod and reel to catch one fish at a time – to research the ocean ecosystems and coastal economies.

It is part of a national overhaul of the way NOAA collects and reports recreational fishing data. The goal of the initiative – known as the Marine Recreational Information Program, or MRIP – is to provide the most accurate information possible that can be used to determine the health of fish stocks. Reliable, universally trusted data will in turn aid anglers, fisheries managers and other stakeholders in their combined efforts to effectively and fairly set the rules that will ensure the longterm sustainability of recreational fishing.

(NJ Fish & Wildlife, n.d., p. 1)[70]

For emergency managers in environmental protection departments, and for community-based emergency managers, understanding the long-term roles of environmental impacts and sustainability are both critical to restoring environmental and economic capabilities, after a disaster which impacts nearby seawater.

13.2.6 Washington State's Quinault Indian Nation – and Sea Level Rise

The Quinault Indian Nation is moving many of its villages away from the shoreline, as a significant concern from both sea level rise and the potential adverse impacts from a tsunami wave. The village of Taholah is being relocated – a process which may take more than ten years – approximately a mile (1.6 km) upland, and further above sea level. And while this project is projected to cost upward of $450 million (€415 million), the Quinault government does not have enough financial resources from its timber production and casino operations to fully support the move.

> Coastal property owners often prefer higher sea walls or engineering solutions to try to hold back a rising sea over relocating or abandoning valuable waterfront real estate.
>
> "There are lots of examples of places that no longer exist on the map, so this has happened before", said oceanographer Ian Miller with Washington Sea Grant at the University of Washington. "There are not too many instances of it happening in this kind of planned, managed, coordinated way".
>
> Taholah, about three hours west of Seattle, has done more to confront that hotter future than most places in the U.S. and it is gradually becoming one of the first American communities to move inland as sea levels rise.
>
> **(Ryan, 2024, p. 1)**[71]

13.3 IMPACTS TO THE DISASTER PHASE CYCLES

As shown in Table 13.1, there are specific impacts to all four of the disaster cycle phases, from **Too Much Seawater**.

13.4 ADVERSE IMPACTS TO THE INCIDENT COMMAND SYSTEM

There may be unique hazards directly impacting Responses and Responders themselves, which should be considered adverse impacts to the Incident Command System (ICS). For too much seawater, Table 13.2 details what may be occurring:

TABLE 13.1
Impacts to the Disaster Phase Cycles from Too Much Seawater

Preparedness/Protection/Prevention
Work toward a continuous cycle of messaging about increasing the public's capacity to be better prepared to Respond themselves to seawater flooding hazards (including its corrosive effects), as well as protecting their homes and places of business. Government will need to support flooding protection work, such as sea walls and dunes, while supporting and sustaining local marine ecology.
Response
Flooding incidents, including infrastructure damage/destruction – whether from tidal seawater water; notice (weather-related) or no-notice (sea wall collapses, ice dam hazards) have massive Response phase work. There are workforce safety impacts – and many of the emergency responders who live in your community may be directly adversely impacted themselves. Communities may be overstressed from repeated flooding incidents and mitigation work takes years to complete. Resource requests, including staffing for the "disasters within the disaster" such as home fires which will occur while flooding is still in the community, can overtax first responder groups.

(Continued)

TABLE 13.1 ***(Continued)***

Impacts to the Disaster Phase Cycles from Too Much Seawater

Recovery

Recovery work is also on a large scale – yet may not reach the level of assistance from the U.S. Federal Government (i.e., Presidentially declared emergency or disaster). The thresholds to reach this vary by state/tribal/territory and adjust over time. A single seawall or bridge collapse are two examples where the recovery costs will probably need to be supported locally and they will probably not rate an "ask" from the governor for Federal assistance. Without this assistance, states, territories, and tribal nations are left to finance any recovery work themselves – and private companies, individuals, etc. must have insurance coverage – especially flood insurance – in order to properly recover.

Mitigation

There are many opportunities and options for Mitigation work, from flooding incidents of any kind. Infrastructure – including personal homes – elevation is one aspect. Purchasing homes in recurrent seawater-based flood areas, by government, to demolish the home and restrict the land's use against further building is another option. The Part 4 section text has a series of "Mitigation Paradoxes" associated with flooding of any kind. Chances are that the higher the economic value of coastal property, the higher the propensity for it to become flooded.

Source: Author.

TABLE 13.2

Adverse Impacts to the Incident Command System

Command (including SO, PIO, and LNOs)

Safety Officers (SOs) need emergency management intelligence about this threat, as soon as possible – especially for hazardous open-water situations (rip tides, ice dams, water rescues, river/street overruns, near utility drainage, etc.). These workforce safety elements are paramount. While there are always risks to responders (including additional threats in oceanwater, such as pathogens and marine life), there may be additional hazards such as oil and hazardous materials in the water, and uncharted risks for search and rescue, as well as recovery work. Dive teams, for example, will need counseling after recovering any drowning victims – so their own life safety from a mental health/wellness is sustained.

Public Information Officers (PIOs) need to urgently activate accurate templated messages to the public which provide critical crisis communications in the appropriate languages and modalities for oversaturation. If an "avoid the area" or "Turn Around, Don't Drown®"[72] warning is needed, for example, it needs to be communicated far and wide, and quickly. It is highly likely that geographical areas which experience recurrent tidal flooding, for example, already understand the risks associated with this hazard – but it bears repeating for new residents and with Climate Change, the intensity and recurrences will only increase over time.

Liaison Officers (LNOs) may be in place on this type of hazard, crossing all of the various groups which operate their own incident command structures/systems. For example, on major flooding incidents, the American Red Cross established their own NIMS-like structure and provides an LNO as the conduit between the two systems. The same will be true for the U.S. Coast Guard.

TABLE 13.2 ***(Continued)***

Adverse Impacts to the Incident Command System

Intelligence
Sources of Emergency Management Intelligence for incidents involving too much seawater will need to include the topography of the water source (inlet, sound, bay, piers, ports, etc.), especially if there are **now** *new* underwater hazards (oil/chemical discharges, bridge collapses, debris, sunken or displaced boats, etc.) as well as cascading human-made hazards (power outages shutting down traffic lights near flooded areas where people are evacuating, fires, etc.). The current flow, tidal impact, temperature, and fluctuating water depth are but some of the factors which help with a Common Operating Picture. Knowledge of mutual-aid support capabilities for both search and rescue, as well as recovery is needed.
Finance/Administration
Generally, there are no adverse impacts to the Finance/Administration branch, specifically associated with a seawater flooding hazard, if the community is properly prepared for them. Overtime and other staffing costs may exceed what is covered in normal fiscal budgets (Search and Rescue efforts from flash flooding may not reach the level of a state-declared emergency and Federal support) and/or emergency assistance reimbursements from higher governmental entities on Declared Emergencies or Disasters (Search and Rescue efforts during a Tropical Storm, for example).
Logistics
Generally, there are no adverse impacts to the Logistics branch from this hazard, again if the jurisdiction is prepared with rescue equipment, sandbags, trained staff and/or mutual aid agreements to support all of this.
Operations
Generally, there are no adverse impacts to the Operations branch – if it is properly staffed and equipped for this Response and Recovery work.
Planning
There are no specific adverse impacts to the Planning section, from this hazard. There is a plethora of planning resources available globally,[73] for flood risk management,[74] local Emergency Management entities,[75] corporations, non-governmental organizations, and more.

Source: Author.

13.4.1 POETE Process Elements for This Hazard

Table 13.3 shows the specifics for **too much seawater**.

13.5 CHAPTER SUMMARY/KEY TAKEAWAYS

Please review the other chapters in this Water Quantity Part, as well as Chapter 14. In many cases, it is not the type of water (fresh vs. salt vs. other), but the fact that there is any type or amount of water in places where there should not be water. That is what causes the adverse impacts. While there is some repetition between these chapters, there are specifics in each, which can add to the overall intelligence curation for an emergency manager.

Further Emergency Management Intelligence research conducted on a continuous basis – including participating in the annual THIRA/SPR and Natural Hazard

TABLE 13.3
POETE Process Elements for This Hazard

Planning	There very well may be a separate annex or appendix to any jurisdiction's emergency operations plan for this specific hazard, since flooding is universal. There should also be elements of other main plan portions and annexes/appendices which cover the planning needed for a seawater flooding incident, which can include extraordinary measures, such as search and rescue. Flooding can occur in all types of weather, and during any season of the year. From this chapter, there are adverse impacts of death and near-death for both the public and responders, historically unique references of past incidents to learn from, and elements of exercises which can be performed, in order to facilitate positive changes and modifications to preparedness, response, recovery, mitigation, and continuity of operations plans.
Organization	The trained staff needed to effect plans, using proper equipment – especially first responder life safety equipment[76] – should be exercised and evaluated on a regular basis each year. Actual incident response and recovery can help determine elements of POETE, including how these types of incidents are organized for staffing. Also, as part of both deployments and exercises, staff involved should be evaluated for both further training needed and also the possibility of advancement in their disaster state roles to leadership positions.
Equipment	Equipment needed for this hazard specifically includes anything needed to rescue someone from any type of water source (ocean, floodwater, quarry, frozen lake, etc.). This should also include decontamination equipment, as saltwater can corrode metal and electronics. Also included for this hazard are flood protection devices such as floodwalls and sandbags, which may be needed to temporarily replace sections of seawalls and dunes, for example.
Training	In addition to standardized Incident Command System training, specific awareness training on building capacity for open-water rescue should be taken by emergency management officials, from an all-hazards perspective.
Exercises and evaluation	A full cycle of discussion and performance exercises should be conducted each year for flooding incidents: from human-made hazards of damage to dunes and sea walls, to massive tidal flooding/sea level rise, etc. These exercises should stand on their own and be part of various cascading incident scenarios – and cover continuity of operations/continuity of government, as well. Exercises should involve the whole community in an exercise planner's invitations for both additional exercise planners/evaluators and actual players, especially since many non-local organizations (USCG, USACE, etc.) are separate legal entities from the local governments they support. In the same way, there are adversely impacted ESFs, RSFs, and CLs as noted in the overview under Part 4, those elements should be exercised and evaluated so that the elements of the POETE cycle are reviewed, and revised as needed (i.e., agreed upon for change, as part of the Improvement Plan components). The AAR/IP is a critical part of this process step, and this chapter's References/Additional Reading section contains a few examples of open-source seawater flooding incident reports and research, which can be the basis for improvement planning everywhere.

Source: Author.

Mitigation Plan efforts (see Chapter 2 for specifics) in your jurisdiction – should also be performed.

NOTES

1 Dinesen, I., & Blixen, K. (1972). *Seven Gothic Tales*. Vintage Books.
2 Kennedy, J. F. (1960). *Remarks of Senator John F. Kennedy, Picnic, Muskegon, MI*. The American Presidency Project. See www.presidency.ucsb.edu/documents/remarks-senator-john-f-kennedy-picnic-muskegon-mi
3 See https://thehill.com/newsletters/energy-environment/4310054-federal-report-projects-widespread-climate-consequences/
4 See www.routledge.com/Marine-Extremes-Ocean-Safety-Marine-Health-and-the-Blue-Economy/Techera-Winter/p/book/9780367662769
5 Barron, M. G., Vivian, D. N., Heintz, R. A., & Yim, U. H. (2020). Long-Term Ecological Impacts from Oil Spills: Comparison of *Exxon Valdez*, *Hebei Spirit*, and Deepwater Horizon. *Environmental Science & Technology*, *54*(11), 6456–6467. https://doi.org/10.1021/acs.est.9b05020
6 See www.cbsnews.com/news/rogue-wave-cuts-power-norwegian-cruise-ship-ms-maud-deadly-storm-northern-europe/
7 See https://skift.com/2013/01/11/is-cruising-safer-one-year-after-the-costa-concordia-disaster/
8 See www.aol.com/bridge-snaps-half-deadly-cargo-054655582.html
9 Wei, K., Zhou, C., Zhang, M., Ti, Z., & Qin, S. (2020). Review of the Hydrodynamic Challenges in the Design of Elevated Pile Cap Foundations for Sea-Crossing Bridges. *Advances in Bridge Engineering*, *1*. https://doi.org/10.1186/s43251-020-00020-9
10 Li, X., Meng, Q., Wei, M., Sun, H., Zhang, T., & Su, R. (2023). Identification of Underwater Structural Bridge Damage and BIM-Based Bridge Damage Management. *Applied Sciences*, *13*(3), 1348. https://doi.org/10.3390/app13031348
11 Zhu, D., Li, Y., Dong, Y., & Yuan, P. (2021). Long-Term Loss Assessment of Coastal Bridges from Hurricanes Incorporating Overturning Failure Mode. *Advances in Bridge Engineering*, *2*(1), 10. https://doi.org/10.1186/s43251-020-00030-7
12 Eyl-Mazzega, M., & Cassignol, É. (2022). The Geopolitics of Seawater Desalination. *Études de l'Ifri*. IFRI. See www.policycenter.ma/sites/default/files/2023-01/Eyl-Mazzega_Cassignol_Desalination_US_2022.pdf
13 Panagopoulos, A., & Haralambous, K.-J. (2020). Environmental Impacts of Desalination and Brine Treatment: Challenges and Mitigation Measures. *Marine Pollution Bulletin*, *161*, 111773. https://doi.org/10.1016/j.marpolbul.2020.111773
14 Roberts, D. A., Johnston, E. L., & Knott, N. A. (2010). Impacts of Desalination Plant Discharges on the Marine Environment: A Critical Review of Published Studies. *Water Research*, *44*(18), 5117–5128. https://doi.org/10.1016/j.watres.2010.04.036
15 United Nations Environment Programme. (2019). *Towards Sustainable Desalination*. United Nations. See www.unep.org/news-and-stories/story/towards-sustainable-desalination
16 Mignone, A. R., Jr. (2013). *Dunes and Ocean Front Structures Under Wave Attack*. NOAA/National Weather Service. See www.weather.gov/media/erh/ta2013-02.pdf
17 See www.iwr.usace.army.mil/Missions/Coasts/Tales-of-the-Coast/Corps-and-the-Coast/Shore-Protection/Beach-Nourishment/
18 USACE. (n.d.). *Beach Nourishment*. U.S. Army. Accessed February 13, 2024. See www.iwr.usace.army.mil/Missions/Coasts/Tales-of-the-Coast/Corps-and-the-Coast/Shore-Protection/Beach-Nourishment/

19 Hays, C. (2023). *How to Navigate the Intracoastal Waterway*. American Nautical Services. See www.amnautical.com/blogs/news/understanding-how-to-navigate-the-intracoastal-waterway
20 See https://oceanservice.noaa.gov/facts/springtide.html
21 See https://oceanservice.noaa.gov/facts/perigean-spring-tide.html
22 National Ocean Service – NOAA. (n.d.). *What Is a King Tide*? U.S. Chamber of Commerce. Accessed February 14, 2024. See https://oceanservice.noaa.gov/facts/kingtide.html
23 See www.scientificamerican.com/article/as-seas-rise-king-tides-increasingly-inundate-the-atlantic-coast/
24 Shore Steward News. *Tides, King Tides & Storm Events*. Washington State University. See https://s3.wp.wsu.edu/uploads/sites/2144/2017/01/SS-NEWS_January-2017_Tides-King-Tides_Word-for-ONLINE_v2.pdf
25 See www.weather.gov/marine/
26 See www.weather.gov/marine/usamz
27 See www.weather.gov/greatlakes/
28 See www.weather.gov/marine/wrdoffmz
29 See www.weather.gov/marine/hsmz
30 NOAA. (2023). *Rip Currents*. U.S. Department of Commerce. See www.noaa.gov/jetstream/ocean/rip-currents
31 See https://oceanservice.noaa.gov/facts/roguewaves.html
32 See https://ocean.weather.gov/gulf_stream.php
33 See www.aoml.noaa.gov/phod/trinanes/AGULHAS/agulhas.html
34 NOAA. (2023). *What Is a Rogue Wave*? U.S. Department of Commerce. See https://oceanservice.noaa.gov/facts/roguewaves.html
35 National Snow and Ice Data Center. (n.d.). *Why Is Sea Ice Important*? CIRES at the University of Colorado Boulder. Accessed February 16, 2024. See https://nsidc.org/learn/parts-cryosphere/sea-ice/quick-facts-about-sea-ice
36 Battles, J. B. (2001). Disaster Prevention: Lessons Learned from the *Titanic*. *Proceedings (Baylor University. Medical Center)*, *14*(2), 150–153. https://doi.org/10.1080/08998280.2001.11927752
37 See www.navcen.uscg.gov/north-american-ice-service-products
38 See https://oceantoday.noaa.gov/fullmoon-seaice-whyisitshrinking/welcome.html
39 Schmidt, C. W. (2011). Out of Equilibrium? The World's Changing Ice Cover. *Environmental Health Perspectives*, *119*(1), A20–A28. https://doi.org/10.1289/ehp.119-a20
40 See https://oceanservice.noaa.gov/hazards/sealevelrise/
41 NOAA. (n.d.). *Sea Level Rise Viewer*. U.S. Chamber of Commerce. Accessed February 13, 2024. See https://coast.noaa.gov/slr/
42 See www.acua.com/Projects/Seawall.aspx
43 See www.newyorker.com/news/annals-of-a-warming-planet/can-seawalls-save-us
44 See www.nws.noaa.gov/directives/sym/pd01001003a012018curr.pdf
45 Heller, J. (1961). *Catch-22*. Simon and Schuster.
46 See www.thecity.nyc/2023/01/27/sea-wall-army-corps-nyc-coastal-plan/
47 See https://time.com/6242869/new-jersey-town-sea-wall-climate-action/
48 See www.bbc.com/news/science-environment-26034196
49 Miller, J. K., Kerr, L., Bredes, A., & Rella, A. (2022). *Ecoshorelines on Developed Coasts: Guidance and Best Practices*. Stevens Institute of Technology. See www.nj.gov/dep/bcrp/docs/nj-dev-eco.pdf
50 See www.nhc.noaa.gov/surge/slosh.php

51 NOAA Fisheries. (2021). *Standard Seawater? Yes, There Is Such A Thing*! NOAA. See www.fisheries.noaa.gov/feature-story/standard-seawater-yes-there-such-thing
52 NOAA. (n.d.). *Tides & Great Lakes Water Levels.* U.S. Department of Commerce. Accessed February 13, 2024. See https://tidesandcurrents.noaa.gov/water_level_info.html
53 See https://oceanservice.noaa.gov/facts/current.html
54 NOAA. (2023). *How Fast Is the Gulf Stream*? U.S. Department of Commerce. See https://oceanservice.noaa.gov/facts/gulfstreamspeed.html
55 NOAA – Center for Operational Oceanographic Products and Services. (n.d.). *Navigation Services.* Accessed February 13, 2024. See https://tidesandcurrents.noaa.gov/navigation_services.html
56 NOAA – Center for Operational Oceanographic Products and Services. (n.d.). *PORTS® (Physical Oceanographic Real-Time System).* Accessed February 13, 2024. See https://tidesandcurrents.noaa.gov/ports.html
57 See www.usgs.gov/special-topics/water-science-school/science/tsunamis-and-tsunami-hazards
58 Applied Sciences Program – NASA. (n.d.). *Developing Tsunami Early Warning Systems.* Applied Sciences Program, Earth Science Division, NASA. Accessed February 16, 2024. See https://appliedsciences.nasa.gov/what-we-do/disasters/tsunamis
59 See https://apnews.com/article/alaska-glacier-dam-flooding-jokuhlaup-95f06f4e0d82445705d3126aac004f5a
60 See https://apnews.com/article/juneau-tourism-mendenhall-glacier-recedes-281e736286abb62d7cee9c6bb6dc8dc9
61 Taylor, C., Robinson, T. R., Dunning, S., Rachel Carr, J., & Westoby, M. (2023). Glacial Lake Outburst Floods Threaten Millions Globally. *Nature Communications, 14*(1), 487. https://doi.org/10.1038/s41467-023-36033-x
62 Hamlington, B. D., Chambers, D. P., Frederikse, T., Dangendorf, S., Fournier, S., Buzzanga, B., & Nerem, R. S. (2022). Observation-Based Trajectory of Future Sea Level for the Coastal United States tracks Near High-End Model Projections. *Communications Earth & Environment, 3*(1), 230. https://doi.org/10.1038/s43247-022-00537-z
63 See https://oceanservice.noaa.gov/podcast/oct15/dd63-nuisance-flooding.html
64 See https://oceanservice.noaa.gov/hazards/sealevelrise/sealevelrise-tech-report.html
65 Younger, S. (2022). *NASA Study: Rising Sea Level Could Exceed Estimates for U.S. Coasts.* NASA. See https://climate.nasa.gov/news/3232/nasa-study-rising-sea-level-could-exceed-estimates-for-us-coasts/
66 Trenchard, T. (2023). *A Disappearing Island: 'The Water Is Destroying us, One House at a Time'.* National Public Radio. See www.npr.org/sections/goatsandsoda/2023/11/19/1213548231/climate-change-disappearing-island-sierra-leone-africa
67 Bramwell, C. (2016). *Insight: Fighting the Pacific's Rising Seas.* Radio New Zealand. See www.rnz.co.nz/national/programmes/insight/audio/201808104/insight-fighting-the-pacific's-rising-seas
68 See https://zenodo.org/records/8069320
69 Brennan, P. (2023). *NASA-UN Partnership Gauges Sea Level Threat to Tuvalu.* NASA. See https://sealevel.nasa.gov/news/265/nasa-un-partnership-gauges-sea-level-threat-to-tuvalu
70 NJ Fish & Wildlife. (n.d.). *NJ Saltwater Recreational Registry Program.* NJ Department of Environmental Protection. Accessed February 19, 2024. See https://dep.nj.gov/njfw/fishing/marine/saltwater-registry/
71 Ryan, J. (2024). *How a Northwest Tribe Is Escaping a Rising Ocean.* National Public Radio. See www.npr.org/2024/02/19/1228727075/how-a-northwest-tribe-is-escaping-a-rising-ocean

72 Turn Around, Don't Drown® is a Registered Trademark of the National Weather Service. Used with Permission, per see www.weather.gov/safety/flood-turn-around-dont-drown

73 See https://assets.gov.ie/117497/503a0c66-7e86-4ebc-8b1d-02bbcbd02195.pdf

74 See www.iwr.usace.army.mil/Missions/Flood-Risk-Management/Flood-Risk-Management-Program/

75 See https://community.fema.gov/ProtectiveActions/s/article/Flood-Response-Planning-Evacuation

76 See www.fema.gov/sites/default/files/2020-08/fema_swiftwater-flood-sar-team_definition_08-03-2020.pdf

REFERENCES/ADDITIONAL READING

Adetoro, O.-I. O., Iyaomolere, M., & Salami, A. (2022). Coastal Flood Prone Communities and Sustainability. In S. Eslamian & F. Eslamian (Eds.), *Disaster Risk Reduction for Resilience: Disaster and Social Aspects* (pp. 357–391). Springer International Publishing. https://doi.org/10.1007/978-3-030-99063-3_15

Giovando, J. (2023). *Summary of Ice Jams and Mitigation Techniques in Alaska*. US Army Corp of Engineers. See https://apps.dtic.mil/sti/trecms/pdf/AD1201563.pdf

Giustiniano, L., Cunha, M. P. E., & Clegg, S. (2016). The Dark Side of Organizational Improvisation: Lessons from the Sinking of Costa Concordia. *Business Horizons*, *59*(2), 223–232. https://doi.org/10.1016/j.bushor.2015.11.007

Le Sourne, H., & Soares, C. G. (Eds.). (2023). *Advances in the Collision and Grounding of Ships and Offshore Structures: Proceedings of the 9th International Conference on Collision and Grounding of Ships and Offshore Structures (ICCGS 2023), Nantes, France, 11–13 September 2023* (1st ed.). CRC Press. https://doi.org/10.1201/9781003462170

Mileski, J. P., Wang, G., & Lamar Beacham, L. (2014). Understanding the Causes of Recent Cruise Ship Mishaps and Disasters. *Research in Transportation Business & Management*, *13*, 65–70. https://doi.org/10.1016/j.rtbm.2014.12.001

Mimura, N. (2013). Sea-Level Rise Caused by Climate Change and Its Implications for Society. *Proceedings of the Japan Academy. Series B, Physical and Biological Sciences*, *89*(7), 281–301. https://doi.org/10.2183/pjab.89.281

NASA Sea Level Change Team, Adams, K., Blackwood, C., Cullather, R., Heijkoop, E., Hamlington, B., Karnauskas, K., Kopp, R., Larour, E., Lee, T., Nerem, R., Nowicki, S., Piecuch, C., Ray, R., Rounce, D., Thompson, P., Vinogradova, N., Wang, O., & Willis, M. (2023). *Assessment of Sea Level Rise and Associated Impacts for Tuvalu, N-SLCT-2023–01*. Technical Report, p. 18. See https://zenodo.org/records/8069320

National Weather Service. (2018). *Coastal Flood Categories*. U.S. Chamber of Commerce. See www.nws.noaa.gov/directives/sym/pd01001003a012018curr.pdf

NOAA. (n.d.a). *Oceans & Coasts*. U.S. Department of Commerce. Accessed February 21, 2024. See www.noaa.gov/ocean-coasts

NOAA. (n.d.b). *Tides & Currents*. U.S. Department of Commerce. Accessed February 21, 2024. See https://tidesandcurrents.noaa.gov/

Román-Rivera, M. A., & Ellis, J. T. (2018). The King Tide Conundrum. *Journal of Coastal Research*, *34*(4), 769–771, 763.

Small-Lorenz, S. L., Shadel, W. P., & Glick, P. (2017). *Building Ecological Solutions to Coastal Community Hazards*. The National Wildlife Federation. See www.nwf.org/-/media/PDFs/Global-Warming/NWF_FINAL_BESCCH_070517.ashx

Techera, E., & Winter, G. (2019). *Marine Extremes: Ocean Safety, Marine Health and the Blue Economy*. Taylor & Francis.

U.S. Coast Guard. (1981). *Navigation and Vessel Inspection Circular No. 5–81: Literature Concerning Hazardous Cargoes.* Department of Transportation/USCG. See www.dco.uscg.mil/Portals/9/OCSNCOE/References/NVICs/NVIC-5-81.pdf

U.S. Coast Guard. (2023). *Report of the International Ice Patrol in the North Atlantic.* U.S. Department of Homeland Security. See www.navcen.uscg.gov/sites/default/files/images/iip/annual_report_icon.png

BIBLIOGRAPHY

Applied Sciences Program – NASA. (n.d.). *Developing Tsunami Early Warning Systems.* Applied Sciences Program, Earth Science Division, NASA. Accessed February 16, 2024. See https://appliedsciences.nasa.gov/what-we-do/disasters/tsunamis

Barron, M. G., Vivian, D. N., Heintz, R. A., & Yim, U. H. (2020). Long-Term Ecological Impacts from Oil Spills: Comparison of *Exxon Valdez*, *Hebei Spirit*, and Deepwater Horizon. *Environmental Science & Technology*, *54*(11), 6456–6467. https://doi.org/10.1021/acs.est.9b05020

Battles, J. B. (2001). Disaster Prevention: Lessons Learned from the *Titanic. Proceedings (Baylor University. Medical Center)*, *14*(2), 150–153. https://doi.org/10.1080/08998280.2001.11927752

Bramwell, C. (2016). Insight: Fighting the Pacific's Rising Seas. *Radio New Zealand.* See www.rnz.co.nz/national/programmes/insight/audio/201808104/insight-fighting-the-pacific's-rising-seas

Brennan, P. (2023). *NASA-UN Partnership Gauges Sea Level Threat to Tuvalu.* NASA. See https://sealevel.nasa.gov/news/265/nasa-un-partnership-gauges-sea-level-threat-to-tuvalu

Dinesen, I., & Blixen, K. (1972). *Seven Gothic Tales.* Vintage Books.

Eyl-Mazzega, M., & Cassignol, É. (2022). The Geopolitics of Seawater Desalination. *Études de l'Ifri*, IFRI. See www.policycenter.ma/sites/default/files/2023-01/Eyl-Mazzega_Cassignol_Desalination_US_2022.pdf

Hamlington, B. D., Chambers, D. P., Frederikse, T., Dangendorf, S., Fournier, S., Buzzanga, B., & Nerem, R. S. (2022). Observation-Based Trajectory of Future Sea Level for the Coastal United States Tracks Near High-End Model Projections. *Communications Earth & Environment*, *3*(1), 230. https://doi.org/10.1038/s43247-022-00537-z

Hays, C. (2023). *How to Navigate the Intracoastal Waterway.* American Nautical Services. See www.amnautical.com/blogs/news/understanding-how-to-navigate-the-intracoastal-waterway

Heller, J. (1961). *Catch-22.* Simon and Schuster.

Kennedy, J. F. (1960). *Remarks of Senator John F. Kennedy, Picnic, Muskegon, MI.* The American Presidency Project. See www.presidency.ucsb.edu/documents/remarks-senator-john-f-kennedy-picnic-muskegon-mi

Li, X., Meng, Q., Wei, M., Sun, H., Zhang, T., & Su, R. (2023). Identification of Underwater Structural Bridge Damage and BIM-Based Bridge Damage Management. *Applied Sciences*, *13*(3), 1348. https://doi.org/10.3390/app13031348

Mignone, A. R., Jr. (2013). *Dunes and Ocean Front Structures Under Wave Attack.* NOAA/National Weather Service. See www.weather.gov/media/erh/ta2013-02.pdf

Miller, J. K., Kerr, L., Bredes, A., & Rella, A. (2022). *Ecoshorelines on Developed Coasts: Guidance and Best Practices.* Stevens Institute of Technology. See www.nj.gov/dep/bcrp/docs/nj-dev-eco.pdf

National Ocean Service – NOAA. (n.d.). *What Is a King Tide*? U.S. Chamber of Commerce. Accessed February 14, 2024. See https://oceanservice.noaa.gov/facts/kingtide.html

National Snow and Ice Data Center. (n.d.). *Why Is Sea Ice Important*? CIRES at the University of Colorado Boulder. Accessed February 16, 2024. See https://nsidc.org/learn/parts-cryosphere/sea-ice/quick-facts-about-sea-ice

NJ Fish & Wildlife. (n.d.). *NJ Saltwater Recreational Registry Program.* NJ Department of Environmental Protection. Accessed February 19, 2024. See https://dep.nj.gov/njfw/fishing/marine/saltwater-registry/

NOAA. (2023a). *How Fast Is the Gulf Stream*? U.S. Department of Commerce. See https://oceanservice.noaa.gov/facts/gulfstreamspeed.html

NOAA. (2023b). *Rip Currents.* U.S. Department of Commerce. See www.noaa.gov/media/image_download/fa0293ef-3e1b-4a5a-92b4-c92f17dfaf31

NOAA. (2023c). *Rip Currents.* U.S. Department of Commerce. See www.noaa.gov/jetstream/ocean/rip-currents

NOAA. (2023d). *What Is a Rogue Wave*? U.S. Department of Commerce. See https://oceanservice.noaa.gov/facts/roguewaves.html

NOAA. (n.d.a). *Sea Level Rise Viewer.* U.S. Chamber of Commerce. Accessed February 13, 2024. See https://coast.noaa.gov/slr/

NOAA. (n.d.b). *Tides & Great Lakes Water Levels.* U.S. Department of Commerce. Accessed February 13, 2024. See https://tidesandcurrents.noaa.gov/water_level_info.html

NOAA – Center for Operational Oceanographic Products and Services. (n.d.a). *PORTS® (Physical Oceanographic Real-Time System).* Accessed February 13, 2024. See https://tidesandcurrents.noaa.gov/ports.html

NOAA – Center for Operational Oceanographic Products and Services. (n.d.b). *Navigation Services.* Accessed February 13, 2024. See https://tidesandcurrents.noaa.gov/navigation_services.html

NOAA Fisheries. (2021). *Standard Seawater? Yes, There Is Such a Thing*! NOAA. See www.fisheries.noaa.gov/feature-story/standard-seawater-yes-there-such-thing

Panagopoulos, A., & Haralambous, K.-J. (2020). Environmental Impacts of Desalination and Brine Treatment – Challenges and Mitigation Measures. *Marine Pollution Bulletin*, *161*, 111773. https://doi.org/10.1016/j.marpolbul.2020.111773

Roberts, D. A., Johnston, E. L., & Knott, N. A. (2010). Impacts of Desalination Plant Discharges on the Marine Environment: A Critical Review of Published Studies. *Water Research*, *44*(18), 5117–5128. https://doi.org/10.1016/j.watres.2010.04.036

Ryan, J. (2024). *How a Northwest Tribe Is Escaping a Rising Ocean.* National Public Radio. See www.npr.org/2024/02/19/1228727075/how-a-northwest-tribe-is-escaping-a-rising-ocean

Schmidt, C. W. (2011). Out of Equilibrium? The World's Changing Ice Cover. *Environmental Health Perspectives*, *119*(1), A20–A28. https://doi.org/10.1289/ehp.119-a20

Shore Steward News. *Tides, King Tides & Storm Events.* Washington State University. See https://s3.wp.wsu.edu/uploads/sites/2144/2017/01/SS-NEWS_January-2017_Tides-King-Tides_Word-for-ONLINE_v2.pdf

Taylor, C., Robinson, T. R., Dunning, S., Rachel Carr, J., & Westoby, M. (2023). Glacial Lake Outburst Floods Threaten Millions Globally. *Nature Communications*, *14*(1), 487. https://doi.org/10.1038/s41467-023-36033-x

Trenchard, T. (2023). *A Disappearing Island: 'The Water Is Destroying Us, One House at a Time'.* National Public Radio. See www.npr.org/sections/goatsandsoda/2023/11/19/1213548231/climate-change-disappearing-island-sierra-leone-africa

United Nations Environment Programme. (2019). *Towards Sustainable Desalination.* United Nations. See www.unep.org/news-and-stories/story/towards-sustainable-desalination

USACE. (n.d.). *Beach Nourishment.* U.S. Army. Accessed February 13, 2024. See www.iwr.usace.army.mil/Missions/Coasts/Tales-of-the-Coast/Corps-and-the-Coast/Shore-Protection/Beach-Nourishment/

Wei, K., Zhou, C., Zhang, M., Ti, Z., & Qin, S. (2020). Review of the Hydrodynamic Challenges in the Design of Elevated Pile Cap Foundations for Sea-Crossing Bridges. *Advances in Bridge Engineering, 1.* https://doi.org/10.1186/s43251-020-00020-9

Younger, S. (2022). *NASA Study: Rising Sea Level Could Exceed Estimates for U.S. Coasts.* NASA. See https://climate.nasa.gov/news/3232/nasa-study-rising-sea-level-could-exceed-estimates-for-us-coasts/

Zhu, D., Li, Y., Dong, Y., & Yuan, P. (2021). Long-Term Loss Assessment of Coastal Bridges from Hurricanes Incorporating Overturning Failure Mode. *Advances in Bridge Engineering*, 2(1), 10. https://doi.org/10.1186/s43251-020-00030-7

Part 5

Complex Incidents, Including Both Quantity and Quality Hazards

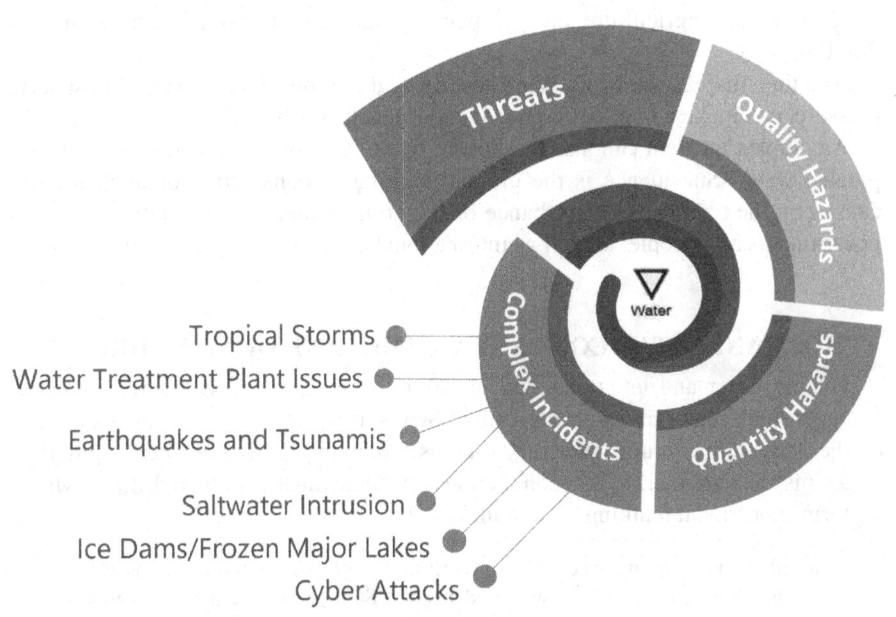

DOI: 10.1201/9781003474685-10

This part covers quantity hazards with water-related threats. It has six chapters:

- Tropical Storms (Chapter 14)
- Earthquakes and Tsunamis (Chapter 15)
- Water Treatment Plant Issues (Chapter 16)
- Saltwater Intrusion into Navigable Freshwater Rivers (Chapter 17)
- Ice Dams/Frozen Major Lakes (Chapter 18)
- Cyberattacks (Chapter 19)

Complex incidents can occur from many different sources and can adversely impact communities with both quantity and quality water-related hazards. Complexity can be defined in scientific terminology as "the study of self organizing complex systems, i.e. systems where macroscopic patterns of behavior emerge spontaneously from the interaction between many simple sub-units" (Finnigan, 2001, p. 241).[1] This definition of complexity has its origins in the work of Per Bak, a Danish scientist who coined the term "self-organized criticality".[2] Complex incidents can be chaotic – community, political, corporate, and other systems can become disorganized during and after disasters, such as hurricanes. The correlation between Chaos Theory[3] and Complex Incidents can easily be made – but this is not causality. Again, emergency managers should not focus as much on the "why did this happen?", as they should on the questions of "how can we fix this?" and "what can we do now, so the adverse impacts do not happen again in the future?"

When it comes to water-related threats and hazards, Complex Incidents may not have their origins in Human-Made ones. Complexity can start with the flow of a river in its bends[4] and turns, and how eventually it will create new pathways or courses. Those new courses can adversely impact homes, governmental infrastructure, businesses, agriculture, etc. – especially during high-water incidents such as flooding.

And then there is the wind. Wind speed and direction, coupled with atmospheric pressure, can make a huge difference in the impacts of storms.[5]

A complex incident can also be created by the lack of any water, or specifically potable water: Subsidence is the gradual sinking or compaction of land, usually caused by the removal or disturbance of the groundwater nearby.[6] Battles between governments and people, as well as international conflicts, have arisen over water.

THE MANHATTAN COMPANY VS. THE BANK OF NEW YORK

Potable water and its connection to public health and fire safety were concerns in all major cities. In the United States, this was very much the case in the early 18th century. Two very famous "founding fathers" ended up pitted against one another, as the connections between water supply and the newly forming national banking system unfolded.

> Manhattan's drinking water went through several privatization schemes as early as the 18th century. These were led by two State Assemblymen, Alexander

Hamilton and Aaron Burr. Burr convinced city officials that public funds would be inadequate to develop a suitable reservoir and aqueduct. He created the Manhattan Company (later to become Chase Bank) to serve the public as the sole supplier of water and took control of the city's water system in 1799. However, hidden in the bill that granted a charter to the Manhattan Company was a clause that stated that the company could use any surplus capital for other purposes. The company was expected to tap into the Bronx River but instead drilled wells into the polluted Collect Pond, which was much cheaper. Additionally, the company only laid 23 miles of pipes and charged an expensive rate of 20 dollars a year, which made it inaccessible to many citizens. Two thirds of the population still relied on polluted wells or buying spring water from expensive private vendors. Because of these failures, in 1816 several Common Council committees were appointed to investigate whether the legislature could grant the city the right to build a public water supply; however, nothing immediate was done about the toxic water and poor service. The first changes began around 1828 after repeated fires destroyed blocks because water mains and fire hydrants had not been extended to all parts of the city.

(Public Water, n.d., p. 1)[7]

While the rivalry between Hamilton and Burr had been festering for years, it is highly likely that Hamilton considered Burr's creation of a company which did banking business in all but name, to be a direct challenge to him.[8]

Excavators for construction work around lower Manhattan, were still finding wooden water pipes as late as 2005[9] under the streets, from this early infrastructure project.

Hamilton also understood the potential of water. Before his untimely death in that famous duel with Burr, Hamilton used his connections and notoriety to help the fledgling United States develop its own industrial capabilities – built on natural waterpower:

He firmly believed that the U.S. had to develop its own industries, and in 1792, he founded Paterson, New Jersey (named for William Paterson), to become the national hub of manufacturing. The Great Falls and the water power they generated were the reasons Hamilton chose this location as the base for his industrial dreams.

(New Jersey Historical Commission, n.d., p. 2)[10]

Global Water Security[11] is a concern for every country's national security:

During the next 10 years, many countries important to the United States will experience water problems – shortages, poor water quality, or floods – that will risk instability and state failure, increase regional tensions, and distract them from working with the United States on important US policy objectives. Between now and 2040, fresh water availability will not keep up with demand absent more effective management of water resources. Water problems will hinder the ability of key countries to

> produce food and generate energy, posing a risk to global food markets and hobbling economic growth. As a result of demographic and economic development pressures, North Africa, the Middle East, and South Asia will face major challenges coping with water problems.
>
> **(Defense Intelligence Agency, 2012, p. iii)**[12]

Complex Incidents will have very different actions for preparedness/prevention/protection, response, recovery, and mitigation, as compared to the other hazard parts described in this book.

Each chapter in this Complex Incidents part will contain the following:

- An overview of the specific incident type, and how its hazards will have adverse impacts to communities
- Case examples for historic reference (many of these will hopefully guide officials as to what ***not*** to do, as lessons learned from other's missteps and misfortunes)
- References/Additional Reading, more specific to the incident

The complex threat and hazards from Climate Change are dispersed throughout this book, since the permutations are great – and in some cases redundant. And since Climate Change is a systemic/chronic disaster, one which should be managed by whole-of-governments[13] – note the multiplier – and ***not*** solely through Emergency Management, this book will not be parsed into "with Climate Change" and "without Climate Change" categories. Emergency managers need to plan for threats and hazards to become more threatening and deadly, sooner rather than later.

The overview section under Part 5 will ***not*** have a common set of adverse impacts to the Disaster Phase Cycles, the Response's ICS, or the U.S.-based ESFs, RSFs, and CLs, as was covered in Parts 3 and 4. Nor will it have a common summary of the POETE process. Each chapter – since they are very different and distinct from one another – will have these elements on their own.

NOTES

1 Finnigan, J. (2001). How Nature Works: The Science of Self-Organized Criticality: Per Bak, Copernicus, an Imprint of Springer-Verlag, New York Inc., New York, 212, pp, $18.00, ISBN 0–387–98738-X. *Agricultural and Forest Meteorology, 108*(3), 241–243. https://doi.org/10.1016/S0168-1923(01)00229-5

2 Bak, P. (1996). *How Nature Works: The Science of Self-Organized Criticality*. Springer-Verlag New York, Inc.

3 Piotrowski, C. (2006). Hurricane Katrina and Organization Development: Part 1. Implications of Chaos Theory. *Organization Development Journal, 24*, 10–19.

4 See www.google.com/maps/@39.5805972,-78.4410891,12.75z?entry=ttu

5 Chavas, D. R., Reed, K. A., & Knaff, J. A. (2017). Physical Understanding of the Tropical Cyclone Wind–Pressure Relationship. *Nature Communications, 8*(1), 1360. https://doi.org/10.1038/s41467-017-01546-9

6 Woodall, T. (2023). Using Satellites to Track Groundwater Depletion in California. *Ohio State News.* Accessed January 8, 2024. See https://news.osu.edu/using-satellites-to-track-groundwater-depletion-in-california/
7 Public Water. (n.d.). *Privatization, Part II: Hamilton, Burr, and the Manhattan Company.* Public Water. Accessed February 22, 2024. See http://public-water.com/story-of-nyc-water/privatization-ii/
8 Rorabaugh, W. J. (1995). The Political Duel in the Early Republic: Burr v. Hamilton. *Journal of the Early Republic, 15*(1), 1–23. https://doi.org/10.2307/3124381
9 See https://aqueduct.org/article/historic-wooden-water-pipes-unearthed
10 New Jersey Historical Commission. (n.d.). *Alexander Hamilton's Dreams of Industry.* The State of New Jersey. See https://nj.gov/state/historical/assets/pdf/it-happened-here/ihhnj-er-hamilton.pdf
11 See www.csis.org/programs/global-food-and-water-security-program
12 Defense Intelligence Agency. (2012). *Global Water Security.* Office of the Director of National Intelligence. See www.dni.gov/files/documents/Special%20Report_ICA%20Global%20Water%20Security.pdf
13 Prasad, M. (2023). Global Pandemics are Extinction-level Events and Should Not Be Coordinated Solely through National or Jurisdictional Emergency Management. *Pracademic Affairs, 3.* See www.hsaj.org/articles/22285

REFERENCES/ADDITIONAL READING

Bak, P. (1996). *How Nature Works: The Science of Self-Organized Criticality.* Springer-Verlag New York, Inc.

United Nations Office for Disaster Risk Reduction. (2023). *Hazards with Escalation Potential: Governing the Drivers of Global and Existential Catastrophes.* UNDRR. See www.undrr.org/publication/hazards-escalation-potential-governing-drivers-global-and-existential-catastrophes

BIBLIOGRAPHY

Bak, P. (1996). *How Nature Works: The Science of Self-Organized Criticality.* Springer-Verlag New York, Inc.

Chavas, D. R., Reed, K. A., & Knaff, J. A. (2017). Physical Understanding of the Tropical Cyclone Wind-Pressure Relationship. *Nature Communications, 8*(1), 1360. https://doi.org/10.1038/s41467-017-01546-9

Defense Intelligence Agency. (2012). *Global Water Security.* Office of the Director of National Intelligence. See www.dni.gov/files/documents/Special%20Report_ICA%20Global%20Water%20Security.pdf

Finnigan, J. (2001). How Nature Works; The Science of Self-Organized Criticality: Per Bak, Copernicus, an Imprint of Springer-Verlag, New York Inc. New York, 212 pp, $18.00, ISBN 0–387–98738-X. *Agricultural and Forest Meteorology, 108*(3), 241–243. https://doi.org/10.1016/S0168-1923(01)00229-5

Piotrowski, C. (2006). Hurricane Katrina and Organization Development: Part 1. Implications of Chaos Theory. *Organization Development Journal, 24*, 10–19.

Prasad, M. (2023). Global Pandemics Are Extinction-level Events and Should Not Be Coordinated Solely Through National or Jurisdictional Emergency Management. *Pracademic Affairs, 3.* See www.hsaj.org/articles/22285

Public Water. (n.d.). *Privatization, Part II: Hamilton, Burr, and the Manhattan Company*. Public Water. Accessed February 22, 2024. See http://public-water.com/story-of-nyc-water/privatization-ii/

Rorabaugh, W. J. (1995). The Political Duel in the Early Republic: Burr v. Hamilton. *Journal of the Early Republic*, *15*(1), 1–23. https://doi.org/10.2307/3124381

Woodall, T. (2023). Using Satellites to Track Groundwater Depletion in California. *Ohio State News*. Accessed January 8, 2024. See https://news.osu.edu/using-satellites-to-track-groundwater-depletion-in-california/

14 Tropical Storms

Some people love the ocean. Some people fear it. I love it, hate it, fear it, respect it, cherish it, loathe it, and frequently curse it. It brings out the best in me and sometimes the worst.

– Roz Savage, MBE. First woman to solo row three oceans (Atlantic, Pacific, Indian).[1]

Tropical Storms – whether they are called Tropical Depressions, Hurricanes, Typhoons, etc. – are a combination of threats – too much seawater and freshwater. The U.S National Weather Service, who analyzed Atlantic Ocean tropical cyclones from 1963 to 2012, found that 88% of the deaths can be attributed to water hazards (see Figure 14.1). Tropical Storms are measured on a scale[2] which is rated only on wind speed,[3] but other factors such as storm surge, torrential rainfall, and the potential for tornadoes can make most Tropical Storms extremely damaging and deadly. The current 1–5 scale of intensity is being advocated for adjustment, to add a sixth level so that the current level 5 would be from 157 miles to 192 miles per hour (253–309 km/h) in maximum sustained winds; and the new proposed level 6 would be for any storm above 192 miles per hour. At least five storms since 2013 reached this higher threshold.[4]

There can also be concurrent and/or cascading threats and hazards from tropical storms:

- Sea Surface temperatures and oceanic heat content can intensify tropical storms.[5]
- Tropical Storms can generate/active/relocate Harmful Algal Blooms.[6]

There are many books and peer-reviewed articles about hurricanes, typhoons, tropical cyclones, etc. Some which have pertinence to Emergency Management, are noted in the Additional Reading section at the end of this chapter. This chapter will touch upon key elements of tropical storms, which can help an emergency manager gauge the severity of the adverse impacts to their communities. What is vitally important for emergency managers to learn is that tropical storms will most likely be the maximum of a notice-type water-related complex incident for their personal careers. If the POETE elements for a tropical storm can be solved completely, as part of the Preparedness phase, the solutioning for almost every other water-related disaster will flow much smoother.

DOI: 10.1201/9781003474685-19

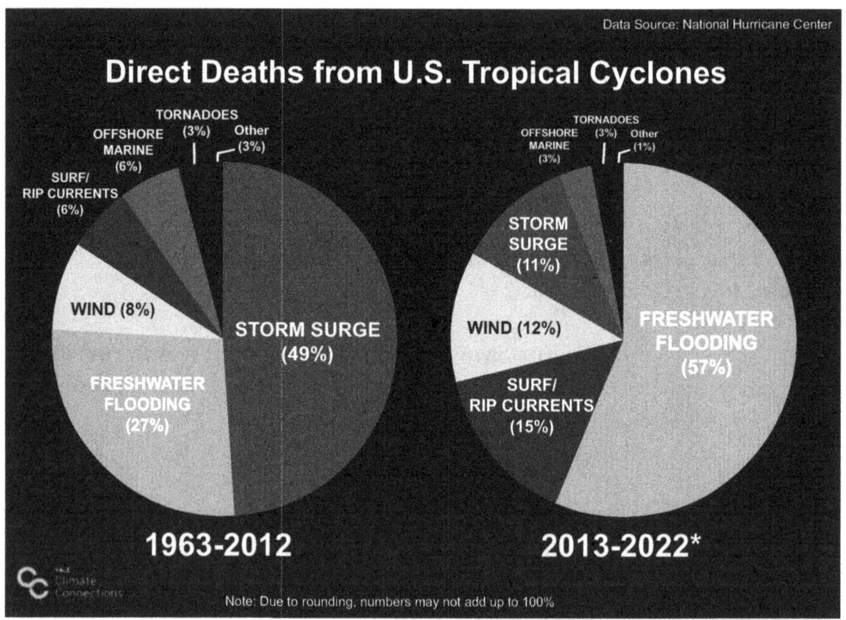

FIGURE 14.1 Direct Deaths from the U.S. Tropical Cyclones.

Source: Lowry, M. (2024). What New Data Reveal about How Hurricanes Kill. *Yale Climate Connections*. Accessed January 27, 2024. See https://yaleclimateconnections.org/2023/08/what-new-data-reveal-about-how-hurricanes-kill/#:~:text=In%20addition%20to%20freshwater%20(rainfall,from%206%25%20in%20earlier%20findings. CC BY-NC-ND 2.5 DEED.

TROPICAL STORMS AND METEOROLOGY

There is much to learn about tropical storms, from a scientific perspective, and in 2010 FEMA collaborated with the National Hurricane Center (NHC) and the University Corporation for Atmospheric Research's (UCAR) Cooperative Program for Operational Meteorology, Education, and Training (COMET). They created a Community Hurricane Preparedness online training and web platform, which is still available today.

> NHC meteorologists provided updated information on NHC products, which have changed greatly during the past 10 years. FEMA representatives discussed the tools provided by their agency to assist in response and evacuation decision-making during hurricane threats. COMET used their training and education expertise to develop a student-friendly on-line distance learning module. At the conclusion of the course, the students should be able to:
>
> - Identify how tropical cyclones form, their climatology, typical tracks, and active season in the Atlantic and Gulf of Mexico
> - Describe the various impacts from tropical cyclone hazards including high winds, storm surge, tornadoes, and heavy precipitation

- Demonstrate familiarity with the tropical cyclone forecast process and terminology
- Explain the use and limitations of products and tools provided by the FEMA, NHC, and the National Weather Service Weather Forecast Offices (NWS WFOs) for hurricane preparedness, planning, response, and operations
- Analyze various sources of information and formulate a plan for dealing with a possible hurricane

The module can be found at: www.meted.ucar.edu/hurrican/chp/[7]
The COMET/MetEd website can be found at: www.meted.ucar.edu[8]

(NHC, 2010, p. 1)[9]

This chapter as well as Chapter 2 has a number of websites and URLs, which are tropical storm related.

COASTAL WATER TEMPERATURE

Coastal Water Temperature, including on the Great Lakes, is something that NOAA's National Centers for Environmental Information (NCEI) monitors in the United States.

> The NCEI Coastal Water Temperature Guide (CWTG) provides recent ocean and Great Lakes temperatures and average water temperatures collected from buoys, tide gauges, and other monitoring stations in the United States and its territories. In addition to water temperature, users have access to station pages that collect data on water levels, wave heights, wind speed, air temperature and pressure. The CWTG also includes a daily average of sea surface temperature to allow users to see water temperatures between physical stations.
>
> **(NCEI, n.d., p. 1)[10]**

Another group at NOAA – the Office of Satellite and Product Operations – monitors Sea Surface Temperatures (SST) on the high seas through SST Contour Charts.[11] Warm water (at least 80°F/27°C) is one of two factors – the other being a weather disturbance, such as a thunderstorm, which brings in warm air from many directions – which are needed to generate a tropical storm.

> Hurricanes start simply with the evaporation of warm seawater, which pumps water into the lower atmosphere. This humid air is then dragged aloft when converging winds collide and turn upward. At higher altitudes, water vapor starts to condense into clouds and rain, releasing heat that warms the surrounding air, causing it to rise as well. As the air far above the sea rushes upward, even more warm moist air spirals in from along the surface to replace it.
>
> As long as the base of this weather system remains over warm water and its top is not sheared apart by high-altitude winds, it will strengthen and grow. More and more

heat and water will be pumped into the air. The pressure at its core will drop further and further, sucking in wind at ever increasing speeds. Over several hours to days, the storm will intensify, finally reaching hurricane status when the winds that swirl around it reach sustained speeds of 74 miles per hour or more.

(NOAA, n.d.a, p. 1)[12]

MARINE LAYER

The Marine Layer is a result of temperature inversion between air masses of different densities and temperatures. This can be like the Sea Breeze.

However, unlike the east coast sea breeze, which reforms almost every day due to the rise and fall of heated and cooled air in the summer atmosphere, the west coast marine layer can persist for days or weeks due to a process called a temperature inversion. This is particularly seen along the coasts of Central and Southern California.

While the eastern Gulf Stream brings warmer tropical water north, water along the west coast of North America moves south from the Gulf of Alaska, bringing much cooler surface temperatures. The water off the California coast can be as much as 30°F (17°C) lower than water at the same latitude on the east coast.

This colder water then cools the air in contact with it, increasing the air's density. When the lower layers of air are cooled from the bottom up, the air higher up remains relatively warmer and less dense, a reverse of what is typically seen. As a result, there is no temperature or pressure imbalance to drive the convection found in a sea breeze. Instead, the air masses remain layered in what is called a temperature inversion.

(NOAA, n.d., p. 1)[13]

MEOWS AND MOMS: STORM SURGE

No other weather-related acronyms are as eye-catching as these two: Starting with MEOWs – which are Maximum Envelopes of Water and ending with MOMs – which is the Maximum of Maximums. Both are related to storm surge,[14] which is inland forward movement of water, from the combination of increased sea level (including tidal rise), wave action, wind speed and direction, and heavy rain.

The Maximum Envelope of Water (MEOW) provides a worst case basin snapshot for a particular storm category, forward speed, trajectory, and initial tide level, incorporating uncertainty in forecast landfall location. These products are compiled when a SLOSH basin is developed or updated.

(National Hurricane Center and Central Pacific Hurricane Center, n.d.a, p. 1)[15]

The Maximum of the Maximum Envelope of High Water (MEOW), or MOM, provides a worst case snapshot for a particular storm category under "perfect" storm conditions. Each MOM considers combinations of forward speed, trajectory, and initial tide level. These products are compiled when a SLOSH basin is developed or updated. As with MEOWs, MOMs are not storm specific and are available to view in the SLOSH display program for all operational basins. No single hurricane will produce the regional

> flooding depicted in the MOMs. Instead, the product is intended to capture the worst case high water value at a particular location for hurricane evacuation planning. The MOMs are also used to develop the nation's eva[c]uation zones.
>
> **(National Hurricane Center and Central Pacific Hurricane Center, n.d., p. 1)[16]**

> Storm surge is a very complex phenomenon because it is sensitive to the slightest changes in storm intensity, forward speed, size (radius of maximum winds-RMW), angle of approach to the coast, central pressure (minimal contribution in comparison to the wind), and the shape and characteristics of coastal features such as bays and estuaries.
>
> **(NHC, n.d., p. 1)[17]**

MEOWs and MOMs are predominantly used for storm surge *potential* planning, rather than a live or forecasted situation awareness element. Both will help with how far inland the water *could* travel. Urban planners should utilize this intelligence for infrastructure design, and Emergency Management should use this risk-calculated[18] intelligence to help warn communities at the start of annual hurricane seasons for example, and to help them plan for evacuation routing and timing, as well.

NATIONAL HURRICANE CENTER STORM SURGE RISK MAPS

The National Hurricane Center produces storm surge risk maps[19] – not to be used for live real-time intelligence – but rather as a long-term planning tool.

> This national depiction of storm surge flooding vulnerability helps people living in hurricane-prone coastal areas. These maps make it clear that storm surge is not just a beachfront problem, with the risk of storm surge extending many miles inland from the immediate coastline in some areas. Storm Surge Risk Maps are provided for the US Gulf and East Coasts, Hawaii, Southern California, US territories – Puerto Rico, US Virgin Islands, Guam and American Samoa. Additional mapped areas include Hispaniola and parts of the Yucatan Peninsula.
>
> **(National Hurricane Center, n.d.b, p. 1)[20]**

NEGATIVE STORM SURGES

Negative Storm Surges, also called Reverse Storm Surges,[21] occur when the wind from a tropical storm pulls water ***away*** from shorelines. Negative storm surges can significantly reduce water depth, adversely impact navigation of boats and ships, and add to coastal erosion.[22]

SEA BREEZE

Sea Breeze is a localized phenomenon, which can have a significant influence on weather in nearby coastal communities.

> The sea breeze develops as the land heats up, air rises and is replaced by cooler air from over the adjacent water. The sea breeze can have a dramatic effect on weather parameters such as temperatures, winds and humidity along and sometimes well inland of the coast. In addition, the sea breeze can act as a trigger for thunderstorms to develop due to the convergence it produces.
>
> **(NWS, n.d., p. 1)**[23]

Seabreeze descriptions are further categorized in the U.S. North Carolina area as Classic Sea Breeze, Southerly Resultant Sea Breeze with Inflection, Southerly Resultant Segmented Sea Breeze, Southerly Resultant Hybrid Sea Breeze, Northeast Resultant Sea Breeze with Inflection, and Northeast Resultant Segmented Sea Breeze.[24] In Florida, they are organized as numbered Regimes.[25] While the sun's heat does not generally penetrate land beyond its top few inches (centimeters) of soil, because of the transparency of water and the wind/wave action which moves water temperatures around, the temperature of the air over water – especially at nighttime – fluctuates much higher than over land.[26]

SLOSH

SLOSH is an acronym for Sea, Lake, and Overland Surges from Hurricanes. Emergency managers need to understand the potential SLOSH threats and hazards in their jurisdiction. While wind on lakes (including the Great Lakes, which are also subject to tidal elevation of water) is included in SLOSH modeling and prediction,[27] it is typically seawater spun up by a tropical storm which generates actual storm surges. As shown in Figure 14.2, storm surge has greater adverse impact than just storm tide alone.

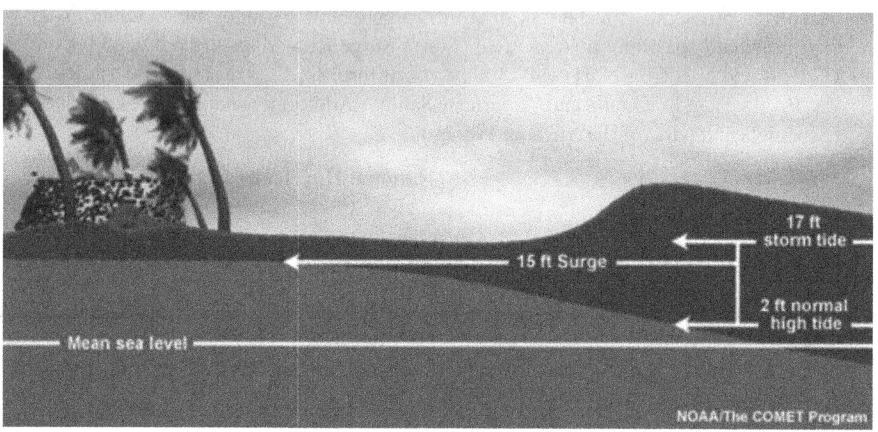

FIGURE 14.2 Storm Surge vs. Storm Tide

Source: National Hurricane Center. (n.d.), *Storm Surge Overview.* U.S. Department of Commerce. Accessed February 23, 2024. See www.nhc.noaa.gov/surge/. Public domain.

The SLOSH model consists of a set of physics equations which are applied to a specific locale's shoreline, incorporating the unique bay and river configurations, water depths, bridges, roads, levees and other physical features.

Modeling Approaches

There are essentially three methods that can be used to estimate surge:

Deterministic Approach – Forecasts surge based on solving physics equations. This approach uses a single simulation based off of a "perfect" forecast which results in a strong dependence on accurate meteorological input. The location and timing of a hurricane's landfall is crucial in determining which areas will be inundated by the storm surge. Small changes in track, intensity, size, forward speed, and landfall location can have huge impacts on storm surge. At the time emergency managers must make an evacuation decision, the forecast track and intensity of a tropical cyclone are subject to large errors, thus a single simulation of the SLOSH model does not always provide an accurate depiction of the true storm surge vulnerability.

Probabilistic Approach – The Probabilistic Surge (P-Surge) product incorporates statistics of past forecast performances to generate an ensemble of SLOSH runs based on distributions of cross track, along track, intensity, and size errors. The latest version explicitly models the astronomical tide.

Composite Approach – Predicts surge by running SLOSH several thousand times with hypothetical hurricanes under different storm conditions. The products generated from this approach are the Maximum Envelopes of Water (MEOWs)[28] and the Maximum of MEOWs (MOMs)[29] which are regarded by NHC as the best approach for determining storm surge vulnerability for an area since it takes into account forecast uncertainty. The MEOWs and MOMs play an integral role in emergency management as they form the basis for the development of the nation's evacuation zones.

(National Hurricane Center and Central Pacific Hurricane Center, n.d., p. 1)[30]

WAVES

Significant Wave Heights – commonly known in maritime terms as "Seas" in the U.S. Marine Forecasts – have their measurements averaged:

This is measured because the larger waves are usually more significant than the smaller waves. For instance, the larger waves in a storm cause the most beach erosion, or the larger waves can cause navigation problems for mariners. Since the Significant Wave Height (Seas) is an average of the largest waves, you should be aware that many individual waves will probably be higher.

If we take a sample forecast of Seas Beyond the Reef of 2 to 4 feet, this implies that the average of the highest one-third waves will have a Significant Wave Height of 2 to 4 feet. But mariners need to keep in mind that roughly one of every ten waves will be greater than 4 feet; one in every one hundred waves will be greater than 5 feet; and one in every 1000 waves will be greater than 6 feet.

As a general rule, the largest individual wave one may encounter is approximately twice as high as the Significant Wave Height (or Seas).

(NWS, n.d., p. 1)[31]

The Douglas Sea Scale – predominantly used by international shipping – measures the size of seas and swells.[32] Both dunes and coastal structures can be undermined by wave action. "Once these coastal defenses are eroded or destroyed by wave action landward buildings are vulnerable to wave attack" (Mignone, Jr., 2013, p. 1).[33]

WIND

Wind speed, direction, and strength are certainly critical factors in tropical storm formation and sustainment. The U.S. National Weather Service uses the Beaufort Wind Scale[34] to measure wind strength. It is named in honor of Sir Francis Beaufort, who devised a method for warship sailing in 1805 for the British Royal Navy. The scale can be used for estimating the power of the wind, without instruments.[35] More depth about wind in general will not be covered by this book.

14.1 CASE EXAMPLES

14.1.1 Babcock Ranch: Florida's First Hurricane-Proof Town?

Babcock Ranch, Florida, is a master-planned community on 18,000 acres (73 km^2) designed to support up to 19,500 homes and 6 million ft^2 (55 hectares) of commercial space.

> Residents zip around in solar-powered golf carts, kayak on the lakes, birdwatch, and congregate at the community pools. But the beautiful aesthetics have a dual purpose: the lakes double up as retaining ponds to protect houses from floods, streets are designed to absorb excess rainfall, and the community hall is reinforced as a storm shelter. A large 870-acre solar panel farm powers the entire development, as well as surrounding communities – making Babcock Ranch America's first solar-powered town.[36]
>
> **(Sherriff, 2023, p. 1)**[37]

After Hurricane Ian struck Florida in 2022,[38] none of Babcock Ranch's homes lost power or access to potable water. They offered their own sports hall as an emergency shelter for one of the surrounding communities who lost power and had greater storm damage.

> Another key component in building resilient towns is location. Babcock Ranch lies inland along Highway 31, around a 45 minute-drive to the region's barrier island beaches, which act as natural buffers to storms. The neighbourhood was also built 30 ft (9.1 m) above sea level, and planners ensured there was plenty of natural land surrounding it which could help buffer storms, particularly excess rainwater.
>
> **(Sherriff, 2023, p. 1)**[39]

Because Babcock Ranch has its own water treatment plant – and its own buried power line connected solar power plant – it did not have a boil-alert or other potable water adverse impacts. There have been questions of affordability and equity in such master-planned developments, but Babcock Ranch had some homes available in the high $200,000 (€185,000) range. It is not a community for Asset Limited, Income Constrained, Employed (ALICE) people, but there are non-governmental groups working in Florida on housing options which are more disaster-resilient, in addition to being affordable.[40]

14.1.2 Caribbean Disaster Emergency Management Agency Upgrades Recovery Tools

The Caribbean Disaster Emergency Management Agency (CDEMA)[41] has upgraded its Comprehensive Disaster Management Audit Tool to include a Sectoral Recovery Capacity Assessment (SRCA):

> To provide a deep analytical dive into the recovery capacity of a given sector, the SRCA spans three areas: governance, competencies, and resources and tools. A country's capacity is rated according to these three areas and a traffic light system is used to present results. These are expected to serve as planning instruments and benefit national governments, sectoral stakeholders, national disaster management agencies, and CDEMA in their efforts to enable a rapid and effective recovery.
>
> Recommendations from the assessment advise the prioritization, design, and implementation of recovery-related capacity-building activities and inform potential investments in recovery by national governments and other development partners.
>
> For example, on the policy side, the tool might support creating rules for different sectors to plan and carry out recovery strategies. It could also recommend making the regulatory environment stronger to manage disasters better by adopting detailed laws and plans about disasters.
>
> **(World Bank, 2024, p. 1)**[42]

This work has started now in three sectors (agriculture, housing, and tourism), targeting in six countries (Antigua and Barbuda, Dominica, Grenada, Saint Lucia, and Saint Vincent and the Grenadines). CDEMA presently comprises 19 Caribbean nations: Anguilla, Antigua and Barbuda, Cayman Islands, Commonwealth of the Bahamas, Barbados, Belize, Commonwealth of Dominica, Grenada, Republic of Guyana, Haiti, Jamaica, Montserrat, St. Kitts & Nevis, Saint Lucia, St. Vincent & the Grenadines, Suriname, Republic of Trinidad & Tobago, Turks & Caicos Islands, and the British Virgin Islands.

14.1.3 Hurricane Daniel in Libya

In 2023, Libya was devastated by Hurricane Daniel. More than 4,000 people were confirmed dead. The major city of Derna suffered dam collapses as well, and international aid is necessary to help restore more than $19 billion (€17.5 billion) in infrastructure damage.

> With a staggering 95% of its population lacking regular water and hygiene access and 60% concerned about water-related risks, immediate interventions for clean water and hygiene essentials are pivotal. Moreover, despite some supermarkets and bakeries

> resuming, food security remains a major concern. A vast 99% of households face challenges accessing daily food due to increased prices, necessitating immediate food relief measures.
>
> **(United Nations Office for the Coordination of Humanitarian Affairs, 2023, p. 1)**[43]

The International Committee of the Red Cross[44] is assisting the Libyan Red Crescent in support of humanitarian aid, as is the United Nations[45] and other worldwide relief groups.[46]

TROPICAL CYCLONE FORECASTS

The U.S. National Weather Service has several products they offer to the public, the media, and emergency managers. Their reports, graphics, forecasts, etc. can be very detailed and sometimes difficult to make evacuation or other Emergency Management decisions.

- The public will be informed about the storm's forecast by local media and authorities.[47] Both need to be in synch. Authorities have a responsibility to also provide earlier notification and support for people with disabilities and access/functional needs. This includes those who need to evacuate sooner than others, and those for whom English is not their primary language. Government is responsible for crisis communications in American Sign Language, for example.
- Recognize that key words such as "tropical storm", "hurricane", "cyclone", "tropical depression" can be confusing – all can have severe life safety adverse impacts to the public and responders. emergency managers may have to translate these and provide easy to action upon communications.
- There are also private weather prediction and analysis services available to emergency managers, including forecasting models from other countries. Even the U.S. National Weather Service[48] experiments with new prediction tools, graphical outputs, and systems.[49] Cost and budgetary factors may determine how many alternatives are available, but it is always better to have cross-validated and backup systems for any continuity of operational aspects of Emergency Management.

14.1.4 Hurricane Idalia Recovery and Mitigation in Cedar Key, Florida

Hurricane Idalia struck the U.S. state of Florida in August 2023, causing significant damage to the northeastern portion of the state. As part of the recovery effort, toward overall disaster resiliency or readiness, the University of Florida (UF) developed an interactive mapping tool[50] to provide residents with information about all the

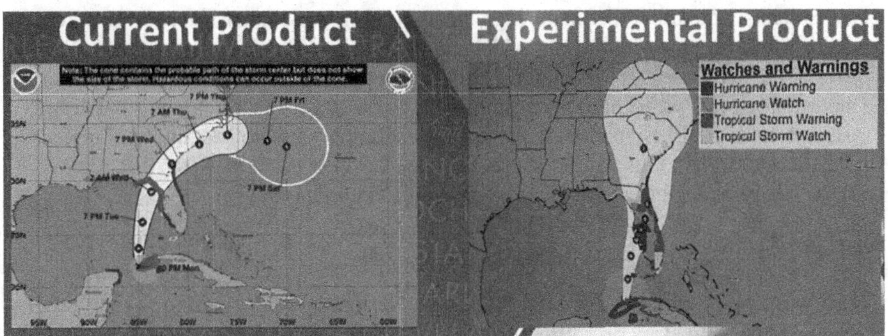

FIGURE 14.3 National Hurricane Center's experimental prediction infographic for 2024.
Source: NWS/NOAA. Public domain.

Recovery Support Function activities impacting them locally, including cultural and natural resources.

> The map tool gives a description of the overview, challenges, and value statement on each map. Statistics in the form of a bar chart and multi-line chart also accompany the map view. The flood vulnerability map is based on future projections, up to the year 2070, of sea level rise and storm and tide events. Also available is information on how residents of Cedar Key can be better prepared for the next disaster. This is the most recent interactive model that UF has produced up to the year 2023, and it is designed for the ease of public use.
>
> **(Resilient Cedar Key, 2023, p. 1)**[51]

14.1.5 Hurricane Otis in Acapulco, Mexico

In 2023, Hurricane Otis struck the Pacific coast of Mexico. It was the "strongest hurricane to in the Eastern Pacific to make landfall in the satellite era"[52] (National Environmental Satellite Data, and Information Service, 2023, p. 1).[53]

> The Category 5 hurricane made landfall near Acapulco, where its heavy rain and 265-kilometer-per-hour (kph) winds unleashed massive landslides and knocked out power lines, killing at least 2 dozen people and causing widespread devastation.
>
> But just 2 days earlier, meteorologists doubted whether Otis – then a tropical storm – would even achieve hurricane status. Forecasters at the U.S. National Hurricane Center expected the storm to undergo "gradual strengthening", with most computer models predicting maximum wind speeds of about 100 kph. Instead, as Otis careened toward Mexico's coastline, its winds increased by 180 kph in 24 hours, a record amount of "rapid intensification".
>
> For meteorologists, it was a tragic reminder that although forecasting methods have drastically improved in recent years, predicting when a minor storm will suddenly explode into a catastrophic hurricane is another matter. "It's difficult to forecast something like that", says John Kaplan, a meteorologist at the National Oceanic and Atmospheric Administration (NOAA).

> Scientists do know the major ingredients driving rapid intensification. Two keys are a warm ocean and moist air, which combine to fuel the convection forces at the storm's center. Sharanya Majumdar, a hurricane researcher at the University of Miami, notes that ocean waters have been "unusually warm" throughout this year's hurricane season, with the El Niño climate pattern channeling even more heat into the tropical Pacific Ocean. As Otis neared shore, it crossed over a patch of water that reached 31°C – several degrees above the average expected for late October.
>
> **(Jacobs, 2023, p. 1)[54]**

There were prior calls[55] to defund Mexico's National Disaster Fund – FONDEN – which was supposed to provide a continuity of disaster funding to impacted communities, based on preemptive bond financing, risk modeling, and an agile design for recovery and reconstruction.[56] Mexico's leaders did request the World Bank to activate multiple tranches of bond funding, to support disaster recovery beyond what private insurance companies were covering.[57]

14.1.6 Island Country of Dominica

Dominica, a small island-nation in the Caribbean, is working toward becoming "hurricane-proof", or more resilient to tropical storms. In 2017, Hurricane Maria impacted all its residents, and damaged 98% of all roof structures on the island.[58]

> The country has 20 targets for resilience by 2030, including an aim for all communities to be self-sufficient for 14 days following a disaster. The aim is for 90% of all housing to be built or retrofitted to comply with resilient building codes.
>
> About two thirds of Dominica is covered in natural vegetation and forest. These plants, and the coral reefs surrounding the island, provide a critical buffer against winds and waves and so need to be protected.
>
> This is part of the resilience plan, which increases protected forest areas and maintains healthy coral reefs around the island through monitoring, restoration, sustainable fishing and by reducing runoff of pesticides from agriculture into the sea.
>
> Dominica's natural assets can also fuel its growth – literally. Dominica aims to become carbon neutral through 100% domestic renewable energy production, which includes investment in a geothermal plant that will produce enough energy to export to neighboring Guadeloupe and Martinique.
>
> **(Wilkinson, 2023, p. 1)[59]**

14.1.7 FEMA Releases Mitigation Assessment Report on Hurricane Ian in Florida

FEMA has a Building Science Disaster Support Program[60] which sends Mitigation Assessment Teams (MATs) with experts, to help assess the performance of buildings, structures, and aspects of the Community Lifelines[61] after disasters like extreme wind, floods, wildfires, and earthquakes. This team was activated for Hurricane Ian, which struck Florida on September 28, 2022. The MAT's report was released in December of 2023.[62]

The 615-page report summarizes the impact of Hurricane Ian and provides an overview of Florida's building codes, standards, and regulations related to floods, wind, floodplain management, evacuation and sheltering. It then outlines the observations of the MAT during the field assessments carried out in Florida, organized into three areas: flooding, wind, and critical facilities. The report then draws 53 conclusions and makes 83 recommendations to improve building, utility, and community resilience on the local and state levels – which can be applicable for many tropical storm impacts.

The recommendations in the report are presented as guidance to the many stakeholders who are involved with the design, construction, and maintenance of infrastructure in Florida, as well as other regions impacted by hurricanes. These include federal, state, and local governments; building officials and floodplain administrators and regulators; the design and construction industry; building code and standard organizations; academia; emergency managers; building owners and operators; and any other stakeholders who can take action to mitigate damage from future natural hazard events.

14.1.8 Palm Beach Gardens, FL: Hurricane Readiness Guides for Residents and Visitors

Palm Beach Gardens in Florida, a community of just over 60,000 people,[63] has a robust and transparent hurricane guide, which can be found at www.pbgfl.com/282/Hurricane-Readiness[64] and includes an accessible version[65] as well. These guides are full-cycle and provide specifics to this community, as well as general life safety suggestions and concepts. Also included in the guides is information about registering for alerts from emergency management and pre-designated shelter sites.

14.1.9 Superstorm Sandy

This hurricane in 2012 was the largest recorded storm – over 1,100 miles wide, in terms of wind span – in the Atlantic Ocean. It also occurred on the astronomically high tide. At least 117 deaths were reported, associated with Superstorm Sandy.[66]

- 24 U.S. states[67] were adversely impacted, generating 13 different U.S. Presidential Disaster Declarations.[68]
- Federal regulations were changed after Sandy, which then allowed for sovereign Tribal Nations to independently request a Presidential Declaration.[69]
- First time in the United States, all six RSFs were activated. Also, the first time that a Federal Recovery Coordinating Officer was assigned, *in addition to* the Federal Coordinating Officer.[70]
- Congressional funding was needed to cover the impacted states' cost-share.[71]
- New Jersey was able to bolster its state-level Voluntary Organizations Active in Disaster (NJVOAD) with paid staff – the first in the United States to do so – and support county/community level groups – including long-term recovery groups[72] – covering all 21 counties from this storm.[73] The

Recovery phase work by VOADs in multiple states, and at the national level, led FEMA to include the National VOAD (NVOAD) as a Federal level support partner to the Community Assistance RSF[74] and also to the Recovery Support Function Leadership Group (RSFLG).[75]

RESEARCHERS WORK TO PREDICT FUTURE URBAN FLOOD RISKS

Scholars at the Massachusetts Institute of Technology (MIT) in 2024 published research on predicting[76] the coastal flooding impacts from tropical storms to urban areas – especially densely populated cities – by studying the damage done in New York City by Superstorm Sandy.

They found that a Climate Change–induced rise in sea level will increase the coastal compounded surge and rainfall flooding hazards, more significantly than just the storm's climatology changes based on warmer temperatures. They "also project that the return period of destructive Sandy-like compound flooding will increase by up to five times by the end of the century" (Sarhadi et al., 2024, p. 1).[77]

There are numerous after-action reports,[78] summary reports[79] on response[80] and recovery,[81] and other documentation on Superstorm Sandy. Long-term individual housing and public assistance[82] recovery has continued, even into 2024.

14.2 ADVERSE IMPACTS TO THE EMERGENCY AND RECOVERY SUPPORT FUNCTIONS AND COMMUNITY LIFELINES

Table 14.1 shows some of the **additional** *adverse impacts of tropical storm hazards on the U.S.-based Emergency Support Functions. Similar tables for* *__all__* *the other Hazard parts of this book should also be reviewed first.*

TABLE 14.1
Water-Related Threats and Hazards to the ESFs

Emergency Support Function (ESF)	Examples of Additional Supporting Actions or Capabilities Related to Tropical Storm-Related Threats or Hazards
#1 – Transportation	Evacuation routes in many coastal communities now involve contra-flow,[83] which must include destination planning (where will everyone go, once they have left a soon-to-be impacted area?) and now account for the increase in electric vehicles, which may not have enough charge to make the journey.[84]

TABLE 14.1 *(Continued)*
Water-Related Threats and Hazards to the ESFs

Emergency Support Function (ESF)	Examples of Additional Supporting Actions or Capabilities Related to Tropical Storm-Related Threats or Hazards
#2 – Communications	There will be significant adverse impacts from tropical storms on this ESF. Communications systems (landlines, cellular, etc.) routinely become disabled during and after these storms. Sustained power outages exhaust backup generators on cell towers. Many times, PSAPs are overwhelmed at a minimum and damaged or destroyed by the storm, at worse. Alert systems may not work (or fully saturate[85] the impacted population), and the prioritization of restoration of communications systems capabilities must be adjudicated at the highest level of operations.[86]
#3 – Public Works and Engineering	Certainly, this ESF has adverse impacts from tropical storms in both the Response and the Recovery phases. Sandbagging and other protective measures will most likely be missions for public works departments locally, as will preemptive storm-drain cleanouts, tree trimming, etc. Engineering groups will need to be ready to provide urgent damage assessment to public infrastructure for usability and potential repair/restoration needs.
#4 – Firefighting	While firefighting is not generally part of the immediate response to a tropical storm, this ESF is also adversely impacted if apparatus is not relocated in advance,[87] out of flood-prone areas. Many local fire departments also have swift-water rescue missions. During the peak of a tropical storm, fire trucks cannot travel; as they are at a high risk of accidents and collisions, in those high-wind conditions.[88] In many tropical storms, gas lines become disconnected, propane tanks float away, and other sources[89] can generate fires – many of which cannot be fought due to inaccessibility.
#5 – Information and Planning	Essential Elements of Intelligence/Information about the impacts to all Community Lifelines are critical for planning the next operational period for tropical storms. This can include capabilities status of all the other ESFs, such as the planned locations for alternative water supply sources (both fixed and mobile), water quality testing results, distribution status, restoration timelines, and more. Status of critical infrastructure/key resources,[90] possibly damaged, needs to be provided by ESF#3. All of this Emergency Management Intelligence is needed by ESF#5 to work through the operational planning cycle (see Chapter 2 for details).
#6 – Mass Care, Emergency Assistance, Temporary Housing, and Human Services	While sheltering, feeding, and family reunification will most likely occur during a tropical storm, advanced notification[91] for people with disabilities and access/functional needs[92] must be part of the planning cycle, in advance of landfall. Consider providing additional doses of methadone[93] in lockboxes, for individuals with drug dependencies who may be evacuating to shelters. Make sure there is family choice support for feeding infants and toddlers.[94] Review material on the National Mass Care Strategy website,[95] for best practices.

(Continued)

TABLE 14.1 *(Continued)*
Water-Related Threats and Hazards to the ESFs

Emergency Support Function (ESF)	Examples of Additional Supporting Actions or Capabilities Related to Tropical Storm-Related Threats or Hazards
#7 – Logistics	Flooding incidents require many of the same logistics support elements as other large-scale incidents (tornados, wildfires, earthquakes, etc.). Staging areas for power restoration, potable water-related distribution support, receiving massive quantities of outside aid materials (including unsolicited/unwanted donations)[96] all will require SLTT warehousing and support. Flooded/damaged railways, roadways, and airports may impact supply-chain integrity.
#8 – Public Health and Medical Services	As previously noted, every incident, emergency, or disaster has a public health and medical services impact. Tropical storms should be declared as a public health emergency. Licensed care facilities (nursing homes, assisted living centers, hospitals, hospices, etc.) all need to be monitored for capabilities and continuity. Consider expanding bed limits in locations which can take additional patients away from sites which could be storm-impacted. Warn providers against patient dumping.[97] Water quantity incidents – especially tropical storms – can generate stormwater hazards, deadly mold in homes, sewage backups, and other public health hazards. Hospitals may become overwhelmed with injuries associated with people traveling through floodwaters, medical patients need to be rerouted from clinics and services in flooded areas, and backups by other hospitals which are not in the impacted geography need to occur.
#9 – Search and Rescue	For tropical storms which produce flooding, this ESF will be activated. FEMA has specific resource typing associated with swift water/flood search and rescue teams.[98] And this is a resource[99] which any U.S. community can request help from a higher-level government within their state, territory, or tribal nation. And those state or territorial governors can request teams from other jurisdictions – even in advance of a flooding incident – to fill gaps in local capacity. See the EMAC section in Chapter 2 for details
#10 – Oil and Hazardous Materials Response	As noted in the case studies in the chapters in this Quantity Part (also, in the Quality Part), there can be tropical storm damage to industrial and other areas where oil and hazardous materials make their way into the streams, rivers, watershed, etc., and, therefore, adversely impact communities. The adverse health impacts from those human-made hazards, including sewer treatment plant overflows, etc., may require missions for cleanup and remediation from this ESF. The high priority of this cleanup and remediation of the oil and/or hazardous materials is both a life safety and an incident stabilization course of action.
#11 – Agriculture and Natural Resources	The adverse impacts to marine wildlife are the primary concern for this ESF. Debris[100] and hazardous materials can be transported into marshlands, estuaries, etc., and harm aquatic and avian life[101] in both the short- and long-term.

TABLE 14.1 *(Continued)*
Water-Related Threats and Hazards to the ESFs

Emergency Support Function (ESF)	Examples of Additional Supporting Actions or Capabilities Related to Tropical Storm-Related Threats or Hazards
#12 – Energy	Energy production[102] and distribution[103] will be significantly impacted. Emergency managers will be tasked with helping to adjudicate restoration priorities – many times, between what appears to be two equally critical key resources. Exercising these decisions in advance can help decision-making processes.
#13 – Public Safety and Security	In addition to additional incident-related Response missions for law enforcement officials, there may be interim Recovery missions as well, such as curfews, no entry zones, etc. There may be investigations related to fatalities attributed to the tropical storm. Public Safety and Security may also be needed to support ESF#6, ESF#7, and ESF#9 in mass care, as well as search and rescue support missions. And of course, Public Safety is never confined to only one incident at a time, so staffing will need to be extended or expanded to cover multiple missions simultaneously. Staff wellness and health is important as well – since some may be impacted by the tropical storm directly themselves. In some SLTTs, separate sheltering[104] for first responders and their families may be implemented.
#14 – Cross-Sector Business and Infrastructure	As tropical storms adversely impact large areas of a jurisdiction, businesses, non-governmental organizations, and other infrastructure will be affected as well. Many of these sectors have homeland security impacts as well, such as banking, oil and gas refining, etc. Stabilizing the Community Lifelines, restoring supply-chain integrity, and coordinating resource needs from the private sector all are missions for this ESF, associated with tropical storms.
#15 – External Affairs	Coordinate with ESF#2 on both internal and external messaging regarding the full disaster cycle of crisis messaging needed for tropical storms. The organization of various PIOs together under a Joint Information Center (JIC). The PIO needs to activate templated messages to the public which provide the LIPER actions they need to take, in the languages and access to crisis communication that the impacted community needs. Emergency Management's public crisis communications messages for tropical storms can also be summarized as "run from the water, hide from the wind".[105] The crisis communications priority of life safety extends to emergency responders as well. The public needs frequent reminders to evacuate early, and that emergency responders may not be able to rescue people, once the storm strikes.

Source: Barton Dunant.

14.2.1 Recovery Support Functions

Similarly, Table 14.2 shows some of the ***additional*** adverse impacts of tropical storm hazards on the U.S. based Recovery Support Functions (RSF). There are similar tables in the other Hazard parts of this book. Those should be reviewed as well.

TABLE 14.2
Recovery Support Functions

Federal Recovery Support Functions (RSFs)
Economic
Most tropical storms will have an economic impact[106] across a community. If there is no federal or international aid, including preparedness/mitigation funding through a potentially competitive Hazard Mitigation Assistance grant program – economic recovery may be slower or nonexistent.[107]
Health and Social Services
As noted in ESF#8, the resultant flooding from tropical storms can have significant adverse impacts to the health of individuals and families, due to hazards associated with contaminated water, mold found in homes after floodwaters recede, etc. Individuals with compromised immune systems, pulmonary issues, and other medical concerns can be significantly harmed after a tropical storm. Social Services programs may also be curtailed, due to site damage and staffing concerns. Special emphasis on recovery and restoring capabilities for adults with disabilities (for example, developmentally disabled group homes), domestic violence shelters, and other community-based residential social service programs must be managed to support housing stability.
Community Assistance
One aspect that FEMA has for Community Assistance in Recovery[108] is to help achieve **equitable** recovery, as incident survivors work toward rebuilding their community. Incorporating community partners, stakeholders, faith-based organizations, etc. is one way to achieve this. There will be considerable economic pressures to rebuild *away* from supporting affordable housing.[109] There will also be massive resistance[110] in any community to reducing or eliminating infrastructure – including rebuilding homes – in coastal areas subject to recurrent flooding.
Infrastructure Systems
As Infrastructure Systems will be damaged from tropical storms, the recovery aspects of restoration of the critical infrastructure/key resources should be prioritized.[111] This will include both public and private sector organizations. This can be a challenge for community leaders: Is a hospital's restoration a higher priority than a school – especially if there were other hospitals in the area which were not damaged? Transparency, collaboration, and coordination on these decisions can be done in advance – via exercises which specifically address these "either this or that" prioritization decisions.
Housing
Housing will most likely be adversely impacted – for example, due to flooding and destruction by wave action against buildings – recovery efforts will need to transition from the ESF#6 – Mass Care/Sheltering through the Recovery phase into temporary housing, interim housing, and possibly permanent housing replacement solutions. In many communities, there is an affordable housing shortage now – before any disasters occur of any kind. The U.S. Department of Housing and Urban Development (HUD)[112] – upon a request from the state or territorial governor – may provide Community Development Block Grant Disaster Recovery (CDBG-DR) funds to help cities, counties, and states recover from disasters, through housing vouchers for Section 8 approved housing units.

TABLE 14.2 *(Continued)*
Recovery Support Functions

Federal Recovery Support Functions (RSFs)
Natural and Cultural Resources
The above noted Response adverse impacts, include potential damage in Recovery aligned to the restoration of recreational activities – including boating and fishing – from the perception that there is ecological harm from the storm.[113] For many coastal communities, their access to the beach, oceans, waterways, etc. is a significant economic factor for them.
Mitigation work – which starts from projects identified through the RSFs – should also be considered by emergency managers, for children-related sites such as sports fields, nature centers, museums, historical sites, etc.; in addition to what has been traditionally focused on government-run K12 schools. The remediation, repair, and future hazard protection work for some NGOs (such as an environmental conservation group or a marine wildlife stranding organization) can count toward Public Assistance[114] projects in the United States.

Source: FEMA.

Generally, adverse impacts from tropical storm threats and hazards target all of the RSFs, but there can be focused attention by Emergency Management for the adverse impacts to Infrastructure Systems, including potable water systems, such as water and wastewater treatment plants (a focus of this book). The lack of potable water can also impact all the RSFs.

14.2.2 Community Lifelines

As noted in Chapter 2, the Community Lifelines (CLs) are a highly useful list of EEIs, which can be visualized as a dashboard. Tropical Storm hazards will have implications for every community lifeline. If roads and power stations and other critical infrastructure are flooded or damaged from waves and wind, almost every CL will become quickly adversely impacted. There are life safety impacts to each CL:

Safety and Security (Law Enforcement/Security, Fire Service, Search and Rescue, Government Service, Community Safety): Tropical storm hazards from raw or ocean water flooding, disruption of water treatment plants, debris in roadways, etc. can block roads, generate additional security and safety missions[115] for law enforcement, and all the other sub-elements within this CL, such as Search and Rescue.

Food, Hydration, Shelter (Food, Hydration, Shelter, Agriculture): Obvious impacts to this Community Lifeline include evacuation missions (including sheltering and feeding) associated with tropical storms. People will hopefully evacuate their homes if the tropical storm is forecasted to strike their area. Some will not, but then after the storm passes, they will still need access to food and water for drinking, bathing, toileting, etc. at their homes. For example, after a tropic storm incident subsides, if that community's water treatment plant was damaged, residents may not return to their homes – and choose to stay at shelters, for feeding and hydration assistance.

Health and Medical (Medical Care, Public Health, Patient Movement, Medical Supply Chain, Fatality Management): Aligns to both Emergency Support Function ESF#8) Public Health missions and the Health and Social Services RSF missions. As noted in the ESFs and RSFs, this Community Lifeline is significantly adversely impacted by any disruption to either site access (roads to hospitals, aid stations, and clinics blocked due to flooding or debris), critical infrastructure damage to health or medical facilities themselves, or damage to their own potable water supply systems, which may happen with tropical storms.

Energy (Power Grid, Fuel): Water and Energy have many nexus points, all of which need emergency management intelligence. As power lines and substations are damaged by ocean water, they may generate hazards of larger power outages and also adversely affect water treatment plants, cellular towers, and other critical infrastructure. The status of this CL is always the most important in any disaster.

Communications (Infrastructure, Responder Communications, Alerts Warnings and Messages, Finance, 911 and Dispatch): One aspect of crisis communications for Public Information Officers, related to tropical

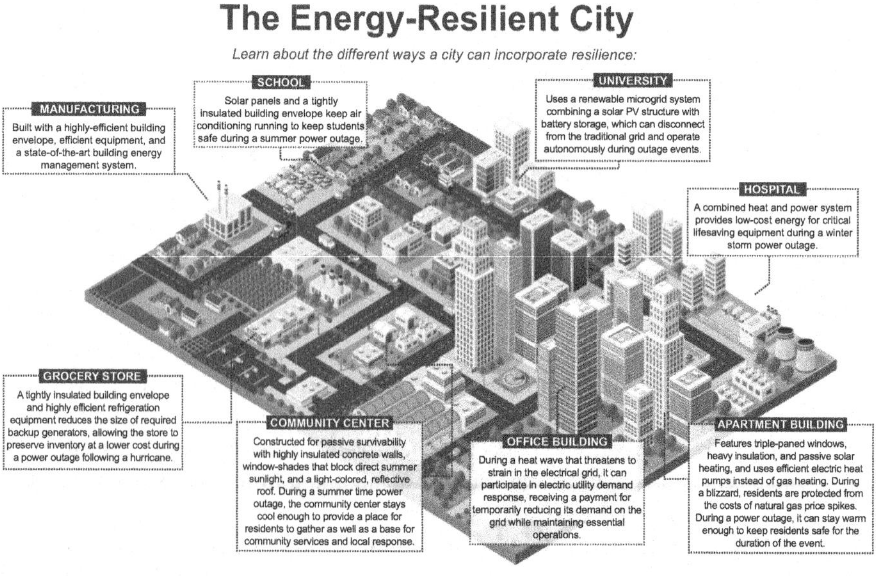

FIGURE 14.4 Energy-resilient city.

Source: Better Buildings Solutions Center/U.S. Department of Energy. (n.d.). *The Energy-Resilient City*. U.S. Department of Energy. See https://betterbuildingssolutioncenter.energy.gov/resilience/communities. Public domain.

storms, is to pre-plan (and exercise) both the notice and no-notice aspects (storms can make landfall quicker than forecasted, additional hazards can occur, etc.) of these storms, utilizing templated public messaging in the languages and delivery methods which have the greatest outreach capabilities for that community. Flooding may be both tidal and "flash flooding" and have little notice to first responders and the impacted jurisdiction, but their adverse impacts – even down to specific streets and buildings – should be known in advance. Chapter 2 has several online tools which can help predict areas which can flood, where sea level rise will occur, etc. A tropic storm is the type of complex incident where a Joint Information Center is needed, throughout the disaster phase cycle.

Transportation (Highway/Roadway/Motor Vehicle, Mass Transit, Railway, Aviation, Maritime): Tropical Storms can have an adverse impact on this CL, as noted in ESF#1 and other areas of this chapter. The negative status of this CL can impact the others as it relates to providing any logistical movement in support of the other CLs in a community. If potable water quantity becomes a hazard in a community, the full capabilities of the Transportation CL will be needed for emergency services Response (police, fire, EMS, etc.), electrical and other utility restoration, and potentially transporting potable water to Points of Distribution, and in bulk transport to critical infrastructure/key resource sites, such as hospitals and dialysis centers. If navigable waterways are blocked or hampered by a tropical storm's impacts, this CL will be impacted as well. Critical repair parts, filters, staff, etc. need to be transported to water and wastewater treatment sites, to make repairs and work toward restoration of the **Water Systems** CL, if they are adversely impacted by a water quantity threat, such as SLOSH-based flooding.

Hazardous Material (Facilities, HAZMAT, Pollutants, Contaminants): Impacts to this Community Lifeline can be for Pollutants and/or Contaminants and the HAZMAT response to the tropical storm and/or any additional cascading incidents, if those commercial and industrial sites are adversely impacted by too much water (flooding), seawall breeches, etc. This could be from natural or human-made hazards. There may be longer-term impacts to facilities, if a contaminated waterway supports those Critical Infrastructure/Key Resources (CIKR), from a needed clean water supply basis. Considerations on how natural freshwater supplies are utilized in dialysis centers and even nuclear power plants, for example.

Water Systems (Potable Water Infrastructure, Wastewater Management):[116] Understanding how flooding conditions can impact a potable water supply system (and the wastewater treatment as well) is critical for Emergency Management.

There may also be longer-term impacts to the food supply if agriculturally used land is flooded – especially by seawater, or experiences damage to infrastructure such as warehouses, transportation vehicles, farm equipment, etc.

14.3 IMPACTS TO THE DISASTER PHASE CYCLES

As shown in Table 14.3, there are specific impacts to all four of the disaster cycle phases from **Tropical Storms**.

TABLE 14.3
Impacts to the Disaster Phase Cycles from Too Much Seawater

Preparedness/Protection/Prevention

Work toward a continuous cycle of messaging about increasing the public's capacity to be better prepared to Respond themselves to the life safety impacts of both tropical storm SLOSH flooding and wave action hazards, as well as protecting their homes and places of business. Governments will need to support flooding protection work, such as sea walls and dunes, while supporting and sustaining local marine ecology. With sea level rise and other risk factors, decisions about building infrastructure in floodplain/flood-prone areas needs to be part of the Prevention discussions and urban planning.

Response

Flooding incidents, including infrastructure damage/destruction – whether from SLOSH seawater inundation, infrastructure destruction, saltwater corrosion, etc. all have massive Response phase work. There are workforce safety impacts – and many of the emergency responders who live in your community may be directly adversely impacted themselves. Communities may be overstressed from repeated tropical storms and mitigation work takes years to complete. Resource requests, including staffing for the "disasters within the disaster" such as home fires, water treatment plant failures, etc., which will occur while flooding is still in the community, can overtax first responder groups.

Recovery

U.S. recovery work is also generally conducted and managed at the State/Territory/Tribal Nation level for tropical storms, not at the local level; especially if the level of adverse impact reaches the threshold assistance from the U.S. Federal Government (i.e., Presidentially declared emergency or disaster). The thresholds to reach this vary by state/tribal/territory and adjust over time. Damage to private homes and commercial entities only is a key aspect where the recovery costs will probably need to be supported through insurance locally, and if local governmental infrastructure is not damaged/destroyed, it will probably not rate an "ask" from the governor for Federal assistance. Without this assistance, states, territories, and tribal nations are left to finance any recovery work themselves – and private companies, individuals, etc. must have insurance coverage – especially flood insurance – in order to properly recover.

Mitigation

There are many opportunities and options for Mitigation work, and tropical storms of any size. Shorelines can be bolstered and strengthened, now more environmentally focused and sustainable.[117] The elevation of infrastructure – including personal homes – is another aspect. Purchasing homes and property in recurrent seawater-based flood areas, by government, to demolish the home and restrict the land's use against further building, is another option. The Part 4 section text has a series of "Mitigation Paradoxes" associated with flooding of any kind. Chances are the higher the economic value of coastal property, the higher the propensity for it to become flooded.

Source: Author.

14.4 ADVERSE IMPACTS TO THE INCIDENT COMMAND SYSTEM

There may be unique hazards directly impacting Responses and Responders themselves, which should be considered adverse impacts to the Incident Command System (ICS). For **Tropical Storms**, Table 14.4 details what may be occurring:

TABLE 14.4
Adverse Impacts to the Incident Command System

Command (including SO, PIO, and LNOs)

Safety Officers (SOs) need continuous emergency management intelligence about this threat. Decisions about holding resources in place during high winds are critical for responder life safety. The post-landfall missions, especially for hazardous swift-water situations in flooded streets, near utility drainage, etc., also need geospatial intelligence on a three-dimensional basis. These workforce safety elements are paramount. While there are always risks to responders (including additional threats in oceanwater, such as pathogens and marine life), there may be additional hazards such as oil and hazardous materials in the water, and uncharted risks for search and rescue, as well as recovery work. Dive teams, for example, will need counseling after recovering any drowning victims – so their own life safety from a mental health/wellness is sustained.

Public Information Officers (PIOs) need to urgently activate accurate templated messages to the public which provide critical crisis communications in the appropriate languages and modalities for oversaturation. If evacuation orders are in place – clear descriptions of mandatory versus recommended must be made. If localized "avoid the area" or "Turn Around, Don't Drown®"[118] warnings are needed, for example, they need to be communicated far and wide, and quickly. It is highly likely that geographic areas which experience recurrent tropical storm adverse impacts such as flooding, for example, already understand the risks associated with this hazard – but it bears repeating for new residents, visitors/tourists, etc. And with Climate Change, the intensity and recurrences will only increase over time.

Liaison Officers (LNOs) may be in place on this type of hazard, crossing all of the various groups which operate their own incident command structures/systems. For example, on tropical storms, the American Red Cross establishes their own NIMS-like organizational structure and provides an LNO as the conduit between the two systems. The same will be true for the U.S. DoD elements, such as the U.S. Coast Guard, US Army Corps of Engineers, etc.

Intelligence

Sources of Emergency Management Intelligence for incidents involving tropical storms will need to include the regular refreshed/curated intelligence about the topography of the water source (inlet, sound, bay, piers, ports, etc.), especially if there are **now** *new* underwater hazards (oil/chemical discharges, bridge collapses, debris, sunken or displaced boats, etc.) as well as cascading human-made hazards (power outages shutting down traffic lights near flooded areas where people are evacuating, fires, etc.). The wind speed and direction, radar imagery of the storm, and SLOSH modeling (MEOWs and MOMs) are just some of the factors which help with a Common Operating Picture. Knowledge of mutual-aid support capabilities for both search and rescue – including EMAC requests and capabilities for interstate Response support – as well as Recovery assistance is needed as well. Emergency Management Intelligence supports all of the general staff leads, not just Operations.

(Continued)

TABLE 14.4 ***(Continued)***
Adverse Impacts to the Incident Command System

Finance/Administration

There will be significant adverse impacts to the staffing, support, systems, etc. associated with the existing local Finance/Administration resources, even if that community is properly prepared for a tropical storm. Overtime and other staffing costs will exceed what is covered in normal fiscal budgets (Emergency contracting for debris removal alone, is something most municipalities cannot support themselves – and economies of scale are usually achieved when this is performed at the County/Parish level or above). The requirements for documentation, proper financial controls, staffing logs, etc. are key when there is a state-declared emergency and/or Federal support. This is needed for emergency assistance reimbursements from higher governmental entities on Declared Emergencies or Disasters (Search and Rescue efforts during a Tropical Storm, for example).

Logistics

Generally, supporting a Tropical Storm logistically is beyond the capabilities of a local jurisdiction's (and in many states, even their county/parish level of government's) Logistics branch. The state/territory/tribal nation will most likely be providing logistical support on a consolidated basis to receive and distribute federal aid/material support, provide warehousing, staging areas, etc.

Operations

There are two historic adverse impacts from Tropical Storms on the Operations Branch: First, if workforce members *and their families* are not supported themselves with evacuations, sheltering, etc., they may prioritize their family's life safety above work.[119] And second, the missions for Operations in a Tropical Storm must be suspended for life safety purposes, due to high winds and other hazards, at the peak of need.

Planning

There are no specific adverse impacts to the Planning section, from this hazard. There is a plethora of planning resources available globally,[120] for tropical storms,[121] local Emergency Management entities,[122] corporations, non-governmental organizations, and more.

Source: Author.

14.4.1 POETE Process Elements for This Hazard

Table 14.5 shows the specifics for **Tropical Storms**.

TABLE 14.5
POETE Process Elements for This Hazard

Planning	There very well may be a separate annex or appendix to any jurisdiction's emergency operations plan for tropical storms, especially in coastal communities. There should also be elements of other main plan portions and annexes/appendices which cover the planning needed for a tropical storm/sea level rise/flooding incident, which can include extraordinary measures, such as search and rescue and contra-flow. Tropical Storms in the Atlantic have a season (June 1 through November 30), but they can – and do – occur outside of this generalized window of warmer seawater. From this chapter, there are adverse impacts of death and near-death for both the public and responders, historically unique references of past incidents to learn from, and elements of exercises which can be performed, in order to facilitate positive changes and modifications to preparedness, response, recovery, mitigation, and continuity of operations plans.

TABLE 14.5 ***(Continued)***

POETE Process Elements for This Hazard

Organization	The **trained** staff needed to effect plans, using proper equipment – especially first responder life safety equipment[123] – should be exercised and evaluated on a regular basis each year. This also includes training with partners who may only be activated at the declared emergency/disaster incident levels. Actual incident response and recovery can help determine elements of POETE, including how these types of incidents are organized for staffing. Also, as part of both deployments and exercises, staff involved should be evaluated for both further training needed and also the possibility of advancement in their disaster state roles to leadership positions.
Equipment	Equipment needed for this hazard specifically includes anything needed to rescue someone from any type of water source (ocean, floodwater, quarry, frozen lake, etc.). This should also include decontamination equipment, as saltwater can corrode metal and electronics.[124] Also included for this hazard are flood protection devices such as floodwalls and sandbags, which may be needed to temporarily replace sections of seawalls and dunes, for example. The geospatial intelligence systems used to track, monitor, and evaluate the emergency management intelligence around tropical storms, is also a form of equipment.
Training	In addition to standardized Incident Command System training, specific awareness training on building capacity for the unique complex aspects of Tropical Storms, should be taken by emergency management officials, from a full disaster phase cycle perspective.
Exercises and evaluation	A full series of discussion and performance exercises should be conducted each year for tropical storms, if there is even a possibility of them impacting the jurisdiction: from human-made hazards of illegal dumping and building against zoning laws, to massive SLOSH and wave impacts, plus future sea level rise, etc. These exercises should stand on their own and be part of various cascading incident scenarios – and cover continuity of operations/continuity of government, as well. Exercises should involve the whole community in an exercise planner's invitations for both additional exercise planners/evaluators and actual players, especially since many non-local organizations (USCG, USACE, etc.) are separate legal entities from the local governments they support. In the same way there are adversely impacted ESFs, RSFs, and CLs as noted, those elements should be exercised and evaluated so that the elements of the POETE cycle are reviewed, and revised as needed (i.e., agreed upon for change, as part of the Improvement Plan components). The AAR/IP is a critical part of this process step, and this chapter's References/Additional Reading section, as well as the footnotes throughout, contains a few examples of open-source tropical storm incident reports and research, which can be the basis for improvement planning everywhere.

Source: Author.

NOTES

1 Savage, R. (2009). *The Ocean Rower's Perspective on Climate Change.* Rozsavage.com. See www.rozsavage.com/the-ocean-rowers-perspective-on-climate-change/
2 See www.nhc.noaa.gov/aboutsshws.php
3 See www.weather.gov/tbw/beaufort

4 See www.npr.org/2024/02/06/1229440080/scientists-explore-whether-to-add-a-category-6-designation-for-hurricanes

5 See www.nbcnews.com/science/environment/hurricane-idalia-went-category-1-category-4-overnight-rcna102545

6 See https://oceanservice.noaa.gov/news/july21/gulf-mexico-hab-forecasts.html

7 See www.meted.ucar.edu/hurrican/chp/

8 See www.meted.ucar.edu

9 NHC. (2010). *COMET's Community Hurricane Preparedness Module Updated for 2010.* U.S. Commerce Department. See www.nhc.noaa.gov/comet_update.shtml

10 NCEI. (n.d.). *Coastal Water Temperature Guide.* U.S Commerce Department. Accessed February 22, 2024. See www.ncei.noaa.gov/products/coastal-water-temperature-guide

11 See www.ospo.noaa.gov/Products/ocean/sst/contour/

12 NOAA. (n.d.). *How Does the Ocean Affect Hurricanes*? U.S. Commerce Department. Accessed February 22, 2024. See https://oceanexplorer.noaa.gov/facts/hurricanes.html

13 NOAA. (n.d.). *The Marine Layer.* U.S. Commerce Department. Accessed February 25, 2024. See www.noaa.gov/jetstream/ocean/marine-layer

14 See https://media.bom.gov.au/social/blog/24/from-sea-to-shore-a-story-of-storm-surges-in-australia/

15 National Hurricane Center and Central Pacific Hurricane Center. (n.d.). *Storm Surge Maximum Envelope of Water (MEOW).* U.S. Department of Commerce. Accessed February 22, 2024. See www.nhc.noaa.gov/surge/meowOverview.php

16 National Hurricane Center and Central Pacific Hurricane Center. (n.d.). *Storm Surge Maximum of the Maximum (MOM).* Accessed February 22, 2024. See www.nhc.noaa.gov/surge/momOverview.php

17 National Hurricane Center. (n.d.). *Storm Surge Overview.* Accessed February 23, 2024. See www.nhc.noaa.gov/surge/

18 See https://noaanhc.wordpress.com/tag/meow/

19 See https://experience.arcgis.com/experience/203f772571cb48b1b8b50fdcc3272e2c/page/Category-5/

20 National Hurricane Center. (n.d.). *National Hurricane Center Storm Surge Risk Maps.* U.S. Department of Commerce. See https://experience.arcgis.com/experience/203f772571cb48b1b8b50fdcc3272e2c/page/Category-5/

21 See www.usatoday.com/story/news/nation/2022/09/29/reverse-storm-surge-meaning-tampa-bay/10458684002/

22 See www.sciencedirect.com/topics/earth-and-planetary-sciences/storm-surge

23 National Weather Service. (n.d.). *Science and Technology.* U.S. Department of Commerce. Accessed February 23, 2024. See www.weather.gov/ilm/ScienceandTechnology#seabreeze

24 See www.weather.gov/ilm/ScienceandTechnology

25 See www.weather.gov/tbw/SB_RegimesMVF

26 See www.noaa.gov/jetstream/ocean/sea-breeze

27 See https://blog.bartondunant.com/slosh-sea-lake-and-overland-surges/

28 See www.nhc.noaa.gov/surge/meowOverview.php

29 See www.nhc.noaa.gov/surge/momOverview.php

30 National Hurricane Center and Central Pacific Hurricane Center. (n.d.). *Sea, Lake, and Overland Surges from Hurricanes (SLOSH).* U.S. Department of Commerce. Accessed February 23, 2024. See www.nhc.noaa.gov/surge/slosh.php

31 National Weather Service. (n.d.). *Significant Wave Height.* U.S. Department of Commerce. Accessed February 23, 2024. See www.weather.gov/key/marine_sigwave

32 See www.noaa.gov/jetstream/ocean/waves/jetstream-max-wind-and-sea-scales

33 Mignone, A., Jr. (2013). *Dunes and Ocean Front Structures under Wave Attack.* U.S. Department of Commerce. See www.weather.gov/media/erh/ta2013-02.pdf
34 See www.weather.gov/tbw/beaufort
35 See https://education.nationalgeographic.org/resource/beaufort-scale/
36 See https://babcockranch.com/
37 Sherriff, L. (2023). Babcock Ranch: Florida's First Hurricane-Proof Town. *BBC.* See www.bbc.com/future/article/20230904-babcock-ranch-floridas-first-hurricane-proof-town
38 See www.nhc.noaa.gov/data/tcr/AL092022_Ian.pdf
39 Sherriff, L. (2023). Babcock Ranch: Florida's First Hurricane-Proof Town. *BBC.* See www.bbc.com/future/article/20230904-babcock-ranch-floridas-first-hurricane-proof-town
40 See https://unitedforalice.org/florida
41 See www.cdema.org/
42 World Bank. (2024). *Ahead of the Storm: An Innovative Disaster Recovery Planning Tool Is Now Available in the Caribbean.* Reliefweb – United Nations Office for the Coordination of Humanitarian Affairs. See https://reliefweb.int/report/world/ahead-storm-innovative-disaster-recovery-planning-tool-now-available-caribbean
43 United Nations Office for the Coordination of Humanitarian Affairs. (2023). *Libya Hurricane Daniel: Situation Report 2–27/09/2023.* See https://reliefweb.int/report/libya/libya-hurricane-daniel-situation-report-2-27092023
44 See www.icrc.org/en/document/libya-icrc-flood-response-october
45 See https://libya.un.org/en/246233-un-delivers-aid-and-support-people-affected-storm-daniel-eastern-libya
46 See www.aljazeera.com/news/2023/9/14/how-to-support-storm-daniel-flood-victims-in-libya
47 Sahana, M., Patel, P. P., Rehman, S., Rahaman, M. H., Masroor, M., Imdad, K., & Sajjad, H. (2023). Assessing the Effectiveness of Existing Early Warning Systems and Emergency Preparedness towards Reducing Cyclone-Induced Losses in the Sundarban Biosphere Region, India. *International Journal of Disaster Risk Reduction*, *90*, 103645. https://doi.org/10.1016/j.ijdrr.2023.103645
48 See https://wpo.noaa.gov/the-hurricane-analysis-and-forecast-system-hafs/
49 See https://www-foxweather-com.cdn.ampproject.org/c/s/www.foxweather.com/weather-news/hurricane-forecast-cone-experimental-watch-warning.amp
50 See https://storymaps.arcgis.com/stories/6e0018f57f2c4faf98cd603436f8255c
51 Resilient Cedar Key. (2023). *Current and Potential Strategies for Resilience in Cedar Key, FL.* University of Florida. See https://storymaps.arcgis.com/stories/6e0018f57f2c4faf98cd603436f8255c
52 See https://twitter.com/NOAASatellitePA/status/1717198619735118261
53 National Environmental Satellite, Data, and Information Service. (2023). *Hurricane Otis Causes Catastrophic Damage in Acapulco, Mexico.* U.S. Department of Commerce. See www.nesdis.noaa.gov/news/hurricane-otis-causes-catastrophic-damage-acapulco-mexico
54 Jacobs, P. (2023). Hurricane Otis Smashed into Mexico and Broke Records. Why Did No One See It Coming? *Science.* See www.science.org/content/article/hurricane-otis-smashed-mexico-and-broke-records-why-did-no-one-see-it-coming
55 See www.independent.co.uk/climate-change/news/mexico-disaster-relief-hurricane-otis-b2437147.html
56 See https://rodrigonietogomez.substack.com/p/defunding-fonden-dismantling-emergency?r=bxhl9&utm_campaign=post&utm_medium=web
57 See https://latinfinance.com/daily-brief/2023/10/29/mexico-triggers-cat-bond-after-hurricane/
58 See https://reliefweb.int/sites/reliefweb.int/files/resources/IB25092017.pdf

59 Wilkinson, E. (2023). How a Small Caribbean Island Is Trying to Become Hurricane-Proof. *The Conversation*. See https://theconversation.com/how-a-small-caribbean-island-is-trying-to-become-hurricane-proof-217999
60 See www.fema.gov/emergency-managers/risk-management/building-science/disaster-support
61 See www.fema.gov/emergency-managers/practitioners/lifelines
62 See www.fema.gov/sites/default/files/documents/fema_rm-hurriance-ian-mat-report-12-2023.pdf
63 See www.census.gov/quickfacts/fact/table/palmbeachgardenscityflorida/PST045223
64 See www.pbgfl.com/282/Hurricane-Readiness
65 See www.pbgfl.com/DocumentCenter/View/13765
66 See www.cdc.gov/mmwr/preview/mmwrhtml/mm6220a1.htm
67 See www.fema.gov/blog/remembering-hurricane-sandy-10-years-later
68 See www.cbo.gov/publication/58840
69 See www.fema.gov/disaster/tribal-declarations
70 See www.fema.gov/sites/default/files/2020-07/elizabeth-zimmerman_sandy-recovery-one-year-later_testimony_11-14-2013.pdf
71 See www.fema.gov/disaster/sandy-recovery-improvement-act-2013
72 See www.fema.gov/blog/long-term-recovery-groups-help-new-jersey-sandy-survivors
73 See www.njvoad.org/about-us/
74 See www.fema.gov/emergency-managers/national-preparedness/frameworks/recovery/recovery-support-functions/community-assistance-rsf
75 See www.fema.gov/emergency-managers/national-preparedness/frameworks/national-disaster-recovery/rsflg
76 See www.homelandsecuritynewswire.com/dr20240125-predicting-flood-risk-from-hurricanes-in-a-warming-climate?page=0,1
77 Sarhadi, A., Rousseau-Rizzi, R., Mandli, K., Neal, J., Wiper, M. P., Feldmann, M., & Emanuel, K. (2024). Climate Change Contributions to Increasing Compound Flooding Risk in New York City. *Bulletin of the American Meteorological Society* (published online ahead of print 2024). https://doi.org/10.1175/BAMS-D-23-0177.1
78 See www.nyc.gov/assets/em/downloads/pdf/hurricane_sandy_aar.pdf
79 See www.redcross.org/about-us/our-work/disaster-relief/hurricane-relief/sandy-response.html
80 See www.redcross.org/local/new-jersey/about-us/news-and-events/news/10-years-later-a-look-back-at-superstorm-sandy.html
81 See www.fema.gov/sites/default/files/2020-11/fema_hurricane-sandy-recovery-collaborating-to-build-resilience_case-study.pdf
82 See www.nyc.gov/content/sandytracker/pages/fema-pa
83 See https://ops.fhwa.dot.gov/publications/fhwahop20026/fhwahop20026.pdf
84 See https://cee.illinois.edu/news/can-we-evacuate-hurricanes-electric-vehicles
85 Raza, M., Awais, M., Ali, K., Aslam, N., Paranthaman, V. V., Imran, M., & Ali, F. (2020). Establishing Effective Communications in Disaster Affected Areas and Artificial Intelligence Based Detection Using Social Media Platform. *Future Generation Computer Systems*, *112*, 1057–1069. https://doi.org/10.1016/j.future.2020.06.040
86 See www.gao.gov/assets/gao-21-297.pdf
87 See https://kmph.com/news/local/naples-fire-station-under-water-yet-firefighters-still-rescue-people
88 See www.hsdl.org/?view&did=692748#:~:text=Cease%20operation%20of%20fire%20apparatus,of%2040%20mph%20or%20greater.
89 See www.nytimes.com/video/climate/100000009065186/idalia-hudson-florida-durst-fire.html

90 See www.dhs.gov/xlibrary/assets/nipp_srtltt_guide.pdf
91 See https://jicrcr.com/index.php/jicrcr/article/view/70/71
92 See www.fema.gov/pdf/about/odic/fnss_guidance.pdf
93 See www.samhsa.gov/medications-substance-use-disorders/statutes-regulations-guidelines/methadone-guidance
94 See https://domesticpreparedness.com/healthcare/challenges-with-pediatric-mass-care-feeding
95 See https://nationalmasscarestrategy.org/
96 See https://hazards.colorado.edu/news/research-counts/avoiding-the-second-disaster-of-unwanted-donations
97 See www.nytimes.com/2012/11/10/nyregion/queens-nursing-home-is-faulted-over-care-after-storm.html
98 See www.fema.gov/sites/default/files/2020-08/fema_swiftwater-flood-sar-team_definition_08-03-2020.pdf
99 Treinish, S. (2017). *Water Rescue: Principles and Practice to NFPA 1006 and 1670: Surface, Swiftwater, Dive, Ice, Surf, and Flood.* International Association of Fire Chiefs/National Fire Protection Association. Jones & Barrett Learning
100 NOAA. (2013). *Severe Marine Debris Event Report: Superstorm Sandy.* U.S. Department of Commerce. See https://marinedebris.noaa.gov/sites/default/files/publications-files/Electronic_2013SuperstormSandy_SMDE_report_to_Congress.pdf
101 See https://blog.nwf.org/2012/10/hurricane-sandys-impact-on-fish-and-wildlife/
102 Gargani, J. (2022). Impact of Major Hurricanes on Electricity Energy Production. *International Journal of Disaster Risk Reduction*, *67*, 102643. https://doi.org/10.1016/j.ijdrr.2021.102643
103 Feng, K., Ouyang, M., & Lin, N. (2022). Tropical Cyclone-Blackout-Heatwave Compound Hazard Resilience in a Changing Climate. *Nature Communications*, *13*(1), 4421. https://doi.org/10.1038/s41467-022-32018-4
104 See www.njsendems.org/van-drew-coastal-evacuation-bill-now-law/
105 See https://oceantoday.noaa.gov/hurricanestormsurge/
106 Strauss, B. H., Orton, P. M., Bittermann, K., Buchanan, M. K., Gilford, D. M., Kopp, R. E., . . . & Vinogradov, S. (2021). Economic Damages from Hurricane Sandy Attributable to Sea Level Rise Caused by Anthropogenic Climate Change. *Nature Communications*, *12*(1), 2720. https://doi.org/10.1038/s41467-021-22838-1
107 See www.un.org/en/chronicle/article/economic-recovery-after-natural-disasters
108 See www.fema.gov/sites/default/files/documents/fema_equitable-recovery-post-disaster-guide-local-officials-leaders.pdf
109 Mitch, N. (2022). *Sea Level Rise and Housing Affordability in Small Coastal Communities: A Case Study in Maine.* Harvard Graduate School of Design. See https://nrs.harvard.edu/URN-3:HUL.INSTREPOS:37371655
110 Parton, L. C., & Dundas, S. J. (2020). Fall in the Sea, Eventually? A Green Paradox in Climate Adaptation for Coastal Housing Markets. *Journal of Environmental Economics and Management*, *104*, 102381. https://doi.org/10.1016/j.jeem.2020.102381
111 See www.fema.gov/pdf/emergency/nrf/nrf-support-cikr.pdf
112 See www.hud.gov/disaster_resources
113 Burger, J. (2015). Ecological Concerns Following Superstorm Sandy: Stressor Level and Recreational Activity Levels Affect Perceptions of Ecosystem. *Urban Ecosystems*, *18*(2), 553–575. https://doi.org/10.1007/s11252-014-0412-x
114 See www.fema.gov/assistance/public
115 See https://hazards.colorado.edu/news/research-counts/looting-or-community-solidarity-reconciling-distorted-posthurricane-media-coverage

116 See www.fema.gov/sites/default/files/documents/fema_p-2181-fact-sheet-4-1-drinking-water-systems.pdf

117 See www.npr.org/2024/02/15/1227741074/nature-shoreline-protection-hurricane-season?sc=18&f=1001

118 Turn Around, Don't Drown® is a registered trademark of the National Weather Service. Used with permission. See www.weather.gov/safety/flood-turn-around-dont-drown

119 Rojek, J., & Smith, M. R. (2007). Law Enforcement Lessons Learned from Hurricane Katrina. *The Review of Policy Research*, *24*(6), 589+. See https://link.gale.com/apps/doc/A173020469/AONE

120 See https://community.wmo.int/en/tropical-cyclone-operational-plans

121 See www.fema.gov/emergency-managers/risk-management/hurricanes

122 See www.fema.gov/sites/default/files/documents/fema_nhp_operational-decision-report_fact-sheet_022024.pdf

123 See www.fema.gov/sites/default/files/2020-08/fema_swiftwater-flood-sar-team_definition_08-03-2020.pdf

124 See www.dco.uscg.mil/Portals/9/DCO%20Documents/5p/CG-5PC/INV/Alerts/USCGSA_0123.pdf

REFERENCES/ADDITIONAL READING

Das, H. P. (2012). *Agrometeorology in Extreme Events and Natural Disasters*. CRC Press.

Elsner, J. B., & Jagger, T. H. (2009). *Hurricanes and Climate Change*. Springer.

FEMA. (2023). *Hurricane Ian in Florida: Building Performance Observations, Recommendations, and Technical Guidance (FEMA P-2342)*. FEMA. See www.fema.gov/sites/default/files/documents/fema_rm-hurriance-ian-mat-report-12-2023.pdf

Fink, S. (2013). *Five Days at Memorial: Life and Death in a Storm-ravaged Hospital*. Atlantic Books Ltd.

International Committee of the Red Cross. (2023). *Libya Flood Response*. ICRC. See www.icrc.org/en/download/file/277963/icrc_flood_response_october_english.pdf

Knutson, T. R., McBride, J. L., Chan, J., Emanuel, K., Holland, G., Landsea, C., . . . & Sugi, M. (2010). Tropical Cyclones and Climate Change. *Nature Geoscience*, *3*(3), 157–163. https://doi.org/10.1038/ngeo779

Michener, W. K., Blood, E. R., Bildstein, K. L., Brinson, M. M., & Gardner, L. R. (1997). Climate Change, Hurricanes and Tropical Storms, and Rising Sea Level in Coastal Wetlands. *Ecological Applications*, *7*, 770–801. https://doi.org/10.1890/1051-0761(1997)007[0770:CCHATS]2.0.CO;2

NOAA National Centers for Environmental Information. (n.d.). *Global Ocean Heat and Salt Content: Seasonal, Yearly, and Pentadal Fields*. U.S. Department of Commerce. Accessed February 26, 2024. See www.ncei.noaa.gov/access/global-ocean-heat-content/

NOAA National Data Buoy Center. (n.d.). *National Data Buoy Center*. U.S. Department of Commerce. Accessed February 26, 2024. See www.ndbc.noaa.gov/

NOAA Weather Program Office. (2023). *Next Generation Hurricane Modeling with the Hurricane Analysis and Forecast System (HAFS)*. NOAA. See https://wpo.noaa.gov/the-hurricane-analysis-and-forecast-system-hafs/

Reynolds, R. (2003). *Weather Rage* (Vol. 5). CRC Press.

Sahana, M., Patel, P. P., Rehman, S., Rahaman, M. H., Masroor, M., Imdad, K., & Sajjad, H. (2023). Assessing the Effectiveness of Existing Early Warning Systems and Emergency Preparedness Towards Reducing Cyclone-Induced Losses in the Sundarban Biosphere Region, India. *International Journal of Disaster Risk Reduction*, *90*, 103645. https://doi.org/10.1016/j.ijdrr.2023.103645

Voyer, J., Dean, M., & Pickles, C. (2015). *Understanding Humanitarian Supply Chain Logistics with System Dynamics Modeling.* System Dynamics Society. See https://proceedings.systemdynamics.org/2015/papers/P1164.pdf

Wehner, M. F., & Kossin, J. P. (2024). The Growing Inadequacy of an Open-Ended Saffir–Simpson Hurricane Wind Scale in a Warming World. *Proceedings of the National Academy of Sciences, 121*(7), e2308901121. https://doi.org/10.1073/pnas.2308901121

Wolf, E. L. (2023). *Physics and Future of Hurricanes.* CRC Press.

BIBLIOGRAPHY

Anthony, R. Mignone, Jr. (2013). *Dunes and Ocean Front Structures Under Wave Attack.* U.S. Department of Commerce. See www.weather.gov/media/erh/ta2013-02.pdf

Better Buildings Solutions Center/U.S. Department of Energy. (n.d.). *The Energy-Resilient City.* U.S. Department of Energy. See https://betterbuildingssolutioncenter.energy.gov/resilience/communities

Burger, J. (2015). Ecological Concerns Following Superstorm Sandy: Stressor Level and Recreational Activity Levels Affect Perceptions of Ecosystem. *Urban Ecosystems, 18*(2), 553–575. https://doi.org/10.1007/s11252-014-0412-x

Feng, K., Ouyang, M., & Lin, N. (2022). Tropical Cyclone-Blackout-Heatwave Compound Hazard Resilience in a Changing Climate. *Nature Communications, 13*(1), 4421. https://doi.org/10.1038/s41467-022-32018-4

Gargani, J. (2022). Impact of Major Hurricanes on Electricity Energy Production. *International Journal of Disaster Risk Reduction, 67,* 102643. https://doi.org/10.1016/j.ijdrr.2021.102643

Jacobs, P. (2023). Hurricane Otis Smashed into Mexico and Broke Records. Why Did No One See It Coming? *Science.* See www.science.org/content/article/hurricane-otis-smashed-mexico-and-broke-records-why-did-no-one-see-it-coming

Lowry, M. (2024). What New Data Reveal about How Hurricanes Kill. *Yale Climate Connections.* Accessed January 27, 2024. See https://yaleclimateconnections.org/2023/08/what-new-data-reveal-about-how-hurricanes-kill/#:~:text=In%20addition%20to%20freshwater%20(rainfall,from%206%25%20in%20earlier%20findings

Mitch, N. (2022). *Sea Level Rise and Housing Affordability in Small Coastal Communities: A Case Study in Maine.* Harvard Graduate School of Design. See https://nrs.harvard.edu/URN-3:HUL.INSTREPOS:37371655

National Environmental Satellite, Data, and Information Service. (2023). *Hurricane Otis Causes Catastrophic Damage in Acapulco, Mexico.* U.S. Department of Commerce. See www.nesdis.noaa.gov/news/hurricane-otis-causes-catastrophic-damage-acapulco-mexico

National Hurricane Center. (n.d.a). *Storm Surge Overview.* U.S. Department of Commerce. Accessed February 23, 2024. See www.nhc.noaa.gov/surge/

National Hurricane Center. (n.d.b). *National Hurricane Center Storm Surge Risk Maps.* U.S. Department of Commerce. See https://experience.arcgis.com/experience/203f772571cb48b1b8b50fdcc3272e2c/page/Category-5/

National Hurricane Center. (n.d.c). *Storm Surge Overview.* Accessed February 23, 2024. See www.nhc.noaa.gov/surge/

National Hurricane Center and Central Pacific Hurricane Center. (n.d.a). *Storm Surge Maximum Envelope of Water (MEOW).* U.S. Department of Commerce. Accessed February 22, 2024. See www.nhc.noaa.gov/surge/meowOverview.php

National Hurricane Center and Central Pacific Hurricane Center. (n.d.b). *Storm Surge Maximum of the Maximum (MOM).* Accessed February 22, 2024. See www.nhc.noaa.gov/surge/momOverview.php

National Hurricane Center and Central Pacific Hurricane Center. (n.d.c). *Sea, Lake, and Overland Surges from Hurricanes (SLOSH)*. U.S. Department of Commerce. Accessed February 23, 2024. See www.nhc.noaa.gov/surge/slosh.php

National Weather Service. (n.d.a). *Science and Technology*. U.S. Department of Commerce. Accessed February 23, 2024. See www.weather.gov/ilm/ScienceandTechnology#seabreeze

National Weather Service. (n.d.b). *Significant Wave Height*. U.S. Department of Commerce. Accessed February 23, 2024. See www.weather.gov/key/marine_sigwave

NCEI. (n.d.). *Coastal Water Temperature Guide*. U.S Commerce Department. Accessed February 22, 2024. See www.ncei.noaa.gov/products/coastal-water-temperature-guide

NHC. (2010). *COMET's Community Hurricane Preparedness Module Updated for 2010*. U.S. Commerce Department. See www.nhc.noaa.gov/comet_update.shtml

NOAA. (2013). *Severe Marine Debris Event Report: Superstorm Sandy*. U.S. Department of Commerce. See https://marinedebris.noaa.gov/sites/default/files/publications-files/Electronic_2013SuperstormSandy_SMDE_report_to_Congress.pdf

NOAA. (n.d.a). *How Does the Ocean Affect Hurricanes*? U.S. Commerce Department. Accessed February 22, 2024. See https://oceanexplorer.noaa.gov/facts/hurricanes.html

NOAA. (n.d.b). *The Marine Layer*. U.S. Commerce Department. Accessed February 25, 2024. See www.noaa.gov/jetstream/ocean/marine-layer

Parton, L. C., & Dundas, S. J. (2020). Fall in the Sea, Eventually? A Green Paradox in Climate Adaptation for Coastal Housing Markets. *Journal of Environmental Economics and Management*, *104*, 102381. https://doi.org/10.1016/j.jeem.2020.102381

Raza, M., Awais, M., Ali, K., Aslam, N., Paranthaman, V. V., Imran, M., & Ali, F. (2020). Establishing Effective Communications in Disaster Affected Areas and Artificial Intelligence Based Detection Using Social Media Platform. *Future Generation Computer Systems*, *112*, 1057–1069. https://doi.org/10.1016/j.future.2020.06.040

Resilient Cedar Key. (2023). *Current and Potential Strategies for Resilience in Cedar Key, FL*. University of Florida. See https://storymaps.arcgis.com/stories/6e0018f57f2c4faf98cd603436f8255c

Rojek, J., & Smith, M. R. (2007). Law Enforcement Lessons Learned from Hurricane Katrina. *The Review of Policy Research*, *24*(6), 589+. See https://link.gale.com/apps/doc/A173020469/AONE

Sahana, M., Patel, P. P., Rehman, S., Rahaman, M. H., Masroor, M., Imdad, K., & Sajjad, H. (2023). Assessing the Effectiveness of Existing Early Warning Systems and Emergency Preparedness Towards Reducing Cyclone-Induced Losses in the Sundarban Biosphere Region, India. *International Journal of Disaster Risk Reduction*, *90*, 103645. https://doi.org/10.1016/j.ijdrr.2023.103645

Sarhadi, A., Rousseau-Rizzi, R., Mandli, K., Neal, J., Wiper, M. P., Feldmann, M., & Emanuel, K. (2024). Climate Change Contributions to Increasing Compound Flooding Risk in New York City. *Bulletin of the American Meteorological Society* (published online ahead of print 2024). https://doi.org/10.1175/BAMS-D-23-0177.1

Savage, R. (2009). *The Ocean Rower's Perspective on Climate Change*. Rozsavage.com. See www.rozsavage.com/the-ocean-rowers-perspective-on-climate-change/

Sherriff, L. (2023). Babcock Ranch: Florida's First Hurricane-Proof Town. *BBC*. See www.bbc.com/future/article/20230904-babcock-ranch-floridas-first-hurricane-proof-town

Strauss, B. H., Orton, P. M., Bittermann, K., Buchanan, M. K., Gilford, D. M., Kopp, R. E., . . . & Vinogradov, S. (2021). Economic Damages from Hurricane Sandy Attributable to Sea Level Rise Caused by Anthropogenic Climate Change. *Nature Communications*, *12*(1), 2720. https://doi.org/10.1038/s41467-021-22838-1

Treinish, S. (2017). *Water Rescue: Principles and Practice to NFPA 1006 and 1670: Surface, Swiftwater, Dive, Ice, Surf, and Flood. International Association of Fire Chiefs/ National Fire Protection Association.* Jones & Barrett Learning.

United Nations Office for the Coordination of Humanitarian Affairs. (2023). *Libya Hurricane Daniel: Situation Report 2–27/09/2023.* See https://reliefweb.int/report/libya/libya-hurricane-daniel-situation-report-2-27092023

Wilkinson, E. (2023). How a Small Caribbean Island Is Trying to Become Hurricane-proof. *The Conversation.* See https://theconversation.com/how-a-small-caribbean-island-is-trying-to-become-hurricane-proof-217999

World Bank. (2024). Ahead of the Storm: An Innovative Disaster Recovery Planning Tool Is Now Available in the Caribbean. *Reliefweb – United Nations Office for the Coordination of Humanitarian Affairs.* See https://reliefweb.int/report/world/ahead-storm-innovative-disaster-recovery-planning-tool-now-available-caribbean

15 Earthquakes and Tsunamis

> Whenever an earthquake or tsunami takes thousands of innocent lives, a shocked world talks of little else.
>
> **– Anne M. Mulcahy**[1]

Generally, it is an earthquake in the ocean floor (submarine) threat which generates tsunami hazards:

> Earthquakes are commonly associated with ground shaking that is a result of elastic waves traveling through the solid earth.
>
> However, near the source of submarine earthquakes, the seafloor is "permanently" uplifted and down-dropped, pushing the entire water column up and down. The potential energy that results from pushing water above mean sea level is then transferred to horizontal propagation of the tsunami wave (kinetic energy). For the case shown above, the earthquake rupture occurred at the base of the continental slope in relatively deep water. Situations can also arise where the earthquake rupture occurs beneath the continental shelf in much shallower water.
>
> **(Pacific Coastal and Marine Science Center, n.d., p. 1)**[2]

As noted in Chapter 14, tsunamis are different from tidal waves.[3] And there are a number of countries (five, at last count) which have effective tsunami early warning systems.[4]

There are many books and peer-reviewed articles about both earthquakes and tsunamis; some which have relevance to Emergency Management are noted in the Additional Reading section at the end of this chapter. This chapter will not gauge the differences in severity of the adverse impacts to communities from tsunamis, but rather focus on the maximum of maximums. What is vitally important for emergency managers to learn is that a tsunami will most likely be the maximum of a **no-notice-type** water-related complex incident, for their personal careers. If the POETE elements for a tsunami can be solved completely, as part of the Preparedness phase, the solutioning for almost every other water-related disaster will flow much smoother. A reminder that the Preparedness phase includes missions of Protection and Prevention – and that both are *against the adverse impacts* from threats and hazards. Humans cannot prevent an earthquake or a tsunami, but they can prevent the adverse impacts to other humans from them.

In the United States, one of the greatest geographical areas of concern for tsunamis is the U.S. West Coast, including Alaska and their Aleutian Islands. This is due to at least two major subduction zones near populated coastlines. And while this

DOI: 10.1201/9781003474685-20

book will not cover earthquakes themselves, but rather note that they can be one significant factor[5] in generating a tsunami wave. It is also important to note that the U.S. west coast is not the only part of the country[6] to experience earthquakes – and there are subduction zones everywhere around the world.

CONFLICTS ON RISK CALCULATIONS IN WASHINGTON STATE

The question is not "if" but rather "when and how bad", for residents of the state of Washington, in the northwest United States, especially communities on the coastline.

> That's because about 70 miles offshore, a jammed-up 800-mile tectonic seam called the Cascadia Subduction Zone is approaching a shattering shakeout. The odds that it will unleash an earthquake in the next 50 years are estimated at 1 in 4. The odds of an 8.7-plus megaquake that would send a tsunami washing 30 ft or more over those communities is 1 in 6. At the tiny Hoh tribal reservation on Washington's Olympic Peninsula, the waves could reach 100 ft and put the tribal center 45 ft underwater.
>
> Washington State's Emergency Management Division calculates that nearly 90,000 people live or work in the outer coast's inundation zone, and there are another 86,000 more along inner waterways that the waves will take longer to reach. On a summer day, they could be joined in the danger zone by up to 248,000 sightseers, clam diggers and other visitors. Western Washington University's Resilience Institute has calculated that as many as 28 percent would be unable to reach higher ground in time to escape the tsunami and 18 percent – up to 60,000 people – would be crushed or swept out to sea.

Open-source media outlets[7] have reported that there is a disconnect between what groups[8]/entities[9] in the U.S. Federal government view as high levels of risk, and what the state[10] perceives these risks to be.

SUBDUCTION ZONES

A Subduction Zone is where two tectonic plates of the earth's crust collide with each other, specifically one under the other. When that happens, "the most powerful earthquakes, tsunamis, volcanic eruptions, and landslides occur" (Subduction Zone Science, 2020, p. 1).[11]

15.1 CASE EXAMPLES

15.1.1 Cascadia Subduction Zones

The probability of a major earthquake striking the U.S. west coast in the 50 years is worth noting:

> The Cascadia Subduction Zone is a 700-mile fault that runs from northern California up to British Columbia and is about 70–100 miles off the Pacific coast shoreline. There have

been 43 earthquakes in the last 10,000 years within this fault. The last [recorded] earthquake that occurred in this fault was on Jan. 26, 1700, with an estimated 9.0 magnitude. This earthquake caused the coastline to drop several feet and a tsunami to form and crash into the land. Evidence for this great earthquake came from Japan. Japanese historical records indicate that a destructive distantly-produced tsunami struck their coast on Jan. 26, 1700. By studying the geological records and the flow of the Pacific Ocean, scientists have been able to link the tsunami in Japan with the great earthquake from the Pacific Northwest. Native American legends also support the timing of this last event.

Oregon has the potential for a 9.0+ magnitude earthquake caused by the Cascadia Subduction Zone and a resulting tsunami of up to 100 feet in height that will impact the coastal area. There is an estimated five to seven minutes of shaking or rolling that will be felt along the coastline with the strength and intensity decreasing the further inland you are.

The Cascadia Subduction Zone has not produced an earthquake since 1700 and is building up pressure where the Juan de Fuca Plate is subsiding underneath the North American plate. Currently, scientists are predicting that there is about a 37% chance that a megathrust earthquake of 7.1+ magnitude in this fault zone will occur in the next 50 years. This event will be felt throughout the Pacific Northwest.

(Oregon Department of Emergency Management, n.d., p. 1)[12]

Multiple states along the west coast, along with FEMA Region 10, are working through the POETE of this threat, including on a whole-of-government/whole-community[13] basis. Annually, there are international self-run exercises conducted for increased earthquake awareness through the "Great Shakeout",[14] which occurs every October.

15.1.2 Indian Ocean/Indonesian Tsunami of 2004

The world's third largest in magnitude earthquake (at 9.1 magnitude), since 1900, generated a tsunami in the Indian Ocean off the coast of Sumatra, one of the islands of Indonesia.

It occurred 18.6 miles (30 kilometers) below the ocean floor along a reverse fault in the Sunda trench where the Indian plate (part of the Indo-Australian plate) subducts beneath the Burma plate (a minor tectonic plate or microplate). The length of the rupture was roughly 800 miles (1,300 kilometers), similar in length to California.

The shaking was felt not only in Indonesia, but also in Bangladesh, India, Malaysia, the Maldives, Myanmar, Singapore, Sri Lanka, and Thailand. The earthquake caused severe damage and casualties in Indonesia (Northern Sumatra) and in India (the Andaman and Nicobar Islands).

The tsunami that followed had catastrophic impacts throughout the Indian Ocean. Once generated, the tsunami radiated outward in all directions, striking the coasts of Indonesia and India's Andaman and Nicobar Islands within 20 minutes of the earthquake and the northeastern coast of Somalia in Africa seven hours later.

Wave heights and inundation distances varied throughout the region based on location relative to the source earthquake as well as the depth of the ocean, the elevation of the coast, and other features of the land both above and below the ocean. In Indonesia's Aceh province in Northern Sumatra, waves reached 167 feet (51 meters) and caused flooding up to three miles (five kilometers) inland.

On the other side of the ocean, in Somalia, waves ranged in height from 11 to 31 feet (3.4 to 9.4 meters). The tsunami was also observed on over 100 coastal water-level stations in the Atlantic and Pacific Oceans, making it a global tsunami.

While deadly tsunamis had happened in the Indian Ocean region (5% of the tsunamis between 1900 and 2017 occurred in the Indian Ocean), the last major tsunami was in 1883. So at the time of the event, there was little public awareness about tsunamis, and there was no official tsunami warning system.

Natural tsunami warnings included the earthquake (close to its source), withdrawal of the sea, and unusually large but not damaging initial waves. Unfortunately, these were not widely understood as warnings, and many witnesses to the event rushed to the water to explore the exposed ocean floor or continued with their daily activities.

In some cases, where tsunami knowledge existed and was used, like in small island communities where stories about tsunamis had been passed down from generations before them, thousands of lives were saved.

The tsunami, not the earthquake, was responsible for most of the impacts, which were observed in 17 countries in Southeastern and Southern Asia and Eastern and Southern Africa. An astonishing roughly quarter million people (227,899) were killed or missing and presumed dead, including tourists, making this the deadliest tsunami in history.

About 1.7 million people were displaced. Total damage was estimated at roughly $13 billion (2017 dollars). Indonesia was the hardest hit country, with over 167,000 lives lost and nearly $6 billion (2017 dollars) in damage.

Waves and wave-carried debris devastated once-thriving communities, destroying homes, businesses, basic services, critical infrastructure, the environment, livelihoods, and entire economies. And the inundation of saltwater damaged soils, vegetation, and crops.

Together, the earthquake and tsunami changed the landscape of many Indian Ocean coastal communities. Coastal erosion and subsidence caused some shorelines to disappear into the ocean while, in some areas, uplift forced coral reefs to rise above its surface.

(NOAA, n.d., p. 1)[15]

15.1.3 The Great East Japan Earthquake of 2011

As is known to many, the Fukushima Dai-ichi nuclear disaster started with an offshore earthquake (thus far the largest magnitude one in Japan), which not only damaged the Fukushima Dai-ichi nuclear power plant in the Tohuku region on the island of Honshu in Japan but also generated a massive tsunami wave.

While the catastrophic impacts from the core meltdown and hydrogen explosion are topics of other books and articles,[16] the "Tohoku tsunami produced waves up to 40 m (132 ft) high. More than 450,000 people became homeless as a result of the tsunami. More than 15,500 people died. The tsunami also severely crippled the infrastructure of the country" (National Geographic, n.d., p. 1).[17]

15.1.4 Valdez Tsunami Inundation Map and the Aleutian Islands in Alaska

The state of Alaska is also subject to tsunami risk, especially in the Aleutian Islands[18] communities, and near the port city of Valdez. While the human populations in these areas are significantly less than the continental U.S. (CONUS) west coast, one interesting facet of this subduction zone is higher frequencies in tsunamis when there is a creeping megathrust between tectonic plates, as compared to a locked one:

> Earthquake and tsunami hazards at subduction zones, according to prevailing models, are thought to be highest where geodetic observations indicate a locked megathrust (the gently dipping fault between converging tectonic plates) and lowest where the megathrust is creeping. However, the presently creeping part of the eastern Aleutian Subduction Zone, which extends over 400 km from Unalaska to the Shumagin Islands, has been proposed as the source area for tsunami scenarios that potentially result in devastating consequences to coastal communities around the Pacific Ocean. The scenarios build upon historical precedents. Disastrous impacts in Hawaii ensued when a tsunami associated with the 1946 Unimak earthquake (Mw 8.6) arose in this creeping region and prompted the advent of the Pacific Tsunami Warning System. The Pacific-wide reach of the 1946 tsunami resulted from a "tsunami earthquake" that involved slow rupture (1.12 km/s average seismic velocity) and tectonic displacement near the Aleutian trench. However, it was a submarine landslide dislodged by earthquake shaking that generated the 42 m high wave that destroyed the Scotch Cap lighthouse on Unimak Island.
>
> **(Witter et al., 2016, p. 1)**[19]

Valdez saw a major earthquake in 1964 – still the largest recorded one in North America, at 9.2 on the Richter Scale – and subsequent tsunami. The wave height from the tsunami reached 27 ft at points and killed 128 people.

> Of the tsunami-related deaths, approximately two-thirds were from the local waves generated during or immediately after the earthquake. Earth movement was so dramatic that seiches were created as far away as Louisiana, where a number of fishing boats were sunk.
>
> The destruction of property was unprecedented. Nearly every coastal community in Alaska was affected, with damage also occurring in British Columbia, and the states of Washington, Oregon and California. The entire town of Valdez, which experienced a tsunami caused by a submarine landslide, had to be relocated due to the resulting ground instability in the area. In some areas, the residents lost their entire livelihoods because of the total destruction of business and industrial facilities.
>
> **(Western States Seismic Policy Council, n.d., p. 1)**[20]

A significant set of maps for the potential tsunami inundation area around Valdez has been produced, and a reference is available at the end of this chapter.

15.2 ADVERSE IMPACTS TO THE EMERGENCY AND RECOVERY SUPPORT FUNCTIONS AND COMMUNITY LIFELINES

Table 15.1 shows some of the **additional** adverse impacts of tsunami threats or hazards on the U.S.-based Emergency Support Functions. Similar tables for ***all*** the other Hazard parts of this book should also be reviewed first.

TABLE 15.1
Water-Related Threats and Hazards to the ESFs

Emergency Support Function (ESF)	Examples of Additional Supporting Actions or Capabilities Related to Tsunami Threats or Hazards
#1 – Transportation	As a no-notice incident of massive adverse impacts, most likely transportation routes into and out of the area where the tsunami flooding occurred will be impassable. National and possibly international aid will be needed for debris removal, bridge repair, etc. Clearing space for helicopter landing areas is a good start to restoring this ESF.
#2 – Communications	One of the first items to be transported into the impacted area should be multiple cell-on-wheels units (COWs), with a priority to restore first responder cellular voice and data capability.[21]
#3 – Public Works and Engineering	National Guard, USACE, private sector (for water companies, etc.) and other public infrastructure missions should be coordinated centrally for synergies, resource request deconfliction, etc.
#4 – Firefighting	Fires can be generated from the adverse impacts of a tsunami.[22] This ESF is also adversely impacted if apparatus is not relocated in advance, out of flood-prone areas. In many tsunamis, gas lines become ruptured, propane tanks float away, and other sources can generate fires[23] – many of which cannot be fought due to inaccessibility.
#5 – Information and Planning	Essential Elements of Intelligence/Information about the impacts to all Community Lifelines are critical for planning the next operational period for tsunamis. This can include capabilities status of all the other ESFs, such as the planned locations for evacuation sheltering, reunification status, water quality testing results, distribution status, restoration timelines, and more. Status of critical infrastructure/key resources,[24] especially those which were damaged, needs to be provided by ESF#3. All of this Emergency Management Intelligence is needed by ESF#5 to work through the operational planning cycle (see Chapter 2 for details).
#6 – Mass Care, Emergency Assistance, Temporary Housing, and Human Services	While sheltering, feeding, and family reunification will most likely occur during a tsunami, inclusion for people with disabilities and access/functional needs[25] must be part of the planning cycle, as soon as possible and integrated into existing sheltering. Make sure there is family choice support for feeding infants and toddlers.[26] Review material on the National Mass Care Strategy website,[27] for best practices.
#7 – Logistics	Tsunamis require many of the same logistics support elements as other large-scale no-notice incidents (tornados, earthquakes, etc.). Staging areas outside of the impact zones, for power restoration, potable water-related distribution support, receiving massive quantities of outside aid materials (including unsolicited/unwanted donations)[28] all will require SLTT warehousing and support. Flooded/damaged railways, roadways, and airports may impact supply-chain integrity.

(Continued)

TABLE 15.1 ***(Continued)***
Water-Related Threats and Hazards to the ESFs

Emergency Support Function (ESF)	Examples of Additional Supporting Actions or Capabilities Related to Tsunami Threats or Hazards
#8 – Public Health and Medical Services	As previously noted, every incident, emergency, or disaster has a public health and medical services impact. Tsunamis should be declared as a public health emergency. Licensed care facilities (nursing homes, assisted living centers, hospitals, hospices, etc.) away from the impacted areas, all need to be monitored for capabilities and continuity themselves. Consider expanding bed limits in locations which can take additional patients away from sites which could be tsunami-impacted. Water quantity incidents – especially extremely destructive ones, such as tsunamis – can generate stormwater hazards, deadly mold in homes, sewage backups, and other public health hazards. Hospitals will become overwhelmed[29] with injuries from the tsunami and need backup support from other hospitals which are outside of the impacted geography.
#9 – Search and Rescue	For tsunamis which produce flooding, this ESF will be activated *after the fact*. Because of the no-notice aspect, there is no real opportunity to perform swift-water rescues[30] during the tsunami wave – nor should they be attempted: the hazards are too great for the life safety of the responders. This may become more of a recovery operation, than a rescue one.
#10 – Oil and Hazardous Materials Response	As noted in the case studies in the chapters in this Quantity Part (also, in the Quality Part), there can be tsunami flooding damage to industrial and other areas where oil and hazardous materials make their way into the streams, rivers, watershed, etc., and, therefore, adversely impact communities. The adverse health impacts from those human-made hazards including sewer treatment plant overflows, etc., may require missions for cleanup and remediation from this ESF. The high priority of this cleanup and remediation of the oil and/or hazardous materials is both a life safety and an incident stabilization course of action.
#11 – Agriculture and Natural Resources	The adverse impacts to marine wildlife are the primary concern for this ESF. Debris[31] and hazardous materials can be transported into marshlands, estuaries, etc., and harm aquatic and avian life[32] in both the short- and long-term.
#12 – Energy	Energy production[33] and distribution[34] will be significantly impacted – if not catastrophically damaged (see Great East Japan Earthquake of 2011 case, above). Emergency managers will be tasked with helping to adjudicate restoration priorities – many times, between what appears to be two or more equally critical key resources. Exercising these decisions in advance can help decision-making processes.

TABLE 15.1 ***(Continued)***
Water-Related Threats and Hazards to the ESFs

Emergency Support Function (ESF)	Examples of Additional Supporting Actions or Capabilities Related to Tsunami Threats or Hazards
#13 – Public Safety and Security	In addition to additional incident-related Response missions for law enforcement officials, there may be interim Recovery missions as well, such as curfews, no entry zones, etc. There may be investigations related to fatalities attributed to the tsunami. Public Safety and Security may also be needed to support every other ESF, as well as search and rescue/recovery support missions. And of course, Public Safety is never confined to only one incident at a time, so staffing will need to be extended or expanded to cover multiple missions simultaneously. Staff wellness and health is important as well – since some and/or their families may be impacted by the tsunami directly themselves.
#14 – Cross-Sector Business and Infrastructure	As tsunamis adversely impact large areas of a jurisdiction, businesses, non-governmental organizations, and other infrastructure will be affected as well. Many of these sectors have homeland security impacts as well such as banking, oil and gas refining, etc. Stabilizing the Community Lifelines, restoring supply-chain integrity, and coordinating resource needs from the private sector are all missions for this ESF, associated with tsunamis.
#15 – External Affairs	Coordinate with ESF#2 on both internal and external messaging regarding the full disaster cycle of crisis messaging needed for tsunamis. Preparedness messaging needs to be year-round, as there is no "season" for tsunamis. Upon a tsunami warning, the PIO needs to activate templated messages to the public which provide the LIPER actions they need to take, in the languages and access to crisis communication that the impacted community needs. Emergency Management's public crisis communications messages for tsunamis, can also be summarized as "run from the water, hide from the wind".[35] The crisis communications priority of life safety extends to emergency responders as well. The public needs frequent and constant reminders to evacuate early and to the highest ground possible, and that emergency responders will not be able to rescue people, once the tsunami strikes.

Source: Barton Dunant.

15.2.1 Recovery Support Functions

Similarly, Table 15.2 shows some of the ***additional*** adverse impacts of Tsunami hazards on the U.S.-based Recovery Support Functions (RSF). There are similar tables in the other Hazard parts of this book. Those should be reviewed as well.

Generally, adverse impacts from tsunami threats and hazards target all the RSFs: no geographically located infrastructure system will be spared.

TABLE 15.2
Recovery Support Functions

Federal Recovery Support Functions (RSFs)

Economic

Tsunamis which make landfall will have an economic impact[36] across a community. If there is no federal or international aid, including preparedness/mitigation funding for future incidents, through a potentially competitive Hazard Mitigation Assistance grant program – economic recovery may be slower or nonexistent.[37]

Health and Social Services

As noted in ESF#8, the resultant flooding from a tsunami can have significant adverse impacts to the health of individuals and families, due to hazards associated with contaminated water, mold found in homes after floodwaters recede, etc. Individuals with compromised immune systems, pulmonary issues, and other medical concerns can be significantly harmed after a tsunami – even without direct injuries from the wave damage. Social Services programs may also be curtailed, due to site damage and staffing concerns. Special emphasis on recovery and restoring capabilities for adults with disabilities (for example, developmentally disabled group homes), domestic violence[38] shelters, and other community-based residential social service programs, must be managed to support housing stability.

Community Assistance

One aspect that FEMA has for Community Assistance in Recovery[39] is to help achieve **equitable** recovery, as incident survivors work toward rebuilding their community. Incorporating community partners, stakeholders, faith-based organizations, etc. is one way to achieve this. There will be considerable economic pressures to rebuild *away* from supporting affordable housing.[40] There will also be massive resistance[41] in any community to reducing or eliminating infrastructure – including rebuilding homes – in coastal areas subject to recurrent flooding and with it the potential for tsunami risk.

Infrastructure Systems

As Infrastructure Systems will be significantly damaged from tsunamis, the recovery aspects of restoration of the critical infrastructure/key resources, should be prioritized.[42] This will include both public and private sector organizations. This can be a challenge for community leaders: is a hospital's restoration a higher priority than a school – especially if there were other hospitals in the area which were not damaged? Transparency, collaboration, and coordination on these decisions can be done in advance – via exercises which specifically address these "either this or that" prioritization decisions. As shown in the cases in this chapter, sometimes the decision is to completely relocate the community, because of the continued risk of flooding and tsunamis.

Housing

Housing will be massively impacted, recovery efforts will need to transition from the ESF#6 – Mass Care/Sheltering through the Recovery phase into temporary housing, interim housing, and possibly permanent housing replacement solutions away from the impacted area. In many communities, there is an affordable housing shortage now – before any disasters occur of any kind. The U.S. Department of Housing and Urban Development (HUD)[43] – upon a request from the state or territorial governor – may provide Community Development Block Grant Disaster Recovery (CDBG-DR) funds to help cities, counties, and states recover from disasters, through housing vouchers for Section 8 approved housing units.

TABLE 15.2 *(Continued)*
Recovery Support Functions

Federal Recovery Support Functions (RSFs)
Natural and Cultural Resources
The above noted Response adverse impacts, include potential damage in Recovery aligned to the restoration of recreational activities – including boating and fishing – from the devastation of ports, boats, etc. For many coastal communities, their access to the beach, oceans, waterways, etc. is a significant economic factor for them.
Mitigation work – which starts from projects identified through the RSFs – should also be considered by emergency managers, for children-related sites such as sports fields, nature centers, museums, historical sites, etc., in addition to what has been traditionally focused on government-run K12 schools. The remediation, repair, and future hazard protection work for some NGOs (such as an environmental conservation group or a marine wildlife stranding organization) can count toward Public Assistance[44] projects in the United States.

Source: FEMA.

15.2.2 Community Lifelines

As noted in Chapter 2, the Community Lifelines (CLs) are a highly useful list of EEIs, which can be visualized as a dashboard. As with the RSFs, tsunami hazards will have implications for every community lifeline. If roads and power stations and other critical infrastructure have been destroyed, almost every CL will become quickly adversely impacted. There are significant life safety impacts to each CL:

Safety and Security (Law Enforcement/Security, Fire Service, Search and Rescue, Government Service, Community Safety): Tsunami hazards, from massive ocean water flooding, disruption of water treatment plants, debris in roadways, etc. can block roads, generate additional security and safety missions[45] for law enforcement, and all the other sub-elements within this CL, such as Search and Rescue.

Food, Hydration, Shelter (Food, Hydration, Shelter, Agriculture): Obvious impacts to this Community Lifeline include post-inundation missions (including sheltering and feeding) associated with tsunamis. People will hopefully evacuate shore areas to higher ground if any tsunami warnings go off. Most will be unable to return to their damaged homes. For those who can, they will still need access to food and water for drinking, bathing, toileting, etc. For example, after a tsunami incident subsides, if that community's water treatment plant was damaged, residents may not return to their homes – and choose to stay at shelters, for feeding and hydration assistance.

Health and Medical (Medical Care, Public Health, Patient Movement, Medical Supply Chain, Fatality Management): Aligns to both Emergency

Support Function ESF#8) Public Health missions and the Health and Social Services RSF missions. As noted in the ESFs and RSFs, this Community Lifeline is significantly adversely impacted by any disruption to either site access (roads to hospitals, aid stations, and clinics blocked due to flooding or debris), critical infrastructure damage to health or medical facilities themselves, or damage to their own potable water supply systems, which may happen with tsunamis.

Energy (Power Grid, Fuel): Water and Energy have many nexus points, all of which need emergency management intelligence. Similar adverse impacts can occur, as shown in Chapter 14. Historically,[46] the adverse impacts of the preceding earthquake can exponentially expand the damage to the community from the tsunami. The status of this CL is always the most important in any disaster.

Communications (Infrastructure, Responder Communications, Alerts Warnings and Messages, Finance, 911 and Dispatch): One aspect of crisis communications for Public Information Officers related to tsunamis is to pre-plan (and exercise) the no-notice aspects (literally only minutes between earthquake sensors being triggered and then tsunami alert systems being activated), utilizing templated public messaging in the languages and delivery methods which have the greatest outreach capabilities for that community.

Transportation (Highway/Roadway/Motor Vehicle, Mass Transit, Railway, Aviation, Maritime): Tsunamis will have a catastrophic impact on this CL, as noted in ESF#1 and other areas of this chapter. The negative status of this CL can impact the others as it relates to providing any Recovery-related logistical movement in support of the other CLs in a community. If potable water quantity becomes a hazard in a community, the full capabilities of the Transportation CL will be needed for emergency services Response (police, fire, EMS, etc.), electrical and other utility restoration, and potentially transporting potable water to Points of Distribution, and in bulk transport to critical infrastructure/key resource sites, such as hospitals and dialysis centers. If navigable waterways are still blocked or hampered after a tsunami wave has gone through a community, this CL will be impacted as well. Critical repair parts, filters, staff, etc. need to be transported to water and wastewater treatment sites, to make repairs and work toward restoration of the **Water Systems** CL, if they are adversely impacted by a water quantity threat, such as this one.

Hazardous Material (Facilities, HAZMAT, Pollutants, Contaminants): Impacts to this Community Lifeline can be for Pollutants and/or Contaminants and the HAZMAT response to the tsunami and/or any additional cascading incidents, if those commercial and industrial sites are adversely impacted by too much water (flooding), seawall breeches, etc. This could be from natural or human-made hazards. There may be longer-term impacts to facilities, if a contaminated waterway supports those

Critical Infrastructure/Key Resources (CIKR), from a needed clean water supply basis. Considerations on how natural freshwater supplies are utilized in dialysis centers and even nuclear power plants (even if they themselves are not damaged), for example.

Water Systems (Potable Water Infrastructure, Wastewater Management):[47] Understanding how the flooding conditions from a tsunami can impact a potable water supply system (and the wastewater treatment, as well) is critical for Emergency Management.

There may also be longer-term impacts to the food supply if agriculturally used land is flooded – especially by seawater, or experiences damage to infrastructure such as warehouses, transportation vehicles, farm equipment, etc.

15.3 IMPACTS TO THE DISASTER PHASE CYCLES

As shown in Table 15.3, there are specific impacts to all four of the disaster cycle phases from **Tsunamis**:

TABLE 15.3
Impacts to the Disaster Phase Cycles from Too Much Seawater

Preparedness/Protection/Prevention

Work toward a continuous cycle of messaging about increasing the public's capacity to be better prepared to Respond themselves solely from the life safety impacts of tsunamis. There really is no ability to protect their homes and places of business. Governments will need to bolster tsunami warning systems, as noted in the introduction of this chapter.

Response

The Response efforts by governmental and non-governmental organizations are generally focused on post-landfall support in areas not devastated by the tsunami.

Recovery

U.S. recovery work is also generally conducted and managed at the State/Territory/Tribal Nation level for tsunamis, not at the local level. While the U.S. thresholds to reach this level of federal assistance vary by state/tribal/territory and adjust over time, the devastation of a tsunami will most likely reach Presidentially Declared (emergency or disaster) status. Still, even with this assistance, the states, territories, and tribal nations are left to finance a portion of any recovery work themselves – and private companies, individuals, etc. must have insurance coverage – especially flood insurance – in order to properly recover.

Mitigation

Jurisdictions should have updated mitigation plans, which include applicable subduction zone threat and hazard risk analysis. Mitigating against both the Response Phase hazards (tsunami flooding, earthquake damage, etc.) and the longer-term Recovery Phase stable changes (land lowering to sea level permanently) is key.[48]

Source: Author.

15.4 ADVERSE IMPACTS TO THE INCIDENT COMMAND SYSTEM

There may be unique hazards directly impacting Responses and Responders themselves, which should be considered adverse impacts to the Incident Command System (ICS). For **Tsunamis**, Table 15.4 details what may be occurring.

15.4.1 POETE Process Elements for This Hazard

Table 15.5 shows the specifics for **Tsunamis.**

TABLE 15.4
Adverse Impacts to the Incident Command System

Command (including SO, PIO, and LNOs)

Safety Officers (SOs) need continuous emergency management intelligence about this threat. Decisions about holding resources in place during high winds are critical for responder life safety. The post-landfall missions, especially for hazardous swift-water situations in flooded streets, near utility drainage, etc.) also need geospatial intelligence on a three-dimensional basis. These workforce safety elements are paramount. While there are always risks to responders (including additional threats in oceanwater, such as pathogens and marine life), there may be additional hazards such as oil and hazardous materials in the water, and uncharted risks for search and rescue, as well as recovery work. Dive teams, for example, will need counseling after recovering any drowning victims – so their own life safety from a mental health/wellness perspective, is sustained.

Public Information Officers (PIOs) need to urgently activate accurate templated messages to the public which provide critical crisis communications in the appropriate languages and modalities for oversaturation. If evacuation orders are in place – clear descriptions of mandatory versus recommended, must be made. If localized "avoid the area" or "Turn Around, Don't Drown®"[49] warnings are needed, for example, they need to be communicated far and wide, and quickly. It is highly likely that geographic areas which experience recurrent tropical storm adverse impacts such as flooding, for example, already understand the risks associated with this hazard – but it bears repeating for new residents, visitors/tourists, etc. And with Climate Change, the intensity and recurrences will only increase over time.

Liaison Officers (LNOs) may be in place on this type of hazard, crossing all the various groups which operate their own incident command structures/systems. For example, on tropical storms, the American Red Cross establishes their own NIMS-like organizational structure and provides an LNO as the conduit between the two systems. The same will be true for the U.S. DoD elements, such as the U.S. Coast Guard, U.S. Army Corps of Engineers, etc.

Intelligence

Sources of Emergency Management Intelligence for incidents involving tropical storms will need to include the regular refreshed/curated intelligence about the topography of the water source (inlet, sound, bay, piers, ports, etc.), especially if there are **now** *new* underwater hazards (oil/chemical discharges, bridge collapses, debris, sunken or displaced boats, etc.) as well as cascading human-made hazards (power outages shutting down traffic lights near flooded areas where people are evacuating, fires, etc.). The wind speed and direction, radar imagery of the storm, and SLOSH modeling (MEOWs and MOMs) are just some of the factors which help with a Common Operating Picture. Knowledge of mutual-aid support capabilities for both search and rescue – including EMAC requests and capabilities for interstate Response support– is needed. Recovery assistance is needed as well. Emergency Management Intelligence supports all the general staff leads, not just Operations.

TABLE 15.4 *(Continued)*
Adverse Impacts to the Incident Command System

Finance/Administration

There will be significant adverse impacts to the staffing, support, systems, etc. associated with the existing local Finance/Administration resources, even if that community is properly prepared for a tropical storm. Overtime and other staffing costs will exceed what is covered in normal fiscal budgets. (Emergency contracting for debris removal alone, is something most municipalities cannot support themselves – and economies of scale are usually achieved when this is performed at the County/Parish level or above.) The requirements for documentation, proper financial controls, staffing logs, etc. are key when there is a state-declared emergency and/or Federal support. This is needed for emergency assistance reimbursements from higher governmental entities on Declared Emergencies or Disasters (Search and Rescue efforts during a Tropical Storm, for example).

Logistics

Generally, supporting a Tropical Storm logistically is beyond the capabilities of a local jurisdiction's (and in many states, even their county/parish level of government's) Logistics branch. The state/territory/tribal nation will most likely be providing logistical support on a consolidated basis to receive and distribute federal aid/material support, provide warehousing, staging areas, etc.

Operations

There are two historical adverse impacts from Tropical Storms on the Operations Branch: First, if workforce members *and their families* are not supported themselves with evacuations, sheltering, etc., they may prioritize their family's life safety above work.[50] And second, the missions for Operations in a Tropical Storm must be suspended for life safety purposes, due to high winds and other hazards, at the peak of need.

Planning

There are no specific adverse impacts to the Planning section, from this hazard. There is a plethora of planning resources available globally,[51] for tropical storms,[52] local Emergency Management entities,[53] corporations, non-governmental organizations, and more.

Source: Author.

TABLE 15.5
POETE Process Elements for This Hazard

Planning	There very well may be a separate annex or appendix to any jurisdiction's emergency operations plan for tropical storms, especially in coastal communities. There should also be elements of other main plan portions and annexes/appendices which cover the planning needed for a tropical storm/sea level rise/flooding incident, which can include extraordinary measures, such as search and rescue and contra-flow. Tropical Storms in the Atlantic have a season (June 1 through November 30), but they can – and do – occur outside of this generalized window of warmer seawater. From this chapter, there are adverse impacts of death and near-death for both the public and responders, historically unique references of past incidents to learn from, and elements of exercises which can be performed, in order to facilitate positive changes and modifications to preparedness, response, recovery, mitigation, and continuity of operations plans.

(Continued)

TABLE 15.5 *(Continued)*
POETE Process Elements for This Hazard

Organization	The **trained** staff needed to effect plans, using proper equipment – especially first responder life safety equipment[54] – should be exercised and evaluated on a regular basis each year. This also includes training with partners who may only be activated at the declared emergency/disaster incident levels. Actual incident response and recovery can help determine elements of POETE, including how these types of incidents are organized for staffing. Also, as part of both deployments and exercises, staff involved should be evaluated for both further training needed and also the possibility of advancement in their disaster state roles to leadership positions.
Equipment	Equipment needed for this hazard specifically includes anything needed to rescue someone from any type of water source (ocean, floodwater, quarry, frozen lake, etc.). This should also include decontamination equipment, as saltwater can corrode metal and electronics. Also included for this hazard are flood protection devices such as floodwalls and sandbags, which may be needed to temporarily replace sections of seawalls and dunes, for example. The geospatial intelligence systems used to track, monitor, and evaluate the emergency management intelligence around tropical storms is also a form of equipment.
Training	In addition to standardized Incident Command System training, specific awareness training on building capacity for the unique complex aspects of Tropical Storms should be taken by emergency management officials from a full disaster phase cycle perspective.
Exercises and evaluation	A full series of discussion and performance exercises should be conducted each year for tropical storms, if there is even a possibility of them impacting the jurisdiction: from human-made hazards of illegal dumping and building against zoning laws, to massive SLOSH and wave impacts, plus future sea level rise, etc. These exercises should stand on their own and be part of various cascading incident scenarios – and cover continuity of operations/continuity of government, as well. Exercises should involve the whole community in an exercise planner's invitations for both additional exercise planners/evaluators and actual players, especially since many non-local organizations (USCG, USACE, etc.) are separate legal entities from the local governments they support. In the same way, there are adversely impacted ESFs, RSFs, and CLs as noted, those elements should be exercised and evaluated so that the elements of the POETE cycle are reviewed and revised as needed (i.e., agreed upon for change, as part of the Improvement Plan components). The AAR/IP is a critical part of this process step, and this chapter's References/Additional Reading section, as well as the footnotes throughout, contains a few examples of open-source tropical storm incident reports and research, which can be the basis for improvement planning everywhere.

Source: Author.

NOTES

1 Mulcahy, A. (2017). *The 2004 Indian Ocean Earthquake and Tsunami: The Story of the Deadliest Natural Disaster of the 21st Century.* CreateSpace Independent Publishing Platform.
2 Pacific Coastal and Marine Science Center. (n.d.). *Life of a Tsunami.* USGS. See www.usgs.gov/centers/pcmsc/life-tsunami
3 See www.usgs.gov/faqs/what-difference-between-a-tsunami-and-a-tidal-wave
4 See www.preventionweb.net/news/these-5-countries-have-advanced-tsunami-warning-systems
5 See www.usgs.gov/faqs/what-it-about-earthquake-causes-a-tsunami
6 See https://dnr.mo.gov/land-geology/hazards/earthquakes/science/facts-new-madrid-seismic-zone
7 See www.politico.com/news/magazine/2023/12/14/tsunami-risk-index-fema-washington-00131544
8 See www.federalregister.gov/documents/2023/05/26/2023-11268/community-disaster-resilience-zones-and-the-national-risk-index
9 See www.politico.com/news/magazine/2023/05/07/coast-guard-tsunami-00058594
10 See www.dnr.wa.gov/programs-and-services/geology/geologic-hazards/Tsunamis#tsunami-hazard-maps
11 Subduction Zone Science. (2020). *Introduction to Subduction Zones: Amazing Events in Subduction Zones.* USGS. See www.usgs.gov/special-topics/subduction-zone-science/science/introduction-subduction-zones-amazing-events
12 Oregon Department of Emergency Management. (n.d.). *Cascadia Subduction Zone.* State of Oregon. Accessed February 26, 2024. See www.oregon.gov/oem/hazardsprep/pages/cascadia-subduction-zone.aspx
13 See https://survivingcascadia.com/contact/
14 See www.shakeout.org/
15 NOAA. (n.d.). *JetStream Max: 2004 Indian Ocean Tsunami.* U.S. Department of Commerce. Accessed February 26, 2024. See www.noaa.gov/jetstream/2004tsu_max
16 Tanaka, S. (2012). Accident at the Fukushima Dai-ichi Nuclear Power Stations of TEPCO – Outline & Lessons Learned. *Proceedings of the Japan Academy. Series B, Physical and Biological Sciences, 88*(9), 471–484. https://doi.org/10.2183/pjab.88.471
17 National Geographic. (n.d.). *Mar 11, 2011 CE: Tohoku Earthquake and Tsunami.* National Geographic. Accessed February 26, 2024. See https://education.nationalgeographic.org/resource/tohoku-earthquake-and-tsunami/
18 See www.usgs.gov/centers/alaska-science-center/science/alaska-aleutian-subduction-zone-studies
19 Witter, R. C., Carver, G. A., Briggs, R. W., Gelfenbaum, G., Koehler, R. D., La Selle, S., Bender, A. M., Engelhart, S. E., Hemphill-Haley, E., & Hill, T. D. (2016). Unusually Large Tsunamis Frequent a Currently Creeping Part of the Aleutian Megathrust. *Geophysical Research Letters, 43*, 76–84, https://doi.org/10.1002/2015GL066083
20 Western States Seismic Policy Council. (n.d.). *1965 Alaska Tsunami.* Accessed February 26, 2024. See www.wsspc.org/resources-reports/tsunami-center/significant-tsunami-events/1964-alaska-tsunami/
21 See www.firstnet.com/coverage/coverage-enhancements/compact-rapid-deployable.html
22 See www.sciencedirect.com/science/article/pii/S1877705813012332/pdf
23 See www.sparisk.com/pubs/Scawthorn-2013-SAFRR-FFT.pdf
24 See www.dhs.gov/xlibrary/assets/nipp_srtltt_guide.pdf
25 See www.fema.gov/pdf/about/odic/fnss_guidance.pdf

26 See https://domesticpreparedness.com/healthcare/challenges-with-pediatric-mass-care-feeding
27 See https://nationalmasscarestrategy.org/
28 See https://hazards.colorado.edu/news/research-counts/avoiding-the-second-disaster-of-unwanted-donations
29 Carballo, M., Daita, S., & Hernandez, M. (2005). Impact of the Tsunami on Healthcare Systems. *Journal of the Royal Society of Medicine*, *98*(9), 390–395. https://doi.org/10.1177/014107680509800902
30 See www.fireengineering.com/technical-rescue/tsunami-a-new-concern-for-responders/
31 NOAA. (2013). *Severe Marine Debris Event Report: Superstorm Sandy*. U.S. Department of Commerce. See https://marinedebris.noaa.gov/sites/default/files/publications-files/Electronic_2013SuperstormSandy_SMDE_report_to_Congress.pdf
32 See https://blog.nwf.org/2012/10/hurricane-sandys-impact-on-fish-and-wildlife/
33 Gargani, J. (2022). Impact of Major Hurricanes on Electricity Energy Production. *International Journal of Disaster Risk Reduction*, *67*, 102643. https://doi.org/10.1016/j.ijdrr.2021.102643
34 Feng, K., Ouyang, M., & Lin, N. (2022). Tropical Cyclone-Blackout-Heatwave Compound Hazard Resilience in a Changing Climate. *Nature Communications*, *13*(1), 4421. https://doi.org/10.1038/s41467-022-32018-4
35 See https://oceantoday.noaa.gov/hurricanestormsurge/
36 Strauss, B. H., Orton, P. M., Bittermann, K., Buchanan, M. K., Gilford, D. M., Kopp, R. E., . . . & Vinogradov, S. (2021). Economic Damages from Hurricane Sandy Attributable to Sea Level Rise Caused by Anthropogenic Climate Change. *Nature Communications*, *12*(1), 2720. https://doi.org/10.1038/s41467-021-22838-1
37 See www.un.org/en/chronicle/article/economic-recovery-after-natural-disasters
38 MacDonald, R. (2005). How Women Were Affected by the Tsunami: A Perspective from Oxfam. *PLoS Medicine*, *2*(6), e178. https://doi.org/10.1371/journal.pmed.0020178
39 See www.fema.gov/sites/default/files/documents/fema_equitable-recovery-post-disaster-guide-local-officials-leaders.pdf
40 Mitch, N. (2022). *Sea Level Rise and Housing Affordability in Small Coastal Communities: A Case Study in Maine*. Harvard Graduate School of Design. See https://nrs.harvard.edu/URN-3:HUL.INSTREPOS:37371655
41 Parton, L. C., & Dundas, S. J. (2020). Fall in the Sea, Eventually? A Green Paradox in Climate Adaptation for Coastal Housing Markets. *Journal of Environmental Economics and Management*, *104*, 102381. https://doi.org/10.1016/j.jeem.2020.102381
42 See www.fema.gov/pdf/emergency/nrf/nrf-support-cikr.pdf
43 See www.hud.gov/disaster_resources
44 See www.fema.gov/assistance/public
45 See https://hazards.colorado.edu/news/research-counts/looting-or-community-solidarity-reconciling-distorted-posthurricane-media-coverage
46 See https://world-nuclear.org/information-library/safety-and-security/safety-of-plants/fukushima-daiichi-accident.aspx
47 See www.fema.gov/sites/default/files/documents/fema_p-2181-fact-sheet-4-1-drinking-water-systems.pdf
48 See www.cfsd401.org/our-district/district-office/hazard-mitigation-plan
49 Turn Around, Don't Drown® is a registered trademark of the National Weather Service. Used with permission, per see www.weather.gov/safety/flood-turn-around-dont-drown
50 Rojek, J., & Smith, M. R. (2007). Law Enforcement Lessons Learned from Hurricane Katrina. *The Review of Policy Research*, *24*(6), 589+. See https://link.gale.com/apps/doc/A173020469/AONE

51 See https://community.wmo.int/en/tropical-cyclone-operational-plans
52 See www.fema.gov/emergency-managers/risk-management/hurricanes
53 See www.fema.gov/sites/default/files/documents/fema_nhp_operational-decision-report_fact-sheet_022024.pdf
54 See www.fema.gov/sites/default/files/2020-08/fema_swiftwater-flood-sar-team_definition_08-03-2020.pdf

REFERENCES/ADDITIONAL READING

Aldrich, D. P. (2012). *Building Resilience: Social Capital in Post-Disaster Recovery.* University of Chicago Press.

Gomberg, J. S., Ludwig, K. A., Bekins, B., Brocher, T. M., Brock, J., Brothers, D. S., . . . & Wein, A. M. (2017). *Reducing Risk Where Tectonic Plates Collide – U.S. Geological Survey Subduction Zone Science Plan* [Report](1428). Circular, Issue. U. S. G. Survey. See https://pubs.usgs.gov/publication/cir1428

Marghany, M. (2018). *Advanced Remote Sensing Technology for Tsunami Modelling and Forecasting.* CRC Press. See www.routledge.com/Advanced-Remote-Sensing-Technology-for-Tsunami-Modelling-and-Forecasting/Marghany/p/book/9780367781118

Murty, T. S., Aswathanarayana, U., & Nirupama, N. (Eds.). (2007). *The Indian Ocean Tsunami.* CRC Press. See www.routledge.com/The-Indian-Ocean-Tsunami/Murty-Aswathanarayana-Nirupama/p/book/9781138496330

Nicolsky, D. J., Suleimani, E. N., Haeussler, P. J., Ryan, H. F., Koehler, R. D., Combellick, R. A., & Hansen, R. A. (2013). *Tsunami Inundation Maps of Port Valdez, Alaska: Alaska Division of Geological & Geophysical Surveys Report of Investigation 2013–1.* https://doi.org/10.14509/25055

Pacific Coastal and Marine Science Center. (2020). *Tsunami and Earthquake Research.* USGS. See www.usgs.gov/centers/pcmsc/science/tsunami-and-earthquake-research

Science on a Sphere. (2022). *120 Years of Earthquakes and Their Tsunamis: 1901–2020.* U.S. Department of Commerce. See https://sos.noaa.gov/catalog/datasets/120-years-earthquakes-tsunamis/

Washington State. (2022). *2022 Cascadia Rising Exercise Series Summary of Conclusions: Critical Transportation and Mass Care Services Tabletop Exercises.* State of Washington. See https://mil.wa.gov/asset/6390e374e0f21

BIBLIOGRAPHY

Carballo, M., Daita, S., & Hernandez, M. (2005). Impact of the Tsunami on Healthcare Systems. *Journal of the Royal Society of Medicine*, *98*(9), 390–395. https://doi.org/10.1177/014107680509800902

Feng, K., Ouyang, M., & Lin, N. (2022). Tropical Cyclone-Blackout-Heatwave Compound Hazard Resilience in a Changing Climate. *Nature Communications*, *13*(1), 4421. https://doi.org/10.1038/s41467-022-32018-4

Gargani, J. (2022). Impact of Major Hurricanes on Electricity Energy Production. *International Journal of Disaster Risk Reduction*, *67*, 102643. https://doi.org/10.1016/j.ijdrr.2021.102643

MacDonald, R. (2005). How Women Were Affected by the Tsunami: A Perspective from Oxfam. *PLoS Medicine*, *2*(6), e178. https://doi.org/10.1371/journal.pmed.0020178

Mitch, N. (2022). *Sea Level Rise and Housing Affordability in Small Coastal Communities: A Case Study in Maine.* Harvard Graduate School of Design. See https://nrs.harvard.edu/URN-3:HUL.INSTREPOS:37371655

Mulcahy, A. (2017). *In the 2004 Indian Ocean Earthquake and Tsunami: The Story of the Deadliest Natural Disaster of the 21st Century*. CreateSpace Independent Publishing Platform.

National Geographic. (n.d.). *Mar 11, 2011 CE: Tohoku Earthquake and Tsunami*. National Geographic. Accessed February 26, 2024. See https://education.nationalgeographic.org/resource/tohoku-earthquake-and-tsunami/

NOAA. (2013). *Severe Marine Debris Event Report: Superstorm Sandy*. U.S. Department of Commerce. See https://marinedebris.noaa.gov/sites/default/files/publications-files/Electronic_2013SuperstormSandy_SMDE_report_to_Congress.pdf

NOAA. (n.d.). *JetStream Max: 2004 Indian Ocean Tsunami*. U.S. Department of Commerce. Accessed February 26, 2024. See www.noaa.gov/jetstream/2004tsu_max

Oregon Department of Emergency Management. (n.d.). *Cascadia Subduction Zone*. State of Oregon. Accessed February 26, 2024. See www.oregon.gov/oem/hazardsprep/pages/cascadia-subduction-zone.aspx

Pacific Coastal and Marine Science Center. (n.d.). *Life of a Tsunami*. USGS. See www.usgs.gov/centers/pcmsc/life-tsunami

Parton, L. C., & Dundas, S. J. (2020). Fall in the Sea, Eventually? A Green Paradox in Climate Adaptation for Coastal Housing Markets. *Journal of Environmental Economics and Management*, *104*, 102381. https://doi.org/10.1016/j.jeem.2020.102381

Rojek, J., & Smith, M. R. (2007). Law Enforcement Lessons Learned from Hurricane Katrina. *The Review of Policy Research*, *24*(6), 589+. See https://link.gale.com/apps/doc/A173020469/AONE

Strauss, B. H., Orton, P. M., Bittermann, K., Buchanan, M. K., Gilford, D. M., Kopp, R. E., . . . & Vinogradov, S. (2021). Economic Damages from Hurricane Sandy Attributable to Sea Level Rise Caused by Anthropogenic Climate Change. *Nature Communications*, *12*(1), 2720. https://doi.org/10.1038/s41467-021-22838-1

Subduction Zone Science. (2020). *Introduction to Subduction Zones: Amazing Events in Subduction Zones*. USGS. See www.usgs.gov/special-topics/subduction-zone-science/science/introduction-subduction-zones-amazing-events

Tanaka, S. (2012). Accident at the Fukushima Dai-ichi Nuclear Power Stations of TEPCO – Outline & Lessons Learned. *Proceedings of the Japan Academy. Series B, Physical and Biological Sciences*, *88*(9), 471–484. https://doi.org/10.2183/pjab.88.471

Western States Seismic Policy Council. (n.d.). *1965 Alaska Tsunami*. Accessed February 26, 2024. See www.wsspc.org/resources-reports/tsunami-center/significant-tsunami-events/1964-alaska-tsunami/

Witter, R. C., Carver, G. A., Briggs, R. W., Gelfenbaum, G., Koehler, R. D., La Selle, S., Bender, A. M., Engelhart, S. E., Hemphill-Haley, E., & Hill, T. D. (2016). Unusually Large Tsunamis Frequent a Currently Creeping Part of the Aleutian Megathrust. *Geophysical Research Letters*, *43*, 76–84. https://doi.org/10.1002/2015GL066083

16 Water Treatment Plant Issues

When the well is dry, they know the worth of water.

– Benjamin Franklin[1]

This chapter covers some of the components of potable water and wastewater treatment plans which can be adversely impacted to and from complex incidents. The complex incident could be initiated, incorporate, or result in a hazard at any or even all the water system's functions or elements. This chapter covers the adverse impacts to water processing. Part 3 has the adverse quality impacts to raw water and natural water sources.

While there are examples of the impacts to and from water treatment plants (including wastewater treatment) in other chapters, this chapter will provide more information on the complexities associated with the use of raw water systems, as well as finished water storage and distribution systems.

Water Treatment Plant Issues may start out as Emergency Management concerns (ESF#3 – Public Works/Engineering; ESF#4 – Firefighting; ESF#6 – Mass Care, Distribution of Emergency Supplies; ESF#8 – Public Health; ESF#14 – Cross-Sector Business and Infrastructure to name but a few which are specifically impacted), but the longer they last, the more they become **endemic** and will need to utilize a whole-of-government approach instead of simply an emergency management one.

Water Treatment Plant Issues obviously have quality of water issues for health, but the lack of water is the quantity issue. Low pressure significantly impacts firefighting. And these issues will only get worse, the longer the problem or incident is unresolved.

CHLORINATION

Adding chlorine to a water system is one way to kill harmful bacteria, but it does come with potential longer-term risks to human health.[2] There can also be supply-chain logistical issues, which can generate cascading hazards for water treatment systems. The U.S. Environmental Protection Agency (USEPA) has researched a number of these research reports globally – and also in light of additional undocumented challenges from the COVID-19 pandemic – collected them into a single consolidated report:

DOI: 10.1201/9781003474685-21

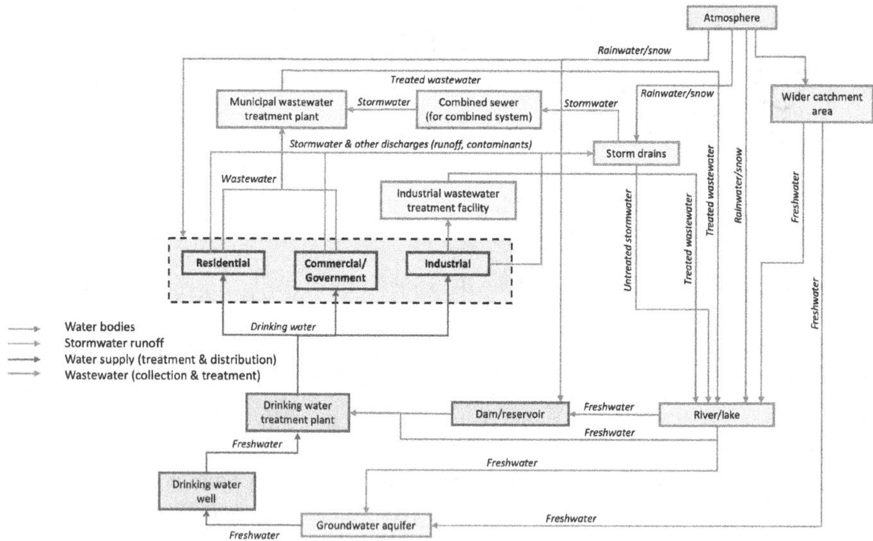

FIGURE 16.1 Typical urban system water flow.

Source: Wan Rosely, W. I. H., & Voulvoulis, N. (2023). Systems Thinking for the Sustainability Transformation of Urban Water Systems. *Critical Reviews in Environmental Science and Technology, 53*(11), 1127–1147. https://doi.org/10.1080/10643389.2022.2131338, p. 1131. CC BY-NC-ND 4.0.

> These studies demonstrate vulnerabilities in production and distribution of water treatment chemicals, and the resulting risk of disruptions in supply of critical water treatment chemicals. However, the studies have been limited in scope and do not capture the severe, and multifaceted supply chain disruptions that started at the beginning of the COVID-19 pandemic. The supply disruptions that have occurred during the pandemic era revealed a range and intensity of supply chains stressors that had not previously been observed in such a short timeframe. While high-impact events such as a pandemic or repeated extreme weather events concentrated on industrial hubs may have been considered low-probability in previous assessments, supply chain risk planning may have to consider greater frequency and cooccurrence of such high-impact events.
>
> **(USEPA, 2023, p. 2)**[3]

Emergency Management staff at water and wastewater treatment systems should become familiar with the steady-state threats and hazards associated with chemical usage in their treatment facilities, including supply-chain integrity/security concerns.

FINISHED WATER STORAGE AND DISTRIBUTION SYSTEMS

Raw water and potable water can both be stored and moved, as they are part of a water system. The locations for these (towers, retention ponds, waterways, channels, aqueducts, etc.) all have hazard risks.

> An interconnection refers to the joining up of two existing water sources or supply systems which facilitates a water transfer between the two. The water resource areas may

> be adjacent to each other and involve only a short transfer or straightforward re-direction of water, or they may be separated by a greater distance with water transferred via a more extensive infrastructure system. The interconnection can be within one water company or provider's resource area or can involve movement of water between separate water companies or providers. Series of interconnections, via displacement chains of supply-demand rebalances, may also be employed to transfer water over longer distances, rather than by using long pipelines.
>
> The type of water involved (i.e. treated or untreated), the distance it is moved and a number of other related factors can all vary significantly.
>
> **(McAlinden, 2015, p. 1)**[4]

Physical and Cyber threats can cascade and/or directly impact these interconnects and water storage systems. Viewing all of the aspects of water supply systems from a risk management perspective is research work being done internationally, as all countries have water supply systems. Hazard and Operability studies are critical components of any water supply system.[5]

THE INTERSECTIONS AND CONFLICTS BETWEEN WATER AND ENERGY

Water in many places is dependent on energy to help make it flow from where it is to where a community needs it to go. And energy in many places is dependent on water, as well. Hydroelectricity is 100% dependent on water. Nuclear power plants need a continuous source of water to cool the reactions to help keep the world safe from radiation.

> The interdependencies of water and energy have gradually been recognized by policy makers, the United Nations, the World Bank, businesses, and many other stakeholders, such as the Union of Concerned Scientists. The International Energy Agency (IEA) paid special attention to the water requirements in the energy sector in its World Energy Outlook in 2012. In 2015, the IEA considered how water scarcity influences the choice of cooling technology in coal-fired power plants in India and China. In 2016, a whole chapter of the World Energy Outlook[6] was devoted to the water-energy nexus.
>
> **(Olsson and Lund, 2017, p. 1)**[7]

16.1 CASE EXAMPLES

Below, please find some case examples from recent history, which exhibit both best practices for disaster readiness and how there can be significant adverse impacts to the LIPER (life safety, incident stabilization, property/asset protection, environmental/economic impacts, and recovery operations) from Complex Incidents impacting Water Treatment Plants.

16.1.1 Albuquerque's Water System

Albuquerque has for years utilized groundwater pumping and aquifer depletion, and was concerned about over-usage of those methods, as well as unsustainability.

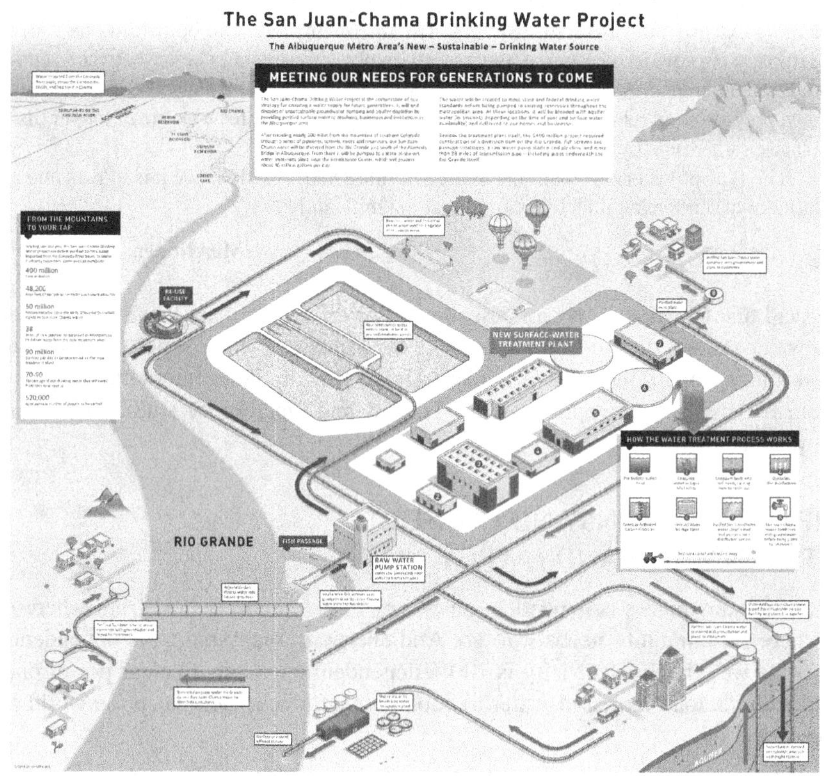

FIGURE 16.2 The San Juan-Chama drinking water project.

Source: Albuquerque Bernalillo County Water Utility Authority. (2021). *Albuquerque's Drinking Water System*. Accessed November 9, 2023. See www.abcwua.org/education-education-el_wsd_2/. Used with permission.

They created their San Juan-Chama Drinking Water Project to ensure a more source-diverse, stable, and sustainable water supply for human use, for generations to come. There are also emergency management elements already built-in to this system besides being the "backup" to their raw water source aquifers and groundwater.

The Albuquerque Bernalillo County Water Utility Authority provides naturally occurring arsenic removal, fish passage, aquifer recharge, surface water treatment, and other elements to benefit this community.

16.1.2 U.S. Virgin Islands Drinking Water Emergency, 2023

As a complete "pendulum swing" shift from the Flint, Michigan Water Supply Switch crisis noted previously, when lead was discovered in routine testing of the potable water supply on the island of St. Croix in the U.S. Virgin Islands, in 2023. Potable water systems on islands are very different from those on the mainland, especially for contingency purposes – there is no ability to interconnect to another nearby water supply system.

A federal emergency was declared by U.S. President Biden, making this incident one which emergency management was activated and supported.[8] There were questions about the testing methods, and whether this level of federal intervention was justified.

Research is currently underway for territory-wide household water storage and treatment, utilizing ultraviolet (UV) light for decontamination.

16.1.3 Jackson, Mississippi, 2022

In 2022, the capital city of Jackson, Mississippi (population of nearly 150,000), lost the functionality of its main water treatment plant, due to both storm damage and years of problems[9] and neglect.[10] For Emergency Management, a water treatment plant failure is very significant – especially if it stops the delivery of pressurized water throughout the communities it serves. The quality issues – which can sometimes be remedied by boiling the water – quickly become outweighed by the low-pressure concerns. The inability to flush toilets becomes a more significant health hazard – one that is more difficult to remedy with only bottled water delivery, as a backup to a water delivery system. The most immediate concern for EM is the lack of or low water pressure for firefighting. *See Chapter 10 for more on the low water pressure threat/hazard.* Massive efforts to deliver potable (i.e., drinkable) water become the operational missions, for months.[11] Outside assistance – including donations from the private sector (*See 'Beer Companies Switch to Canned Water for Disasters' box, which follows*) – but this takes time,[12] funding,[13] and coordination. Failing to solve this problem, through whole of government solutions, not just via Emergency Management courses of action, will adversely impact all the elements of the Recovery Support Functions, including Economic Recovery and resiliency.[14]

BEER COMPANIES SWITCH TO CANNED WATER FOR DISASTERS

Several U.S. breweries[15] annually switch off their beer production in cans and turn their assembly lines into producing canned water instead, for use by non-governmental organizations on disaster responses. Emergency managers globally, should be aware of this Recovery resource, from an Emergency Management Intelligence perspective. This may be an option for longer-term disasters when the potable water supply is adversely affected. The companies[16] donate this water and may even provide logistical support for transportation. On U.S. declared disasters, these donations should be included in the calculation for cost-share match on public assistance projects.

16.1.4 Confederated Tribes of Warm Springs (Oregon) – New Water Treatment Plant

In 2023, two U.S. federal agencies helped fund a water treatment plant to the Confederated Tribes of Warm Springs, in Oregon. The U.S. EPA and the Indian Health Service[17] sent $23.8 million (approx. €22,000) to this community to continue to draw water from the nearby Deschutes River, but through a new plant which will replace one built in the early 1980s.

> The Confederated Tribes of Warm Springs have struggled for years with issues at their water treatment plant, which was built in the early 1980s. People who live on the reservation have faced frequent water outages before and during the pandemic,[18] raising questions about public health safety. The aged water treatment facility also was forced to shut down temporarily last March following a fire,[19] and has faced scrutiny from the EPA for violations.
>
> **(Land, 2022, p. 1)[20]**

16.1.5 Flint, Michigan, Water Supply Switch

This incident is the very big, very bad one – in terms of 21st century disasters in the United States adversely impacting a water supply system. And its cause was human made (a financial only decision), done on purpose, and not in error:

> On April 25, 2014, the City of Flint, Michigan changed their municipal water supply source from the Detroit-supplied Lake Huron water to the Flint River. The switch caused water distribution pipes to corrode and leach lead and other contaminants into municipal drinking water. In October 2016, Flint residents were advised not to drink the municipal tap water unless it had been filtered through an NSF International approved filter certified to remove lead. Although the city reconnected to the original Detroit water system that same month, the potential damage was already done and a state of emergency was declared on January 16, 2016.
>
> **(CASPER, 2020, p. 1)[21]**

In addition to the physical health impacts of lead poisoning in children (brain damage, hearing and speech problems, and learning disabilities) and adults (high blood pressure, heart and kidney disease, and reduced fertility),[22] there were concerns about behavioral impacts such as anxiety, stress, depression, and increased substance use, from the adverse effects of this water system change:

> 66% of households reported one or more adult members reported experiencing at least one behavioral health issue “more than usual”.
>
> 54% of households reported that at least one child experienced at least one behavioral health issue “more than usual”.
>
> 22.5% of households reporting difficulties getting access to behavioral health services.
>
> 34% of individuals self-reported symptoms of anxiety and 29% self-reported symptoms of depression.

> 51% of households felt that the physical health of at least one member had worsened due to Flint water crisis.
>
> (CASPER, 2016, p. 14)[23]

Emergency managers and elected officials should utilize this case as the beacon of what **not** to do, in order to save money or reduce costs. It is also a significant case study, as to how quickly poor crisis communications to the public can further damage physical health of individuals, mental health of populations, as well as increase a general distrust by the public of government.

Also, when reviewing the material on the Flint, Michigan water system crisis, pracademics will find the term "emergency manager" used quite frequently, although in this case in Michigan (and 20 other states in the United States), have states laws which define an emergency manager as someone who takes over **financial** control of a municipality when they have a fiscal crisis. This is not the role this book is using for an emergency manager, and individuals who have this financial-only role, should be retitled as an "Emergency Fiscal Manager" (an option in Michigan, but one not taken yet) or something else to distinguish the major differences between supporting those who manage incidents of scale (disasters, major emergencies, crises, etc.), and managing an insolvent or fiscally challenged municipality. True emergency managers in a community – including their public information officers – need to distance themselves from any "Emergency Fiscal Managers", as to what they do, how they serve the public, etc. The long-term public health impacts from the catastrophic failures in Flint alone[24] signal the strong need for a comprehensive emergency management program to be implemented in every major community – before, during, and after incidents of scale.

16.1.6 Water System Quality Issues in the UK

Lead pipes in buildings, contaminating the drinking water supply, remains a challenge in the UK. Even brass fittings, such as those used in water fountains can be soldered with lead solder, and therefore leech lead over time. Chlorine added to the drinking water for waterborne pathogens bacterial elimination has a side effect of corroding the pipes as well. Lead is a toxin shown to adversely impact brain development in young children. The process of checking and replacing pipes in nearly 8,000,000 buildings in the UK is long-term, systemic disaster response and mitigation work.[25]

Research on contamination of river water in the UK was conducted, and published in 2022:

- Water quality in British rivers has changed substantially since the industrial revolution.
- Between 1760 and 1940, point-source pressures are likely to have increased.
- From 1940, pressures from nutrients and pesticides have increased in many areas.
- The current picture is mixed: urban quality has improved, rural quality has declined.
- Diffuse-source pollution and novel pollutants remain as significant water quality threats.[26]

PORTLAND, OR, REGIONAL WATER PROVIDERS CONSORTIUM

The Portland, Oregon Regional Water Providers Consortium has developed a guide to encourage residents to store 14 gallons (53 liters) of potable water *per person*, in order to support them for up to two (2) weeks, or a gallon (3.75 liters) of water per person, per day.

Much of this material was designed for the health protection for the public, from a catastrophic incident – such as an earthquake – which would take water systems and other infrastructure offline. Emergency water is needed for cooking, hygiene, and drinking. Families with individuals who have special needs – as well as any pets – will need more stored water. Their online guidance includes ways to treat non-potable water, how to properly store potable water, how to get additional water in an emergency, and provides additional disaster preparedness tips. The information is available in English and 13 other languages.[27]

FIGURE 16.3 Get ready. Get water.
Source: Regional Water Providers Consortium. Public domain.

16.2 IMPACTS TO THE DISASTER PHASE CYCLES

As shown in Table 16.1, there are specific impacts to all four of the disaster cycle phases, related to complex threats and hazards to Water Treatment Plans.

16.3 ADVERSE IMPACTS TO THE INCIDENT COMMAND SYSTEM

Please see Chapter 2 and other chapters in Parts 3 and 4 for definitions and more details for the adverse impacts to the Incident Command System, for *additional community-based* response capabilities associated with complex incidents involving Water Treatment Plants and Water/Wastewater systems. Table 16.2 has impacts for the Incident Command System of the Water/Wastewater plant itself.

TABLE 16.1
Impacts to the Disaster Phase Cycles from Human-Made Hazards to Water Sources

Preparedness/Protection/Prevention

In addition to the information in both quality and quantity threats noted in Parts 3 and 4, all the elements of a community's water system need to be incorporated in that community's emergency planning. Besides the obvious inclusion of the water/wastewater treatments plant sites themselves, other components such as pumping stations, aqueducts, watershed/reservoir areas, etc. need to be considered Critical Infrastructure/Key Resources (CIKR)[28] and therefore included in overall protection and prevention plans.

Response

Emergency Management's response courses of action for any complex threat or hazard adversely impacting the local water sources and water supply system, should include providing alternative sources of potable water, crisis communications about the status of the water, and public health syndromic surveillance. This is not a comprehensive list, but a reminder that there are significant response actions needed for any hazard impacting any part of a community's water supply system, including their water and wastewater treatment plants. These Response phase actions should all be coordinated through the local emergency management agency, as much as possible (each jurisdiction may be different). Do not rely on the water systems authority to provide all the resources and response capability – in fact, their representative should be a liaison officer (or possibly part of unified command) to the local jurisdiction's incident command system, not the other way around.

Recovery

There will be cleanup and remediation work, possible infrastructure repair work, supply-chain restoration, financial reimbursement to government by culpable companies and other entities, continued health monitoring for the public, continued crisis communications, cooperation with investigators, including state/tribal/territorial and even federal level officials, and more. The more standardized the key metrics and/or dashboards, the better. For example, use of the Community Lifelines for monitoring the essential elements of intelligence in that community. Also using the CLs helps with the transition to and from outside support, such as Incident Management Teams (IMTs). And most likely, in the longer-term time frame for recovery, another incident of scale will occur and consume the time and staffing of emergency managers and others. *"When it rains, it pours"*. Self-care and wellness are critical for not only first responders, but everyone involved in this disaster phase cycle.

Mitigation

It will depend on the cascading disaster type, as to whether the costs to Respond and Recover from this incident (or incidents) will generate any U.S. Federal Governmental subsidy (i.e., cost-share match) through the *Stafford Act*. There may be other funding available from other Federal[29] or SLTT sources, as well. State/Tribal/Territorial legislation may mandate mitigation actions by the organization operating the water supply system, or there may not be such legal mandates. Emergency managers may have to utilize other funding sources to provide much needed mitigation work to reduce the possibility of a similar adverse set of hazards from any type of hazard to water treatment plants. If there was a Declared Disaster in the same jurisdiction which impacted the same water supply system from a natural hazard, it is possible that mitigation work to help prevent and protect against the adverse impacts of that natural hazard *and also* prevent and protect against the adverse impacts of a human-made hazard. In the same way that building an earthen berm around infrastructure at a sewage treatment plant helps prevent nearby river flooding, it may also protect against eco-terrorists from sabotaging the equipment.

Source: Author.

TABLE 16.2
Adverse Impacts to the Incident Command System

Command (including SO, PIO, and LNOs)

Safety Officers (SOs) need emergency management intelligence about the adverse impacts to the water treatment plant(s), as soon as possible. Beyond the critical worker life safety issues on-scene, as the water sources may be contaminated, potable water will be needed for responders, emergency operations center staff, and of course the public. This may extend beyond just drinking water into handwashing, bathing, etc. These workforce safety elements are paramount. While there are always pathogens and waterborne diseases which do not get filtered out of a water treatment process, extra attention and advanced notice to deploying response teams of this **hardship** should be made.

Public Information Officers (PIOs) provide Preparedness messaging in advance, to include the need for home/work water storage. In the same ways individuals and families need to have food supplies on hand to shelter-in-place, they need to have potable water stored, in case their water supply system and/or water treatment plant cannot provide potable water.

See the Shelter-in-Place section in Chapter 2 and also the References/Additional Reading links for more specifics.

Liaison Officers (LNOs) from both the water supply companies/entities and potentially the organizations deemed responsible for the failures in the Water or Wastewater Treatment Plant will need to be prepared to financially commit to the Incident/Unified Command their support quickly, in order to effect solutions on a timely basis. In the United States, there may also be LNOs from federal agencies (such as the U.S. Environmental Protection Agency), who need to represent the full capabilities and capacities of their organizations, absent a Presidentially Declared Disaster or Emergency, where a unified Federal Coordinating Officer is then assigned to support the State (or Tribal) Coordinating Officer

Intelligence

Sources of Emergency Management Intelligence for incidents where any hazards are adversely impacting water and/or wastewater treatment plants will need to include more than just law-enforcement related ones. Additional subject-matter expertise of water systems officials from this jurisdiction, higher levels of government and possibly derived from mutual-aid compacts (i.e., EMAC), may be needed.

Finance/Administration

- Budgets for alternative water storage, delivery, and distribution may not be adequate. Emergency declarations at the local jurisdiction may not be sufficient to support emergency contracts, etc. Overtime and other staffing costs may exceed what is covered in normal fiscal budgets and/or emergency assistance reimbursements from higher governmental entities.
- If it is determined that a private entity is responsible for the damages to a water or wastewater treatment plant, the costs for remediation and cleanup may be borne by that entity. Governmental Finance/Administration officials will need to manage costs and expenses in such a way as to obtain reimbursement on a timely basis. Additional elements of penalties and fines are not normally the responsibility of emergency management, but the organizations who are support partners to a jurisdiction's emergency management organization (such as the governmental environmental protection agency) may have dual responsibilities of both supporting the response/recovery work and issuing penalties and/or fines to the entity responsible for the adverse impacts caused by them. Emergency Management staff at those agencies must clearly divest themselves from any direct missions involved in the investigation or prosecution of malfeasance or criminal activity.

TABLE 16.2 *(Continued)*
Adverse Impacts to the Incident Command System

Logistics

- With adverse impacts to and from water treatment plants, command may designate points of distribution (PODs) for potable water to be donated to the public. This can take many forms (water from a truck where the public must bring their own containers, individual size bottled water cases which have weight and are usually distributed to residents via vehicle-based, gallon bottles of water, etc.). How this water is ordered/requisitioned is critical for logistics to align to the needs of the operation. Many times, there are also donations of water – from both corporations and the general public – which have to be managed as in-kind donations, probably through a non-governmental organization. In the event the incident is Stafford Act qualified for Public Assistance reimbursement of costs – these material donations, as well as hours of donated labor from volunteers, can be applied to public assistance projects[30]
- Logistics may also be tasked with providing parts, filters, supplies, etc. to the water or wastewater treatment plant impacted. Normal supply-chain integrity and/or security may be adversely impacted by the complex incident, as well

Operations

Firefighting may be impacted – for other threats/hazards, for example, responding to a fire in the same jurisdiction as the impacted water supply system. Low or no water pressure adversely affects the water available through fire hydrants.

Planning

There are no specific adverse impacts to the Planning section, from this hazard. There are, however, additional planning activities, aligned planners from other incident commands, etc. associated with this hazard. Full-cycle planning through Mitigation work, will most likely involve large infrastructure projects which may require Planning elements maintain the operational tempo/battle rhythm much longer that what is typically performed on shorter-term incidents, such as winter storms, minor river flooding, etc.

Source: Author.

16.3.1 POETE Process Elements for This Hazard

Table 16.3 shows the POETE elements for **Complex Threats and Hazards to Water Treatment Plants.**

16.4 CHAPTER SUMMARY/KEY TAKEAWAYS

Complex Incidents – including those with both quality and quantity threats to potable water delivery – can adversely impact water and wastewater treatment plants. The subsequent hazards for a community can have significant impacts on hospitals, dialysis centers, nursing homes/assisted living centers, daycare sites, etc. "Boil" alerts mean massive protocol changes at hospitals, for example, longer than the alert itself. Community preparedness messaging can be very effective in minimizing adverse impacts from incidents involving these hazards – but the community may be focused on the other aspects of the complex incident. In other words, taking a shower after an earthquake may not be a priority for many. The crisis communications about the impacts to the water supply system should be accomplished through a

TABLE 16.3
POETE Process Elements for This Hazard

Planning	While there will most likely not be a separate annex or appendix to any jurisdiction's emergency operations plan for this specific hazard (Water Treatment Plants), there should be elements of other main plan portions and annexes/appendices which cover the planning needed for any complex or coordinated attack which includes an adverse impact to water or wastewater treatment plants. Each Water Supplier may have a separate emergency response plan, with separate mandates from the Emergency Plans in the communities they serve. From this chapter, there are several adverse impacts, historical references of past incidents to learn from, and elements of exercises which can be performed, in order to facilitate positive changes and modifications to preparedness, response, recovery, mitigation, and continuity of operations plans to benefit everyone.
Organization	The **trained** staff needed to implement the plans, using proper equipment should be exercised and evaluated on a regular basis each year. Actual incident response and recovery can help determine elements of POETE, including how these types of incidents are organized for staffing. Also, as part of both deployments and exercises, staff involved should be evaluated for both further training needed and the possibility of advancement in their disaster state roles to leadership positions. It is most critical that this work – and organization – be collaborative between the water utility companies and the local emergency management organization/entity.
Equipment	Equipment needed for this hazard specifically includes anything needed to remediate/repair the water or wastewater treatment plant. Also needed is any special equipment (traffic controls, pallet jacks, water tanker trucks, etc.) needed at – and in support of – emergency potable water distribution points. This can also include at warehouses and supply depots.
Training	Consider the elements of water and wastewater systems, as noted in this chapter, for training courses for both jurisdictional emergency management and water treatment plant emergency management staff. This will work best when courses, webinars, conferences, and other training methodologies are conducted collaboratively. In the United States, courseware on the National Infrastructure Protection Plan[31] is an example.
Exercises and evaluation	A full cycle of discussion and performance exercises should be designed for complex hazards to Water Treatment Plants – both on their own and as part of complex coordinated attack scenarios. These exercises should involve the whole community in exercise planner's invitations for both additional exercise planners/evaluators and actual players, especially since many water supply system operators are separate legal entities from the local governments they serve. In the same way, there are adversely impacted ESFs, RSFs, and CLs as noted in this chapter, those elements should be exercised and evaluated so that the elements of the POETE cycle are reviewed, and revised as needed (i.e., agreed upon for change, as part of the Improvement Plan components).

Source: Author.

whole-of-government/whole-community effort. A community cannot ***overcommunicate*** hazards associated with threats to their potable water sources and supply systems, so messaging via public media, community/civic/religious groups, libraries, social media, etc., all should be maximized using consistent messaging before, during, and after any incidents. If the water or wastewater treatment plant was damaged itself, longer-term remediation, repair, restoration, **and additional** elements which protect it from harm/prevent adverse impacts in the future, should be incorporated into Mitigation work. On U.S. Presidentially declared emergencies and disasters, the partial (usually 75%) Federal funding for this work, may be available.

16.4.1 What to Read Next

The next chapter will continue the theme of Complex Incidents, including Both Quantity and Quality Hazards, but with a focus on Earthquakes and Tsunamis. Cross-check each of the chapters in this part, since some hazard impacts may be the same and some may differ. Also, be prepared for the possibility of any combination of threats and/or hazards when it comes to complex or cascading incidents: It may start out as a human-made hazard to a water source, but then may produce different hazards, including natural ones, somewhere else.

NOTES

1 Franklin, B. (~1735). The Way to Wealth. *Poor Richard Improved.* See https://oll.libertyfund.org/titles/bigelow-the-works-of-benjamin-franklin-vol-ii-letters-and-misc-writings-1735–1753#lf1438–02_label_111
2 Hattersley, J. G. (2003). *The Negative Health Effects of Chlorine.* Townsend Letter for Doctors and Patients, (238), 60+. See https://link.gale.com/apps/doc/A100767859/AONE?u=nysl_oweb&sid=bookmark-AONE&xid=a27b096d
3 USEPA. (2023). *Understanding Water Treatment Chemical Supply Chains and the Risk of Disruptions.* USEPA. See www.epa.gov/system/files/documents/2023-03/Understanding%20Water%20Treatment%20Chemical%20Supply%20Chains%20and%20the%20Risk%20of%20Disruptions.pdf
4 McAlinden, B. (2015). *What Are Water Transfers and Interconnections*? Institution of Civil Engineers. Accessed January 12, 2024. See www.ice.org.uk/engineering-resources/briefing-sheets/what-are-water-transfers-and-interconnections
5 Kombo Mpindou, G. O. M., Escuder Bueno, I., & Chordà Ramón, E. (2022). Risk Analysis Methods of Water Supply Systems: Comprehensive Review from Source to Tap. *Applied Water Science*, *12*(4), 1–20. https://doi.org/10.1007/s13201-022-01586-7
6 See www.iea.org/reports/world-energy-outlook-2016
7 Olsson, G., & Lund, P. D. (2017). Water and Energy – Interconnections and Conflicts. *Global Challenges*, *1*(5), 1700056. https://doi.org/10.1002/gch2.201700056
8 See https://time.com/6339104/us-virgin-islands-lead-water-emergency/
9 See www.npr.org/2022/08/31/1120166328/jackson-mississippi-water-crisis
10 See www.cnn.com/2022/04/19/us/jackson-mississippi-water-crisis/index.html
11 See www.cnn.com/2022/08/30/us/jackson-water-system-failing-tuesday/index.html
12 See www.reuters.com/world/us/floods-knock-out-drinking-water-supply-jackson-mississippi-2022-08-30/
13 See www.npr.org/2022/08/31/1120166328/jackson-mississippi-water-crisis

14 See https://www-foxnews-com.cdn.ampproject.org/c/s/www.foxnews.com/us/jackson-mississippi-preparing-go-without-water-periodically-for-up-10-years-crisis-continues.amp
15 See www.marketingdive.com/news/anheuser-busch-inbev-celebrates-canned-water-program/692106/
16 See www.journal-news.com/news/molson-coors-brewery-cans-water-for-communities-in-crisis/NV6V53ALCNGAXFW4UHS4RFALY4/
17 See www.ihs.gov/
18 See www.opb.org/news/article/water-crisis-returns-to-warm-springs-as-virus-cases-rise/
19 See www.opb.org/article/2022/03/20/reservation-water-treatment-plant-offline-due-to-fire/
20 Land, J. A. (2022). *Warm Springs Will Receive Nearly $24 Million to Replace Problem-Prone Water Treatment Facility*. Oregon Public Broadcasting. Accessed January 23, 2024. See www.opb.org/article/2022/12/21/warm-springs-will-receive-nearly-24-million-to-replace-problem-prone-water-treatment-facility/
21 Community Assessment for Public Health Emergency Response (CASPER). (2020). *Flint Water Crisis. Centers for Disease Control and Prevention*. Accessed November 10, 2023. See www.cdc.gov/nceh/casper/pdf-html/flint_water_crisis_pdf.html
22 Ruckart, P. Z., Ettinger, A. S., Hanna-Attisha, M., Jones, N., Davis, S. I., & Breysse, P. N. (2019). The Flint Water Crisis: A Coordinated Public Health Emergency Response and Recovery Initiative. *Journal of Public Health Management and Practice: JPHMP*, *25*(Suppl 1, Lead Poisoning Prevention), S84–S90. https://doi.org/10.1097/PHH.0000000000000871
23 Community Assessment for Public Health Emergency Response (CASPER). (2016). *After the Flint Water Crisis: May 17–19, 2016*. Centers for Disease Control and Prevention. Accessed November 10, 2023. See www.michigan.gov/documents/flintwater/CASPER_Report_540077_7.pdf
24 *See Emergency Management 101 section, on disaster typology used in this book.*
25 Speight, V. (2021). What Contaminants Lurk in the UK's Drinking Water? An Expert Explains. *TheConversation.com*. Accessed November 10, 2023. See https://theconversation.com/what-contaminants-lurk-in-the-uks-drinking-water-an-expert-explains-167734
26 Whelan, M. J., Linstead, C., Worrall, F., Ormerod, S. J., Durance, I., Johnson, A. C., . . . & Tickner, D. (2022). Is Water Quality in British Rivers "Better Than at Any Time Since the End of the Industrial Revolution"? *Science of The Total Environment*, *843*, 157014. https://doi.org/10.1016/j.scitotenv.2022.157014
27 See www.regionalh2o.org/emergency-preparedness
28 See www.fema.gov/pdf/emergency/nrf/nrf-support-cikr.pdf
29 See https://tribalbusinessnews.com/sections/energy/14669-doi-320m-available-for-tribal-water-infrastructure-projects
30 See www.govinfo.gov/content/pkg/FR-2013-11-01/pdf/2013-26018.pdf
31 See https://training.fema.gov/is/courseoverview.aspx?code=IS-860.c&lang=en

REFERENCES/ADDITIONAL READING

Cybersecurity & Infrastructure Security Agency. (2013). *2013 National Infrastructure Protection Plan*. USDHS. See www.cisa.gov/resources-tools/resources/2013-national-infrastructure-protection-plan

Cybersecurity & Infrastructure Security Agency. (2023). *Infrastructure Resilience Planning Framework: Application to Regional Water & Wastewater Planning*. USDHS. See www.cisa.gov/resources-tools/resources/infrastructure-resilience-planning-framework-application-regional-water-wastewater-planning

Hanna-Attisha, M. (2018). *What the Eyes Don't See: A Story of Crisis, Resistance, and Hope in an American City*. Random House Publishing Group.

Institution of Civil Engineers. (2022). *Water from Source to Tap: A Free Online Exhibition*. Accessed January 12, 2024. See www.ice.org.uk/events/exhibitions/water-from-source-to-tap

Preisner, M., Smol, M., & Szołdrowska, D. (2021). Trends, Insights and Effects of the Urban Wastewater Treatment Directive (91/271/EEC) Implementation in the Light of the Polish Coastal Zone Eutrophication. *Environmental Management*, *67*, 342–354. https://doi.org/10.1007/s00267-020-01401-6

Regional Water Providers Consortium. (n.d.). *Emergency Preparedness*. Regional Water Providers Consortium. Accessed January 5, 2024. See www.regionalh2o.org/emergency-preparedness

Soll, D. (2013). *Empire of Water: An Environmental and Political History of the New York City Water Supply* (1st ed.). Cornell University Press. https://doi.org/10.7591/9780801468070

USEPA. (2023). *Understanding Water Treatment Chemical Supply Chains and the Risk of Disruptions*. USEPA. See www.epa.gov/system/files/documents/2023-03/Understanding%20Water%20Treatment%20Chemical%20Supply%20Chains%20and%20the%20Risk%20of%20Disruptions.pdf

The White House. (2022). *Bipartisan Infrastructure Law Tribal Playbook*. The White House. Accessed January 3, 2024. See www.whitehouse.gov/build/resources/bipartisan-infrastructure-law-tribal-playbook/

BIBLIOGRAPHY

Albuquerque Bernalillo County Water Utility Authority. (2021). *Albuquerque's Drinking Water System*. Accessed November 9, 2023. See www.abcwua.org/education-education-el_wsd_2/

Community Assessment for Public Health Emergency Response (CASPER). (2016). *After the Flint Water Crisis: May 17–19, 2016. Centers for Disease Control and Prevention*. Accessed November 10, 2023. See www.michigan.gov/documents/flintwater/CASPER_Report_540077_7.pdf

Community Assessment for Public Health Emergency Response (CASPER). (2020). *Flint Water Crisis. Centers for Disease Control and Prevention*. Accessed November 10, 2023. See www.cdc.gov/nceh/casper/pdf-html/flint_water_crisis_pdf.html

Franklin, B. (~1735). The Way to Wealth. *Poor Richard Improved*. See https://oll.libertyfund.org/titles/bigelow-the-works-of-benjamin-franklin-vol-ii-letters-and-misc-writings-1735–1753#lf1438-02_label_111

Hattersley, J. G. (2003). *The Negative Health Effects of Chlorine*. Townsend Letter for Doctors and Patients, 238, 60+. See https://link.gale.com/apps/doc/A100767859/AONE?u=nysl_oweb&sid=bookmark-AONE&xid=a27b096d

Kombo Mpindou, G. O. M., Escuder Bueno, I., & Chordà Ramón, E. (2022). Risk Analysis Methods of Water Supply Systems: Comprehensive Review from Source to Tap. *Applied Water Science*, *12*(4), 1–20. https://doi.org/10.1007/s13201-022-01586-7

Land, J. A. (2022). *Warm Springs Will Receive Nearly $24 Million to Replace Problem-Prone Water Treatment Facility*. Oregon Public Broadcasting. Accessed January 23, 2024. See www.opb.org/article/2022/12/21/warm-springs-will-receive-nearly-24-million-to-replace-problem-prone-water-treatment-facility/

McAlinden, B. (2015). What Are Water Transfers and Interconnections? *Institution of Civil Engineers*. Accessed January 12, 2024. See www.ice.org.uk/engineering-resources/briefing-sheets/what-are-water-transfers-and-interconnections

Olsson, G., & Lund, P. D. (2017). Water and Energy – Interconnections and Conflicts. *Global Challenges*, *1*(5), 1700056. https://doi.org/10.1002/gch2.201700056

Ruckart, P. Z., Ettinger, A. S., Hanna-Attisha, M., Jones, N., Davis, S. I., & Breysse, P. N. (2019). The Flint Water Crisis: A Coordinated Public Health Emergency Response and Recovery Initiative. *Journal of Public Health Management and Practice: JPHMP*, *25*(Suppl 1, Lead Poisoning Prevention), S84–S90. https://doi.org/10.1097/PHH.0000000000000871

Speight, V. (2021). What Contaminants Lurk in the UK's Drinking Water? An Expert Explains. *TheConversation.com*. Accessed November 10, 2023. See https://theconversation.com/what-contaminants-lurk-in-the-uks-drinking-water-an-expert-explains-167734

University of Michigan School of Public Health. (2018). *Emergency Manager Law Primer Protecting the Public's Health During Financial Emergencies: Lessons Learned from the Flint Water Crisis*. University of Michigan. Accessed November 10, 2023. See www.networkforphl.org/wp-content/uploads/2020/01/Emergency-Manager-Law-Primer-Protecting-the-Public%E2%80%99s-Health-During-Financial-Emergencies-%E2%80%93-Lessons-Learned-from-the-Flint-Water-Crisis.pdf

USEPA. (2023). *Understanding Water Treatment Chemical Supply Chains and the Risk of Disruptions*. USEPA. See www.epa.gov/system/files/documents/2023-03/Understanding%20Water%20Treatment%20Chemical%20Supply%20Chains%20and%20the%20Risk%20of%20Disruptions.pdf

Wan Rosely, W. I. H., & Voulvoulis, N. (2023). Systems Thinking for the Sustainability Transformation of Urban Water Systems. *Critical Reviews in Environmental Science and Technology*, *53*(11), 1127–1147. https://doi.org/10.1080/10643389.2022.2131338

Whelan, M. J., Linstead, C., Worrall, F., Ormerod, S. J., Durance, I., Johnson, A. C., . . . & Tickner, D. (2022). Is Water Quality in British Rivers "Better Than at Any Time Since the End of the Industrial Revolution"? *Science of the Total Environment*, *843*, 157014. https://doi.org/10.1016/j.scitotenv.2022.157014

17 Saltwater Intrusion into Navigable Freshwater Rivers

> And it is an interesting biological fact that all of us have, in our veins the exact same percentage of salt in our blood that exists in the ocean, and, therefore, we have salt in our blood, in our sweat, in our tears. We are tied to the ocean. And when we go back to the sea – whether it is to sail or to watch it – we are going back from whence we came.
>
> **John F. Kennedy**[1]

Saltwater intrusion into freshwater rivers, especially navigable rivers, will have complex adverse impacts. The initial threat is the lack of water – recurrent and significant drought conditions – which reduce the river levels to extremely low levels – specifically below sea level. When this happens, the ocean water moves upstream into areas where normally freshwater exists. This can adversely affect water treatment plants, shipping (in navigable rivers), recreational boating and swimming, the marine environment of the river, and more. There are human-made interventions which can help Protect and Mitigate against this, including the construction of underwater sills (or levees), which are barriers to slow/stop the movement of saltwater intrusion.

As noted in other chapters, storm surges or high tides can also move saltwater into agricultural areas (fields, farms, etc.), as well.

> Because of its low elevation, land along much of the Northeast seaboard is at risk from saltwater intrusion. In this region, many acres of farmland are lost every year because they are becoming too wet and salty to grow crops. Salt tolerant marsh plants that border farm fields are moving inland and onto the former fields in a process called "marsh migration". Though farmers currently view marsh migration as a problem, it may also provide opportunities. Farmers may be able to develop wetland conservation easements on this land. In this case, planting certain native plants can help prevent or slow the migration of undesirable or invasive species such as Phragmites australis. Native plants can provide other valuable ecosystem services too.
>
> Saltwater can also impact water quality by "unlocking" nutrients from fertilizers in farm fields. This is due to the unique chemistry of saltwater and how it interacts with soil. Once these nutrients become mobile, they can travel through networks of agricultural ditches into larger coastal water bodies such as tidal creeks and marshes. There, the excess nutrients can cause excess algae growth. When the algae die, they are

DOI: 10.1201/9781003474685-22

broken down by bacteria. This process can use up all the oxygen in the water. Depleted oxygen levels can result in fish kills, loss of animal habitat, and other harmful effects on coastal ecosystems and wildlife.

(USDA, n.d., p. 1)[2]

17.1 CASE EXAMPLES

17.1.1 2023 Presidential Emergency Declaration for Mississippi Saltwater Intrusion

In 2023, drought conditions in the midwestern part of the United States caused the Mississippi River water levels to fall below sea level. What is normally gravity-fed from the north to the south, the oceanwater being higher than the freshwater from the river, means saltwater inundation upstream. From a request from the governor of Louisiana for federal assistance, the federal emergency declaration provides financial and technical assistance, under Public Assistance Category B (Protective Measures):

> Under an Emergency Declaration (FEMA – 3600-EM-LA) signed by the President on September 27, 2023, four (4) parishes: Jefferson, Orleans, Plaquemines, and St. Bernard Parishes have been designated adversely affected by the emergency and are eligible for emergency protective measures (Category B), including direct Federal assistance and reimbursement for temporary measures that address reduced water treatment capability due to saltwater intrusion resulting from low water levels of the Mississippi River.
>
> This public notice concerns activities that may affect historic properties, activities that are located within or affect wetland areas or the 100-year floodplain and critical actions within the 500-year floodplain. Such activities may adversely affect the historic property, floodplain, or wetland, or may result in continuing vulnerability to flood damage.

(FEMA, 2023, p. 1)[3]

In the case of the Mississippi River, during this emergency in 2023, the freshwater flow downstream was only 150,000 ft^3/s (4,248 m^3/s), when 300,000 ft^3/s (8,496 m^3/s) is needed to stop the saltwater from moving upstream (Speck, 2023).[4]

The U.S. Army Corps of Engineers already had a project for Mississippi River sill work underway, from 2022:

> To stop the salt water from moving upriver and reduce the risk to freshwater intakes, the New Orleans District began construction of an underwater barrier sill Oct. 11 at river mile 64, which is near Myrtle Grove, Louisiana, to arrest the progression of saltwater intrusion.
>
> The sill is being created using sediment dredged from a designated area just upstream for this purpose. Throughout the construction phase, the Corps will test salinity levels in the river to determine where the saltwater wedge is and how high the sill may need to be built.
>
> Currently, the greatest risk associated with the saltwater intrusion is the appearance of unsafe salinity levels at the intakes of municipal drinking water intakes in Plaquemines Parish. Additionally, the U.S. Coast Guard has issued navigation restrictions around the sill construction site from river mile 63.5 to 64.

> The New Orleans District and U.S. Coast Guard are closely coordinating with the navigation industry to ensure vessel traffic in the river is not significantly impacted. Draft restrictions may be issued as construction of the sill progresses.
>
> **(USACE, 2022, p. 1)**[5]

17.1.2 International Examples

Saltwater intrusion is not a hazard limited to the United States.[6] Researchers[7] in 2023, have reviewed saltwater intrusion incidents in Australia, China, Italy, Mexico, Libya, South Korea, and other countries in Europe. References to those research papers can be found at the end of this chapter.

17.2 ADVERSE IMPACTS TO THE EMERGENCY AND RECOVERY SUPPORT FUNCTIONS AND COMMUNITY LIFELINES

Table 17.1 shows some of the ***possible*** adverse impacts of saltwater inundation into navigable freshwater rivers – both threats and hazards on selected U.S.-based Emergency Support Functions. Similar tables for the other Hazard parts of this book should also be reviewed first.

TABLE 17.1
Water-Related Threats and Hazards to the ESFs

Emergency Support Function (ESF)	Examples of Additional Supporting Actions or Capabilities Related to Intrusion into Navigable Freshwater Rivers
#1 – Transportation	As noted, the lower levels of the river (below sea-level) can impact shipping, as will any dredging work which blocks channels, etc.
#3 – Public Works and Engineering	USACE work in the dredging and sill replacement work will be the prioritized missions, if a declaration is made for emergency protective measures.
#4 – Firefighting	This ESF will not be impacted, unless the firefighting uses raw water from the river, or the water processing plant has issues which reduce hydrant pressure.
#6 – Mass Care, Emergency Assistance, Temporary Housing, and Human Services	If potable water supply is adversely impacted, this will be similar to what is covered in Chapter 14
#8 – Public Health and Medical Services	If potable water supply is adversely impacted, this will be similar to what is covered in Chapter 14
#11 – Agriculture and Natural Resources	Recreational use of the impacted river may be impaired or suspended. Also, if potable water supply is adversely impacted, this will be similar in Response to what is covered in Chapter 14
#12 – Energy	If potable water supply to a power plant is adversely impacted, this will be similar in Response, to what is covered in Chapter 14
#14 – Cross-Sector Business and Infrastructure	Commercial shipping on the river and any other industrial uses will be impacted.

Source: Barton Dunant.

17.2.1 Recovery Support Functions

Similarly, Table 17.2 shows some of the ***additional*** adverse impacts of these saltwater inundation hazards on **selected** U.S. based Recovery Support Functions (RSF). There are similar tables in the other Hazard parts of this book. Those should be reviewed as well.

17.2.2 Community Lifelines

As with the RSFs, saltwater inundation hazards will have implications for only some of the community lifelines.

Food, Hydration, Shelter (Food, Hydration, Shelter, Agriculture): Should only be adversely affected if the water supply in a community is diminished or cut off, due to the saltwater inundation in the river.

Health and Medical (Medical Care, Public Health, Patient Movement, Medical Supply Chain, Fatality Management): Should not be adversely affected, unless the sole water supply to a healthcare provider is from a water processing facility which is on the impacted river. It should be noted that individual health – especially children, people who are immunocompromised, and seniors – will be adversely impacted if saltwater gets into the water supply system, including wells, aquifers, etc.[11] This is a public health concern.

TABLE 17.2
Recovery Support Functions

Federal Recovery Support Functions (RSFs)

Economic

The suspension of commercial maritime traffic on the navigable river will have a significant economic impact. There are cost-factors for agriculture (including fishing)[8] as well.[9] Longer-term economic[10] burden of purchasing water for individuals and families can also be a factor.

Health and Social Services

Only if the healthcare facility is adversely impacted via their water supply coming from the inundated river. The recovery of health of individuals and families potentially impacted by contaminated (with saltwater) potable water is a major concern, which should be monitored.

Infrastructure Systems

The underwater sills, levees, dredging, etc. are all infrastructure systems impacts, and missions during Recovery. Also, all these resources may have been previously allocated to other Mitigation projects, so those will be on hold or postponed.

Natural and Cultural Resources

The above noted Response adverse impacts, include potential damage in Recovery aligned to the restoration of recreational activities – including boating and fishing – from saltwater contamination of the river, streams, and surrounding land. For many river communities, their access to the water is a significant economic factor for them.

Source: FEMA.

Energy (Power Grid, Fuel): Should not be impacted, unless any power plants (including nuclear) utilize the river – and saltwater inundation would be a concern.

Transportation (Highway/Roadway/Motor Vehicle, Mass Transit, Railway, Aviation, Maritime): The maritime aspects of this CL need to be monitored through the dashboard/common operating picture.

Water Systems (Potable Water Infrastructure, Wastewater Management):[12] Understanding how saltwater inundation can impact a potable water supply system (and the wastewater treatment, as well) is critical for Emergency Management.

There may also be longer-term impacts to the food supply if agriculturally used land is impacted by seawater, or experiences damage to infrastructure such as warehouses, transportation vehicles, farm equipment, etc.

17.3 IMPACTS TO THE DISASTER PHASE CYCLES

As shown in Table 17.3, there are specific impacts to all four of the disaster cycle phases, from **Saltwater Inundation into Navigable Freshwater Rivers**.

17.4 ADVERSE IMPACTS TO THE INCIDENT COMMAND SYSTEM

There may be unique hazards directly impacting Responses and Responders themselves, but the Response work for Saltwater Intrusion into Navigable Freshwater Rivers will most likely not be organized under a NIMS/ICS structure. If it did, it would follow other Mitigation-oriented command and control structures, and probably have longer operational periods than a single day.

TABLE 17.3
Impacts to the Disaster Phase Cycles from Too Much Seawater

Preparedness/Protection/Prevention
Monitoring saltwater intrusion through remote sensing.[13] Work on wetlands, coastal land development, shoreline maintenance, etc. can also help.[14]
Response
The Response efforts by governmental and non-governmental organizations are generally focused on underwater sill work, as previously noted.
Recovery
There are agricultural recovery efforts, especially if it is short-term inundation.[15]
Mitigation
Changes in industrial usage near rivers and other human-made interventions which exacerbate climate change droughts, water levels in the river, etc. should be considered.

Source: Author.

17.4.1 POETE Process Elements for This Hazard

Table 17.4 shows the specifics for **Saltwater Inundation into Navigable Freshwater Rivers**:

TABLE 17.4
POETE Process Elements for This Hazard

Planning	If the incident received a U.S. Presidential emergency or disaster declaration, FEMA will invoke is standard planning process.[16] Coastal states with major navigable rivers, such as Louisiana in the United States, will have deliberative plans for saltwater inundation.[17]
Organization	The **trained** staff needed to effect plans, using proper equipment – especially first responder life safety equipment[18] – should be exercised and evaluated on a regular basis each year. This also includes training with partners who may only be activated at the declared emergency/disaster incident levels. Actual incident response and recovery can help determine elements of POETE, including how these types of incidents are organized for staffing. Also, as part of both deployments and exercises, staff involved should be evaluated for both further training needed and also the possibility of advancement in their disaster state roles to leadership positions
Equipment	Equipment needed for this hazard specifically includes anything needed to repair or replace underwater sills. This should also include decontamination equipment, as saltwater can corrode metal and electronics. The geospatial intelligence systems used to track, monitor, and evaluate the emergency management intelligence around saltwater inundation monitoring, is also a form of equipment.
Training	In addition to standardized Incident Command System training, specific awareness training on building capacity for the unique complex aspects of Saltwater Inundation should be taken by emergency management officials, from a full disaster phase cycle perspective.
Exercises and evaluation	A full series of discussion and performance exercises should be conducted each year, for saltwater inundation, if there is even a possibility of them impacting the jurisdiction. These exercises should stand on their own and be part of various cascading incident scenarios – and cover continuity of operations/continuity of government, as well. Exercises should involve the whole community in an exercise planner's invitations for both additional exercise planners/evaluators and actual players, especially since many non-local organizations (USCG, USACE, etc.) are separate legal entities from the local governments they support. In the same way, there are adversely impacted ESFs, RSFs, and CLs as noted, those elements should be exercised and evaluated so that the elements of the POETE cycle are reviewed, and revised as needed (i.e., agreed upon for change, as part of the Improvement Plan components). The AAR/IP is a critical part of this process step.

Source: Author.

NOTES

1 Kennedy, J. F. (1962). *Remarks in Newport at the Australian Ambassador's Dinner for the America's Cup Crews (383).* Public Papers of the Presidents: John F. Kennedy, 1962. See www.jfklibrary.org/learn/about-jfk/life-of-john-f-kennedy/john-f-kennedy-quotations
2 USDA. (n.d.). *Saltwater Intrusion: A Growing Threat to Coastal Agriculture.* USDA. See www.climatehubs.usda.gov/hubs/northeast/topic/saltwater-intrusion-growing-threat-coastal-agriculture
3 FEMA. (2023). *EM-3600-LA Public Notice 001.* FEMA. See www.fema.gov/disaster-federal-register-notice/em-3600-la-public-notice-001
4 Speck, E. (2023). *Biden Approves Louisiana Emergency as Mississippi River Saltwater Intrusion Threatens Drinking Water.* Fox Weather. Accessed December 7, 2023. See www.foxweather.com/extreme-weather/louisiana-federal-help-mississippi-river-drought-drinking-water-supply.amp
5 USACE. (2022). *Corps Begins Constructing Underwater Sill to Halt Saltwater Intrusion in Mississippi River.* U.S. Army. See www.mvn.usace.army.mil/Media/News-Releases/Article/3185907/corps-begins-constructing-underwater-sill-to-halt-saltwater-intrusion-in-missis/
6 See www.washingtonpost.com/climate-environment/2023/09/29/saltwater-intrusion-louisiana-drinking-water/
7 Wang, Z., Guan, Y., Zhang, D., Niyongabo, A., Ming, H., Yu, Z., & Huang, Y. (2023). Research on Seawater Intrusion Suppression Scheme of Minjiang River Estuary. *International Journal of Environmental Research and Public Health, 20*(6), 5211. https://doi.org/10.3390/ijerph20065211
8 See https://dergipark.org.tr/tr/download/article-file/1128247
9 See www.rff.org/publications/journal-articles/the-spread-and-cost-of-saltwater-intrusion-in-the-us-mid-atlantic/
10 Alameddine, R. T., & El-Fadel, M. (2018). Household Economic Burden from Seawater Intrusion in Coastal Urban Areas. *Water International, 43*(2), 217–236. https://doi.org/10.1080/02508060.2017.1416441
11 See www.nbcnews.com/science/environment/salt-water-creeping-mississippi-cause-health-concerns-rcna117360
12 See www.fema.gov/sites/default/files/documents/fema_p-2181-fact-sheet-4-1-drinking-water-systems.pdf
13 See https://sustainability.stanford.edu/news/understanding-saltwater-intrusion-through-remote-sensing
14 See https://eri.iu.edu/erit/strategies/saltwater-intrusion.html
15 White Jr., E., & Kaplan, D. (2017). Restore or Retreat? Saltwater Intrusion and Water Management in Coastal Wetlands. *Ecosystem Health and Sustainability, 3*(1), e01258. https://doi.org/10.1002/ehs2.1258
16 See www.fema.gov/sites/default/files/documents/fema_incident-action-planning-process.pdf
17 See https://gohsep.la.gov/emergency/saltwater
18 See www.fema.gov/sites/default/files/2020-08/fema_swiftwater-flood-sar-team_definition_08-03-2020.pdf

REFERENCES/ADDITIONAL READING

Alcérreca-Huerta, J. C., Callejas-Jiménez, M. E., Carrillo, L., & Castillo, M. M. (2019). Dam Implications on Salt-Water Intrusion and Land Use Within a Tropical Estuarine Environment of the Gulf of Mexico. *Science of the Total Environment, 652*, 1102–1112.

Alfarrah, N., & Walraevens, K. (2018). Groundwater Overexploitation and Seawater Intrusion in Coastal Areas of Arid and Semi-Arid Regions. *Water, 10*(2), 143.

Cunillera-Montcusí, D., Beklioğlu, M., Cañedo-Argüelles, M., Jeppesen, E., Ptacnik, R., Amorim, C. A., . . . & Matias, M. (2022). Freshwater Salinisation: A Research Agenda for a Saltier World. *Trends in Ecology & Evolution, 37*(5), 440–453. https://doi.org/10.1016/j.tree.2021.12.005

Custodio, E. (2010). Coastal Aquifers of Europe: An Overview. *Hydrogeology Journal, 18*(1), 269.

Jeen, S. W., Kang, J., Jung, H., & Lee, J. (2021). Review of Seawater Intrusion in Western Coastal Regions of South Korea. *Water, 13*(6), 761.

Mastrocicco, M., Busico, G., Colombani, N., Vigliotti, M., & Ruberti, D. (2019). Modelling Actual and Future Seawater Intrusion in the Variconi Coastal Wetland (Italy) Due to Climate and Landscape Changes. *Water, 11*(7), 1502.

Werner, A. D. (2010). A Review of Seawater Intrusion and Its Management in Australia. *Hydrogeology Journal, 1*(18), 281–285.

Xu, Z., Ma, J., & Hu, Y. (2019). Saltwater Intrusion Function and Preliminary Application in the Yangtze River Estuary, China. *International Journal of Environmental Research and Public Health, 16*(1), 118.

BIBLIOGRAPHY

Alameddine, R. T., & El-Fadel, M. (2018). Household Economic Burden from Seawater Intrusion in Coastal Urban Areas. *Water International, 43*(2), 217–236. https://doi.org/10.1080/02508060.2017.1416441

FEMA. (2023). *EM-3600-LA Public Notice 001*. FEMA. See www.fema.gov/disaster-federal-register-notice/em-3600-la-public-notice-001

Kennedy, J. F. (1962). *Remarks in Newport at the Australian Ambassador's Dinner for the America's Cup Crews (383). Public Papers of the Presidents: John F. Kennedy, 1962.* See www.jfklibrary.org/learn/about-jfk/life-of-john-f-kennedy/john-f-kennedy-quotations

Speck, E. (2023). *Biden Approves Louisiana Emergency as Mississippi River Saltwater Intrusion Threatens Drinking Water.* Fox Weather. Accessed December 7, 2023. See www.foxweather.com/extreme-weather/louisiana-federal-help-mississippi-river-drought-drinking-water-supply.amp

USACE. (2022). *Corps Begins Constructing Underwater Sill to Halt Saltwater Intrusion in Mississippi River.* U.S. Army. See www.mvn.usace.army.mil/Media/News-Releases/Article/3185907/corps-begins-constructing-underwater-sill-to-halt-saltwater-intrusion-in-missis/

USDA. (n.d.). *Saltwater Intrusion: A Growing Threat to Coastal Agriculture.* USDA. See www.climatehubs.usda.gov/hubs/northeast/topic/saltwater-intrusion-growing-threat-coastal-agriculture

Wang, Z., Guan, Y., Zhang, D., Niyongabo, A., Ming, H., Yu, Z., & Huang, Y. (2023). Research on Seawater Intrusion Suppression Scheme of Minjiang River Estuary. *International Journal of Environmental Research and Public Health, 20*(6), 5211. https://doi.org/10.3390/ijerph20065211

White Jr., E., & Kaplan, D. (2017). Restore or Retreat? Saltwater Intrusion and Water Management in Coastal Wetlands. *Ecosystem Health and Sustainability, 3*(1), e01258. https://doi.org/10.1002/ehs2.1258

18 Ice Dams and Frozen Major Lakes

> Constant kindness can accomplish much. As the sun makes ice melt, kindness causes misunderstanding, mistrust, and hostility to evaporate.
>
> **– Attributed to Albert Schweitzer[1]**

Ice forming in large rivers and major lakes can be a threat which generates cascading hazards (ice dams or jams), adversely impacting the transportation sector. Emergency Management may become activated/tasked with supporting missions to reduce or eliminate the hazards. In addition to smaller ice dams, which can form around bridge structures over rivers and even ocean waterways, ice dams can cause other damages (to ports, as ice jams[2] in navigable waterways themselves, etc.) in communities. Also, the thawing of ice downstream before it thaws upstream may also cause hazards, including creating ice dams.

Ice roads, which are human-made roads[3] carved through the snow and onto frozen major lakes, are being challenged globally[4] by climate change.[5] As the temperatures rise, the stability and consistency of thick ice upon which truck transportation is driven have become riskier. Also, the length of the season, which these critical and sometimes sole transportation routes into isolated communities – including sovereign tribal nations,[6] is getting shorter.[7]

The hazard of ice dams on homes and businesses (where ice accumulates on roofs, eaves, etc.) is covered in Chapter 10.

18.1 CASE EXAMPLES

18.1.1 Denison's Ice Road: Canada

Each year in the 1950s–1970s, John Denison and his team would carve a 520 km (320 mile) road over frozen lakes in northern Canada.

> Winter roads are seasonal roads that only exist during the winter – they run over frozen land and frozen lakes and rivers. Many northern communities in Canada rely on them for their yearly supplies of bulk goods, including fuel and building supplies, which are too costly to ship by air. Because of a warming climate, a progressive shortening of the operational time windows is observed, and is predicted to continue based on climate model projections. Compared to all-season roads, winter roads are less well understood; they are also unevenly managed across Canada. This state of affairs represents a liability for Northerners and could be addressed via the systematic characterization of

DOI: 10.1201/9781003474685-23

individual roads. This would help the assessment of community vulnerability and costs for remediation measures. It would also guide decision-making and prioritization.

(Barrette et al., 2022, p. 842)[8]

18.1.2 Mackinac Island, Michigan

While year-round residents on this island are around 500, during peak tourist season it can reach over 16,000 people. There is no bridge to the island from either the upper peninsula or the mainland portion of Michigan. While it has the same name as the island, the Mackinac bridge connects these two major parts of Michigan, but not this island.

There is a ferry service to and from the mainland, and a small private airfield, but until recently no plans to have any bridges. Recently, changes in their master plan[9] noted the need for an emergency bridge to be temporarily constructed, as they did not have one. In the event of heavy ice flow or a complete icing over of Lake Michigan (which occurs regularly enough for "ice bridges"[10] to be constructed), the ferry service is very seasonal. There are no cars, trucks, or other motorized vehicles allowed on Mackinac Island.

18.2 ADVERSE IMPACTS TO THE EMERGENCY AND RECOVERY SUPPORT FUNCTIONS AND COMMUNITY LIFELINES

Table 18.1 shows some of the ***possible*** adverse impacts of ice dams and frozen major lakes – both as threats and hazards on selected U.S.-based Emergency Support Functions. Similar tables for all the other Hazard parts of this book should also be reviewed first.

FIGURE 18.1 Mackinac Island.

Source: Google Maps. © 2024. www.google.com/maps/place/Mackinac+Island,+MI+49757. Used with permission.

TABLE 18.1
Water-Related Threats and Hazards to the ESFs

Emergency Support Function (ESF)	Examples of Additional Supporting Actions or Capabilities Related to Ice Dams and Frozen Major Lakes
#1 – Transportation	As climate change is impacting temperatures, transportation safety over frozen lakes is a concern.[11] Transportation routes on and over rivers can be adversely impacted by ice dams, especially if they damage bridge infrastructure.
#3 – Public Works and Engineering	See above, as it relates to ice dam infrastructure damage. Also, where there are ice roads, there may be ESF work needed for repair.
#8 – Public Health and Medical Services	Only concern for Emergency Management is medical transportation routes may be blocked, if roadways are impacted by these hazards.
#11 – Agriculture and Natural Resources	Ice dams can cause rivers to overflow. Recreational aspects of frozen lakes (i.e., ice fishing) may be adversely impacted by early thawing/lack of thick ice.
#14 – Cross-Sector Business and Infrastructure	Commercial transportation predominately impacted by either reduced river traffic due to ice dams and/or ice roads becoming unusable.

Source: Barton Dunant.

TABLE 18.2
Recovery Support Functions

Federal Recovery Support Functions (RSFs)

Economic
The suspension of commercial maritime traffic on the navigable river will have a significant economic impact. The same is true for ground transportation over ice roads.

Health and Social Services
Only if the routes to healthcare facilities are damaged/blocked, there are significant individual health risks[12] to ice road truck transportation.

Infrastructure Systems
The ice roads themselves are infrastructure systems, and any damage to bridges, piers, etc. from ice dams will also need support from this Recovery Support Function.

Natural and Cultural Resources
The above noted Response adverse impacts, include potential damage in Recovery aligned to the restoration of recreational activities – including ice fishing – from ice dams and/or issues with frozen major lakes. For many river communities, their access to continuously navigable water is a significant economic factor for them.

Source: FEMA.

18.2.1 Recovery Support Functions

Similarly, Table 18.2 shows some of the ***additional*** adverse impacts of these ice dams and frozen major lakes hazards on **selected** U.S. based Recovery Support Functions (RSF). There are similar tables in the other Hazard parts of this book. Those should be reviewed as well.

18.2.2 Community Lifelines

As with the RSFs, ice dams and frozen major lakes have hazards with implications for only some of the community lifelines.

Health and Medical (Medical Care, Public Health, Patient Movement, Medical Supply Chain, Fatality Management): Should not be adversely affected, unless the primary route to a healthcare provider is compromised due to an ice dam. Ice roads are not normally used for medical transportation. It should be noted that individual health – especially that of the truck drivers – will be adversely impacted if ice roads fail. This is a public health concern.

Energy (Power Grid, Fuel): Should not be impacted, except in locations where fuel transported by truck over ice roads, is the only way to deliver it.[13]

Transportation (Highway/Roadway/Motor Vehicle, Mass Transit, Railway, Aviation, Maritime): The maritime aspects of this CL need to be monitored through the dashboard/common operating picture, as do the land-based transportation over ice roads.

18.3 IMPACTS TO THE DISASTER PHASE CYCLES

As shown in Table 18.3, there are specific impacts to all four of the disaster cycle phases, from **Ice Dams and Frozen Major Lakes**.

18.4 ADVERSE IMPACTS TO THE INCIDENT COMMAND SYSTEM

There may be unique hazards directly impacting Responses and Responders themselves, but the Response work for Ice Dams and Frozen Major Lakes may or may not

TABLE 18.3
Impacts to the Disaster Phase Cycles from Too Much Seawater

Preparedness/Protection/Prevention
Monitoring ice in rivers requires the help of the public for management. The U.S. National Weather Service has a "River Ice Spotter"[14] program, where people can voluntarily report ice jams or dams.
Response
The Response efforts by governmental[15] and non-governmental organizations are generally focused on search and rescue, enabling transportation infrastructure restoration, and preventing possible river flooding.
Recovery
There can be infrastructure Recovery phase elements, such as bridge and road repair work.
Mitigation
Changes in transportation practice[16] for ice roads, monitoring[17] and intervening in ice dam formation, and other human-made interventions should be considered, especially with Climate Change impacts in the long term.

Source: Author.

be organized under a NIMS/ICS structure. If it did, it should follow other Response into Recovery-oriented command and control structures – highlighting the need for a Safety Officer[18] in Command – and will probably have longer operational periods than a single day.

18.4.1 POETE Process Elements for This Hazard

Table 18.4 shows the specifics for **Ice Dams and Frozen Major Rivers**.

TABLE 18.4
POETE Process Elements for This Hazard

Planning	If the incident received a U.S. Presidential emergency or disaster declaration, FEMA will invoke is standard planning process.[19] Nations with major frozen lakes near the Arctic Peninsula should have deliberative plans for ice road failures.
Organization	The **trained** staff needed to effect plans, using proper equipment – especially first responder life safety equipment[20] – should be exercised and evaluated on a regular basis each year. This also includes the potential for training and exercising with international partners who may only be activated at large scale cross-border declared emergency/disaster incidents. Actual incident response and recovery missions can help determine elements of POETE, including how these types of incidents are organized for staffing. Also, as part of both deployments and exercises, staff involved should be evaluated for both further training needed and also the possibility of advancement in their disaster state roles to leadership positions.
Equipment	Equipment needed for this hazard specifically includes anything needed to repair bridges, roadways, remove trucks and heavy equipment in bad weather, search and rescue equipment, etc. This should also include decontamination equipment, as saltwater may be a factor, and can corrode metal and electronics. The geospatial intelligence systems used to track, monitor, and evaluate the emergency management intelligence around ice dams and frozen major rivers, is also a form of equipment.
Training	In addition to standardized Incident Command System training, specific awareness training on building capacity for the unique complex aspects of both ice dams and challenges with frozen major rivers, should be taken by emergency management officials, from a full disaster phase cycle perspective.
Exercises and evaluation	A full series of discussion and performance exercises should be conducted each year, for both ice dams/jams and ice road hazards at frozen major lakes, if there is even a possibility of them impacting the jurisdiction. These exercises should stand on their own and be part of various cascading incident scenarios – and cover continuity of operations/continuity of government, as well. Exercises should involve the whole community in an exercise planner's invitations for both additional exercise planners/evaluators and actual players, especially since many commercial and non-governmental organizations are separate legal entities from the local governments they work with. In the same way, there are adversely impacted ESFs, RSFs, and CLs as noted, those elements should be exercised and evaluated so that the elements of the POETE cycle are reviewed, and revised as needed (i.e., agreed upon for change, as part of the Improvement Plan components). The AAR/IP is a critical part of this process step.

Source: Author.

NOTES

1 link.gale.com/apps/doc/A757750822/AONE?u=nysl_oweb&sid=sitemap&xid=5da48696
2 See www.weather.gov/media/dmx/Hydro/DMX_InfoSht_IceJamsAndFlooding.pdf
3 See https://edmontonjournal.com/news/local-news/famous-ice-roads-around-the-world
4 Zhang, R., Zuo, Y., Sun, Z., & Cong, S. (2024). Changes in Accessibility of Chinese Coastal Ports to Arctic Ports under Melting Ice. *Journal of Marine Science and Engineering*, *12*, 54. https://doi.org/10.3390/jmse12010054
5 See www.openaccessgovernment.org/the-end-for-ice-roads-ice-road-truckers/151348/
6 See www.france24.com/en/live-news/20240209-melting-ice-roads-cut-off-indigenous-communities-in-northern-canada
7 See www.theweathernetwork.com/en/news/lifestyle/travel/northern-saskatchewan-residents-concerned-about-supplies-as-ice-roads-delayed-by-warm-winter
8 Barrette, P. D., Hori, Y., & Kim, A. M. (2022). The Canadian Winter Road Infrastructure in a Warming Climate: Toward Resiliency Assessment and Resource Prioritization. *Sustainable and Resilient Infrastructure*, *7*(6), 842–860. https://doi.org/10.1080/23789689.2022.2094124
9 See www.michigan.gov/mdot/-/media/Project/Websites/MDOT/Travel/Mobility/Public-Transportation/SDNT-Reports/Study-2021/Mackinac-Island-Transportation-Master-Plan.pdf
10 See www.freep.com/story/news/local/michigan/2023/02/16/drummond-island-ice-bridge-canada-snowmobile/69893622007/
11 See www.labmanager.com/deteriorating-safety-on-frozen-lakes-in-a-warming-world-28918
12 See https://gitnux.org/ice-road-trucker-death-rate/
13 See www.thekag.com/newsletters/driving-canadas-ice-road/
14 See www.weather.gov/cle/RiverIceSpotters
15 See www.aptnnews.ca/national-news/4-first-nations-in-manitoba-declare-state-of-emergency-because-of-winter-road-issues/
16 See https://oceanskycruises.com/airships-for-cargo-transportation/
17 See www.wired.com/story/climate-change-threatens-ice-roads-satellites-could-help/
18 See https://highways.dot.gov/public-roads/autumn-2021/04
19 See www.fema.gov/sites/default/files/documents/fema_incident-action-planning-process.pdf
20 See www.fema.gov/authorized-equipment-list-item/03wa-02-bord

REFERENCES/ADDITIONAL READING

Beltaos, S. (1995). *River Ice Jams*. Water Resources Publications.

Iglauer, E. (1991). *Denison's Ice Road*. Harbour Publishing. See https://harbourpublishing.com/products/9781550170412

Wilson Center. (n.d.). *Above the Permafrost Line: Winter Roads*. Accessed February 28, 2024. See www.wilsoncenter.org/sites/default/files/media/uploads/documents/Above%20the%20Permafrost_Winter%20Roads.pdf

BIBLIOGRAPHY

Barrette, P. D., Hori, Y., & Kim, A. M. (2022). The Canadian Winter Road Infrastructure in a Warming Climate: Toward Resiliency Assessment and Resource Prioritization. *Sustainable and Resilient Infrastructure*, *7*(6), 842–860. https://doi.org/10.1080/23789689.2022.2094124

Zhang, R., Zuo, Y., Sun, Z., & Cong, S. (2024). Changes in Accessibility of Chinese Coastal Ports to Arctic Ports under Melting Ice. *Journal of Marine Science and Engineering*, *12*, 54. https://doi.org/10.3390/jmse12010054

19 Cyberattacks

America's critical infrastructure is only as strong as its weakest link. And in the United States, water may be the greatest vulnerability. The United States has approximately 52,000 drinking water and 16,000 wastewater systems, most of which service small- to medium-size communities of less than 50,000 residents. Each of these systems operates in a unique threat environment, often with limited budgets and even more limited cybersecurity personnel to respond to these threats.

– Dr. Samantha Ravich, chair of Center on Cyber and Technology Innovation at the Foundation for Defense of Democracies[1]

This chapter covers how a cyberattack – of any kind – can adversely impact the water quality and/or quantity in any community. Cyberattacks can also impact ports, dams, locks, remote sensors, and other aspects of maritime activities. This book will not cover all the possible cyberattack methods, nor will it provide any information or intelligence which is not open-source available. Every nation's potable water supply and navigable waterways both are national security concerns and should be protected as part of a country's national cybersecurity strategy.[2]

> Safe drinking water is a prerequisite for protecting public health and all human activity. Properly treated wastewater is vital for preventing disease and protecting the environment. Thus, ensuring the supply of drinking water and wastewater treatment and service is essential to modern life and the nation's economy.
>
> **(CISA, n.d., p. 1)**[3]

In an environment of limited resources, prioritization is essential. One key aspect of this prioritization is creating a collaborative environment where leadership and IT understand what cybersecurity efforts are integral to the continuing delivery of the most critical services. The following are some non-technical questions for leaders across jurisdictions to ask when prioritizing cybersecurity:

- *Interdependencies:* What IT-enabled business processes and services would most impact our mission if disrupted? What data is most critical to protect? What IT systems do that data reside on? Is there a complete inventory?
- *Security efforts:* How do we protect those processes, services, and data? Are the security controls we have in place working? Have we assessed them? How many of

DOI: 10.1201/9781003474685-24

my systems and business services are currently vulnerable to disruption? Are we engaged in any "Bad Practices"?[4]

- *Response capabilities:* How would we respond in case of a disruption? Do we have a cyber incident management plan? Have we tested it, and is it up to date? Does our continuity plan account for a cybersecurity disruption?
- *Preparedness:* Are we ready for ransomware? Do our backups meet our restoration needs? Are they securely stored offline?
- *Budget:* Given any identified gaps, is our budget adequate for the task of securing our most critical services? Where should we invest more?

(Ballesteros, 2024, p. 1)[5]

CISA SUPPORT IN THE UNITED STATES

The U.S. Cybersecurity & Infrastructure Security Agency (CISA), within the U.S. Department of Homeland Security, is tasked with protecting all the critical infrastructure systems, including the Water and Wastewater Sector. As part of the National Infrastructure Protection Plan,[6] each sector has a specific sub-plan within. Reference to the Water and Wastewater Systems Sector-Specific Plan, last updated in 2015, is included at the end of this chapter.

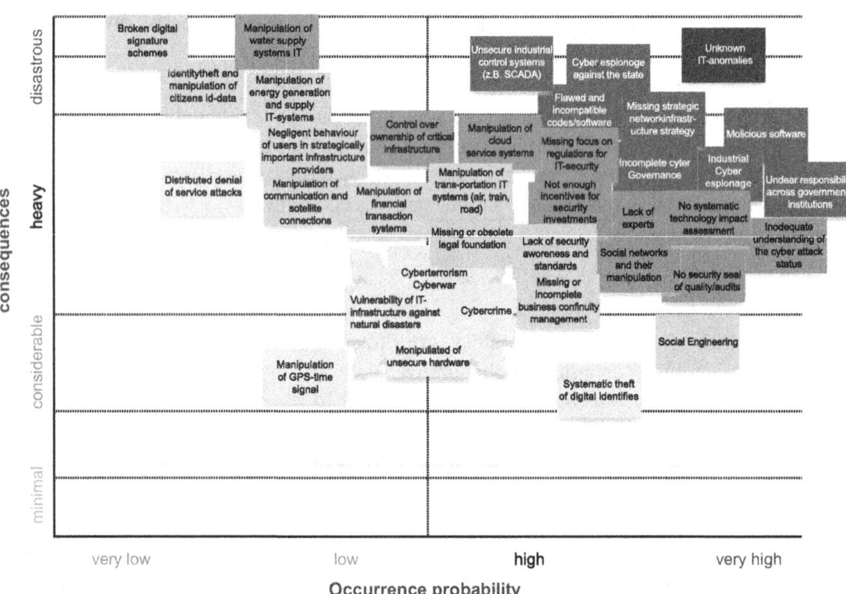

FIGURE 19.1 Cyber Risk Matrix 2011.

Source: Federal Chancellery of the Republic of Austria. (2013). *Austrian Cyber Security Strategy.* See www.enisa.europa.eu/topics/national-cyber-security-strategies/ncss-map/AT_NCSS.pdf. Public domain

> There are approximately 153,000 public drinking water systems and more than 16,000 publicly owned wastewater treatment systems in the United States. More than 80 percent of the U.S. population receives their potable water from these drinking water systems, and about 75 percent of the U.S. population has its sanitary sewerage treated by these wastewater systems.
>
> The Water and Wastewater Systems Sector is vulnerable to a variety of attacks, including contamination with deadly agents; physical attacks, such as the release of toxic gaseous chemicals; and cyberattacks. The result of any variety of attack could be large numbers of illnesses or casualties and/or a denial of service that would also impact public health and economic vitality. The sector is also vulnerable to natural disasters. Critical services, such as firefighting and healthcare (hospitals), and other dependent and interdependent sectors, such as Energy, Food and Agriculture, and Transportation Systems, would suffer negative impacts from a denial of service.
>
> Additionally, both the ability to "supply water" and "manage wastewater" are considered National Critical Functions[7] – functions of government and the private sector so vital to the U.S. that their disruption, corruption, or dysfunction would have a debilitating effect on security, national economic security, national public health or safety, or any combination thereof.
>
> **(CISA, n.d., p. 1)[8]**

Many other nations have similar long-term strategic plans for cybersecurity protection of their water and wastewater sectors; some general (Canada,[9] Poland,[10] and Japan,[11] for example) and others specifically targeted (the United Kingdom[12] and France,[13] as other examples).

USCG TO INSPECT SHIPS AT PORTS FOR CYBERSECURITY

In 2024, the U.S. Federal Government announced it was tasking the U.S. Coast Guard with inspecting the cybersecurity of ports, ships, and other U.S. maritime facilities.

> The order also will allow the Coast Guard to conduct cyber safety inspections of facilities and ships, and even limit the movement of vessels suspected of posing a cyber threat to U.S. ports.
>
> The changes also call for mandatory reporting of cyberattacks on maritime facilities.
>
> "The continuity of their [U.S. ports] operations has a clear and direct impact on the success of our country, our economy and our national security", White House Deputy National Security Adviser Anne Neuberger told reporters. "A cyberattack can cause just as much, if not more, damage than a storm or another physical threat".
>
> There are more than 300 ports in the United States, employing an estimated 31 million Americans. U.S. officials say those ports contribute about $5.4 trillion to the country's economy while serving as the main points of entry for cargo from around the world.
>
> "Any disruption to the MTS [Marine Transportation System], whether manmade or natural, physical or in cyberspace, has the potential to cause cascading impacts to our domestic or global supply chains", said Rear Admiral John Vann, commander of the Coast Guard Cyber Command.
>
> **(Seldin, 2024, p. 1)[14]**

There had already been a concern expressed over foreign-made ship-to-shore cranes, which are designed to be remotely controlled and operated. Those cranes are also vulnerable to exploitation by cyberattacks, including potentially tracking shipment information and damaging ports and port equipment.

FEDERAL OVERSIGHT OR OVERREACH?

There have been questions in several U.S. states, as to whether the federal oversight on cybersecurity, would be considered overreach, from the states' rights and private business independence perspectives. In 2023, the EPA announced it will no longer require cybersecurity audits, due to lawsuits and lobbying efforts of water system trade associations.

> Owners and operators of these systems are struggling to combat the deluge of ransomware and state-backed attacks and infiltration of the nation's most sensitive networks. For critical infrastructure sectors, the consequences for a major cyberattack can be dire, and U.S. water utilities have been identified as particularly lacking in security.
>
> The withdrawal of the rule does not bode well for future efforts to harmonize[15] regulations among the existing 16 critical infrastructure sectors. Many critical infrastructure sectors like water and wastewater lack cybersecurity regulations. Using a voluntary approach to regulate cybersecurity in these industries was described in the National Cybersecurity Strategy as resulting in "inadequate and inconsistent outcomes".
>
> Using the EPA to regulate the cybersecurity of water utilities represented a creative piece of policymaking by the Biden administration, but the effort to do so has been controversial from the start,[16] with the water industry loudly opposing the use of EPA's existing authorities to add cybersecurity regulations. Some experts[17] questioned whether sanitary survey was the right tool to enforce cybersecurity mandates, as the process traditionally does not involve auditors who understand the complex nature of protecting industrial systems.
>
> **(Vasquez, 2023, p. 1)[18]**

19.1 CASE EXAMPLES

19.1.1 Florida Water System Cyberattack

In 2021, a hacker attempted to poison the water system in Oldsmar, Florida, by using a dormant software package, which was still on the water system's network. The amount of sodium hydroxide (lye) was increased 100-fold remotely but was quickly disrupted by the water system's operator and returned to normal amounts for purification use. This is not the only water supply system to be subject to remote cyberattacks. In 2020, a similar attack was made on a water treatment plant in Israel.[19]

In 2023, other water systems were cyberattacked – some, as part of a cascading threat.[20] Some advocate for more regulations – and funding – from the U.S. Environmental Protection Agency (who has oversight at the national level for water systems) is needed to keep up with emerging technology threats to water supply systems.[21]

19.1.2 Port of Houston Cyberattack: 2021

In September of 2021, the Port of Houston announced that it was the subject of an attempted cyberattack the prior month. The hack involved a password management program, used by the port and other organizations. CISA, the FBI, and the USCG issued a joint advisory,[22] "warning that the vulnerability in the software "poses a serious risk" to critical infrastructure companies, defense contractors and others" (Suderman, 2021, p. 1).[23]

19.1.3 Universities Help in Port Security Protection/Prevention Elements

There are collaborative conferences and meetings in the United States, between academia and the federal, state, and local agencies, ports and private industry on maritime security (also known as MARSEC). Cybersecurity has been a prime topic for nearly a decade. In 2017, Texas A&M University – Galveston hosted the annual conference on MARSEC, which had one of the first workshops on Cybersecurity for this industry. "Texas A&M-Galveston is also home to the Texas A&M Maritime Academy, one of six state maritime academies and the only one in the southern United States, which trains over 400 cadets annually for maritime service and employment around the world" (Texas A&M University – Galveston, n.d., p. 1).[24]

19.2 ADVERSE IMPACTS TO THE EMERGENCY AND RECOVERY SUPPORT FUNCTIONS AND COMMUNITY LIFELINES

Table 19.1 shows some of the ***additional*** adverse impacts of a cyberattack hazards on the U.S.-based Emergency Support Functions. Similar tables for **all** the other Hazard parts for this book should also be reviewed first.

19.2.1 Recovery Support Functions

Similarly, Table 19.2 shows some of the ***additional*** adverse impacts of Cyberattack hazards on the U.S.-based Recovery Support Functions (RSF). There are similar tables in the other Hazard parts of this book. Those should be reviewed as well.

Generally, adverse impacts from cyberattack hazards can target all the RSFs. For example, no geographically located infrastructure system is truly impenetrable.

19.2.2 Community Lifelines

As noted in Chapter 2, the Community Lifelines (CLs) are a highly useful list of EEIs, which can be visualized as a dashboard. As with the RSFs, cyberattack hazards will have implications for every community lifeline. A cyberattack can start at one CL and cascade to others – especially along the lines of critical infrastructure

TABLE 19.1
Water-Related Threats and Hazards to the ESFs

Emergency Support Function (ESF)	Examples of Additional Supporting Actions or Capabilities Related to Cyberattack Threats or Hazards
#1 – Transportation	As noted in this chapter, any cyberattack on maritime transportation, would impact this ESF.
#2 – Communications	The same is true for communications systems. This type of cyberattack may or may not be targeting the water and wastewater sectors or the transportation sector but may cascade from a general communications systems outage into a water-related hazard.
#3 – Public Works and Engineering	Impacts to water and wastewater treatment plants, levees, dams, locks, etc. are all infrastructure related.
#4 – Firefighting	If water supply is cut off or curtailed for firefighting purposes, this ESF may be impacted.
#5 – Information and Planning	This ESF is impacted in two possible ways, if the cyberattack disrupts intelligence curation and/or the ability to communicate planning work throughout Emergency Management.
#8 – Public Health and Medical Services	Monitoring and curating intelligence of possible cascading cyberattacks (or adverse impacts of the same attack) on healthcare facilities is critical. If potable water supply systems were impacted, there will be a need for public health monitoring of the water quality and possible contamination risks to the health of individuals.
#10 – Oil and Hazardous Materials Response	As noted in the case studies in the chapters in this Quantity Part (also, in the Quality Part), there can be flooding damage generated by a cyberattack, to industrial and other areas where oil and hazardous materials then make their way into the streams, rivers, watershed, etc., and, therefore, adversely impact communities. The adverse health impacts from those human-made hazards, including sewer treatment plant overflows, etc., may require missions for cleanup and remediation from this ESF. The high priority of this cleanup and remediation of the oil and/or hazardous materials is both a life safety and an incident stabilization course of action.
#11 – Agriculture and Natural Resources	If water systems are impacted from a quantity perspective toward use by agriculture (farming, irrigation, etc.), this ESF can be adversely impacted.
#12 – Energy	The same is true for energy production systems. This type of cyberattack may or may not be targeting the water and wastewater sectors or the energy sector but may cascade from a general cyberattack across all sectors, and then adversely impact the Energy ESF. This can have water sector impacts as well. For example, if the temperature of discharge water rises too high, it can have ecological impacts to the waterways and downstream.[25]
#13 – Public Safety and Security	In addition to any incident-related Response missions for law enforcement officials, there will be investigations related to the criminality aspects attributed to the cyberattack. And of course, Public Safety is never confined to only one incident at a time, so staffing will need to be extended or expanded to cover multiple missions simultaneously. Staff wellness, mental health, and physical health are important as well – since some and/or their families may be impacted by the cyberattack directly themselves.

TABLE 19.1 *(Continued)*
Water-Related Threats and Hazards to the ESFs

Emergency Support Function (ESF)	Examples of Additional Supporting Actions or Capabilities Related to Cyberattack Threats or Hazards
#14 – Cross-Sector Business and Infrastructure	Cyberattacks can cross multiple jurisdictions, businesses, non-governmental organizations, and other infrastructure – all can be adversely impacted to different degrees, for their continuity of operations. Many of these sectors have homeland security impacts as well such as banking, oil and gas refining, etc. Stabilizing the Community Lifelines, restoring supply-chain integrity and security, and coordinating resource needs from the private sector all are missions for this ESF, associated with cyberattacks.
#15 – External Affairs	Coordinate with ESF#2 on both internal and external messaging regarding the full disaster cycle of crisis messaging needed for a cyberattack. Preparedness messaging needs to be year-round, as there is no "season" for cyberattacks. Upon notification of a cyberattack, the PIO needs to activate templated messages to the public as warranted, which provide the LIPER actions they need to take, in the languages and access to crisis communication that the impacted community needs. The crisis communications priority of life safety extends to emergency responders as well. The public needs frequent and constant reminders to be aware of any water supply concerns (quantity and/or quality) to action themselves, as soon as possible.

Source: Barton Dunant.

TABLE 19.2
Recovery Support Functions

Federal Recovery Support Functions (RSFs)

Economic

Many cyberattacks are designed to cause economic harm. The restoration of services and work by governmental and non-governmental organizations may need economic recovery assistance from government.

Health and Social Services

As noted in ESF#8, any excess water quantity or diminished water quality hazards which result from the cyberattack will have a health-related Recovery aspect. Vulnerable populations, including those with compromised immune systems, pulmonary issues, and other medical concerns can be significantly harmed after a cyberattack which harms the water supply system, even if they did not have direct injuries from the cyberattack itself. Social Services programs may also be curtailed, due to any cascading impacts of the cyberattack. Special emphasis on recovery and restoring capabilities for adults with disabilities (for example, developmentally disabled group homes), domestic violence[26] shelters, and other community-based residential social service programs must be managed to support housing, health, and social services stability.

Infrastructure Systems

As Infrastructure Systems may be significantly damaged from a cyberattack, the recovery aspects of restoration of the critical infrastructure/key resources should be prioritized.[27] This will include both public and private sector organizations. This can be a challenge for community leaders: Is a sewage treatment plant's restoration a higher priority than a 9–1–1 dispatch center – especially if there were other dispatch centers in the area which were not damaged? Transparency, collaboration, and coordination on these decisions can be done in advance – via exercises which specifically address these "either this or that" prioritization decisions.

(Continued)

TABLE 19.2 *(Continued)*
Recovery Support Functions

Federal Recovery Support Functions (RSFs)

Natural and Cultural Resources

The above-noted Response adverse impacts include potential damage in Recovery aligned to the restoration of natural waterways, if they were damaged as part of the cyberattack (chemical discharge released, river temperature artificially risen, etc.). For many, some communities on the water, recreational use may be connected to local economies.

Source: FEMA.

elements.[28] There can be significant life safety impacts from a cyberattack, to each CL:

Safety and Security (Law Enforcement/Security, Fire Service, Search and Rescue, Government Service, Community Safety): Cyberattacks can shut down public safety dispatching services, disrupt communications networks, and generate additional security and safety missions[29] for law enforcement, and all the other sub-elements within this CL, such as Community Safety.

Food, Hydration, Shelter (Food, Hydration, Shelter, Agriculture): Obvious impacts to this Community Lifeline include *any other* incident missions (for example, sheltering and feeding), which are then compounded by a cyberattack. If a flooding incident occurs (whether due to a cyberattack or not) and people evacuate to a shelter and then a (or another) cyberattack impacts that shelter site, then this CL will be adversely impacted.

Health and Medical (Medical Care, Public Health, Patient Movement, Medical Supply Chain, Fatality Management): Aligns to both Emergency Support Function ESF#8) Public Health missions and the Health and Social Services RSF missions. As noted in the ESFs and RSFs, this Community Lifeline is significantly adversely impacted by any disruption to a community's water supply system from a cyberattack, as well as any critical infrastructure damage to health or medical facilities themselves, or damage to their own potable water supply systems, which may happen with cyberattacks.

Energy (Power Grid, Fuel): Water and Energy have many nexus points, all of which need emergency management intelligence, especially for cyberattacks. The status of this CL is always the most important in any disaster.

Communications (Infrastructure, Responder Communications, Alerts Warnings and Messages, Finance, 911 and Dispatch): One aspect of crisis communications for Public Information Officers, related to cyberattacks, is to pre-plan (and exercise) the no-notice aspects (especially the possibility that normal communications systems which the PIO uses, may be disabled), utilizing templated public messaging in the languages and ***alternative***delivery methods which have the greatest outreach capabilities for that community.

Transportation (Highway/Roadway/Motor Vehicle, Mass Transit, Railway, Aviation, Maritime): If navigable waterways and/or ports are hampered/disrupted from a cyberattack, this CL will be impacted as well. Critical repair parts, filters, staff, etc. need to be transported to water and wastewater treatment sites, to make repairs and work toward restoration of the **Water Systems** CL, if they are adversely impacted by a water quantity or quality hazards generated by a cyberattack.

Hazardous Material (Facilities, HAZMAT, Pollutants, Contaminants): Impacts to this Community Lifeline can be for Pollutants and/or Contaminants cleanups and restoration, if adversely impacted by the cyberattack. Commercial or industrial sites may be adversely impacted by too much water (flooding), malfeasance in dam openings, discharged water into riverways contamination, etc. This could be from natural or human-made hazards. There may be longer-term impacts to facilities, if a contaminated waterway supports those Critical Infrastructure/Key Resources (CIKR), from a needed clean water supply basis. Considerations on how natural freshwater supplies are utilized in dialysis centers and even nuclear power plants (even if they themselves are not damaged), for example.

Water Systems (Potable Water Infrastructure, Wastewater Management):[30] Understanding how the chemical processing valve controls, alkalinity level monitoring, filtration systems, and possible over-flooding conditions from a cyberattack[31] can impact a potable water supply system (and the wastewater treatment, as well) is critical for Emergency Management.

There may also be longer-term impacts to the food supply if agriculturally used land is cut off from potable water supply or if the cyberattack cascades into farming infrastructure such as warehouses, transportation vehicles, farm equipment, etc.

19.3 IMPACTS TO THE DISASTER PHASE CYCLES

As shown in Table 19.3, there are specific impacts to all four of the disaster cycle phases from **Cyberattacks**.

19.3.1 Adverse Impacts to the Incident Command System

There may be unique hazards directly impacting Responses and Responders themselves, which should be considered adverse impacts to the Incident Command System (ICS). For **Cyberattacks**, Table 19.4 details what may be occurring.

19.3.2 POETE Process Elements for This Hazard

Table 19.5 shows the specifics for **Cyberattacks**.

TABLE 19.3
Impacts to the Disaster Phase Cycles from Too Much Seawater

Preparedness/Protection/Prevention

Work toward a continuous cycle of messaging about increasing the public's capacity to be better prepared to Respond themselves solely from the life safety impacts of cyberattacks, especially ones impacting water systems and waterways. There is some ability to protect their homes and places of business in advance, but the cascading aspects abound. If an organization backs up its datafiles to the cloud, and then the internet service provider is cyberattacked, that organization may not be able to access those files. Governments constantly need to bolster cyberattack protection elements.

Response

The Response efforts by governmental and non-governmental organizations are generally focused both continuity of operations and restoring capability, from what was damaged or destroyed from the cyberattack. This can – and should – be collaborative across different countries,[32] especially on global attacks.

Recovery

U.S. recovery work is also generally conducted and managed at the local level, but there can be some State/Territory/Tribal Nation level and U.S. Federal level support for cyberattacks. While the U.S. thresholds to reach this level of federal assistance vary, historically they have not reached a Presidentially Declared (emergency or disaster) status. Still, even with any assistance from the U.S. Federal Government, the local jurisdictions,[33] states, territories, and tribal nations, local organizations will most likely be left to finance the majority of any recovery work themselves – and private companies, individuals, etc. must have insurance coverage[34] – especially cyberthreat insurance – in order to properly recover.

Mitigation

Jurisdictions should have updated mitigation plans, which include applicable cybersecurity threat and hazard risk analysis. Mitigating against both the Response Phase hazards (critical infrastructure damage, adverse impacts to hospitals and healthcare, potable water supply hazards, etc.) and the longer-term Recovery Phase stable changes is key.[35]

Source: Author.

TABLE 19.4
Adverse Impacts to the Incident Command System

Command (including SO, PIO, and LNOs)

Safety Officers (SOs) need continuous emergency management intelligence about this threat. For example, deciding to move to continency sites for continuity of government/operations from a life safety perspective may happen if life safety systems (fire, security, etc.) are compromised from the cyberattack. While there are always risks to responders, including additional threats from excess water quantity or poor water quality concerns, such as pathogens and marine life, there may be additional hazards such as oil and hazardous materials in the water, and uncharted risks for search and rescue, as well as recovery work. Dive teams, for example, will need counseling after recovering any drowning victims – so their own life safety from a mental health/wellness perspective, is sustained.

Public Information Officers (PIOs) need to urgently activate accurate templated messages to the public which provide critical crisis communications in the appropriate languages and modalities for oversaturation. Backup communications systems should be utilized whenever a cyberattack is the threat. If any health-related warnings are needed, they need to be communicated far and wide, and quickly. It is highly likely that the cyberattack will not be limited to one organization or even one geographical areas. This may be another opportunity to activate and use a Joint Information Center.

TABLE 19.4 *(Continued)*

Adverse Impacts to the Incident Command System

Liaison Officers (LNOs) may be in place on this type of hazard, crossing all the various groups which operate their own incident command structures/systems. For example, on cyberattacks, CISA may establish their own NIMS-like organizational structure and provides an LNO as the conduit between the two systems. The same will be true for the U.S. DoD elements, the FBI, and other groupings involved in the investigation aspects as well.

Intelligence

Sources of Emergency Management Intelligence for incidents involving cyberattacks will need to include the regular refreshed/curated intelligence about status of systems restoration, even in the private sector, especially if the attack cascades across sectors. Knowledge of mutual-aid support capabilities for both systems restoration and continuity of operations for possible interstate/international Response support is needed. Recovery assistance is needed as well. Emergency Management Intelligence supports all the general staff leads, not just Operations.

Finance/Administration

There will be significant adverse impacts to the staffing, support, systems, etc. associated with the existing local Finance/Administration resources, even if that community is possibly prepared somewhat for a cyberattack. Overtime and other staffing costs will exceed what is covered in normal fiscal budgets (Emergency contracting for systems restoration alone, is something most municipalities cannot support themselves – and economies of scale are usually achieved when this is performed at the County/Parish level or above when multiple locations are attacked at the same time). The requirements for documentation, proper financial controls, staffing logs, etc. are key when there is a possibility of cyberthreat insurance[36] coverage and financial support.

Logistics

Generally, there are no additional logistical needs – except for backup communications systems such as satellite telephones/internet, amateur radio, etc. – from a cyberattack.

Operations

Operations needs a whole-of-government and whole-community focus for Response. This can also include international groups, for cross-border incident responses and investigations.[37]

Planning

There are no specific adverse impacts to the Planning section, from this hazard. There have been calls for global[38] action on cybersecurity threats, which will involve collaborative and coordinated planning in the future.

Source: Author.

TABLE 19.5

POETE Process Elements for This Hazard

Planning	There can be a separate annex[39] or appendix to any jurisdiction's emergency operations plan for cyberattacks or cyberthreats. There should also be elements of other main plan portions and annexes[40]/appendices which cover the planning needed for a cyberattack which can include extraordinary measures, such as oil and hazardous materials cleanup. From this chapter, there can be adverse impacts of death and near-death for both the public and responders, historically unique references of past incidents to learn from, and elements of exercises which can be performed, to facilitate positive changes and modifications to preparedness, response, recovery, mitigation, and continuity of operations plans.

(Continued)

TABLE 19.5 *(Continued)*
POETE Process Elements for This Hazard

Organization	The **trained** staff needed to effect plans, using proper equipment – especially technical systems restoration equipment – should be exercised and evaluated on a regular basis each year. This also includes training with partners who may only be activated at the interstate/cross-border incident levels. Actual incident response and recovery can help determine elements of POETE, including how these types of incidents are organized for staffing. Also, as part of both deployments and exercises, staff involved should be evaluated for both further training needed and also the possibility of advancement in their disaster state roles to leadership positions.
Equipment	Equipment needed for this hazard specifically includes anything needed to recover data files and support continuity of operations for computer systems and electronic devices. This should also include replacing valves, filters, etc., *at a higher level of protection against future cyberattacks.* and electronics. The geospatial intelligence systems used to track, monitor, and evaluate the emergency management intelligence around cyberattacks[41] is also a form of equipment.
Training	In addition to standardized Incident Command System training, specific awareness training on building capacity for the unique complex aspects of cyberattacks should be taken by emergency management officials, from a full disaster phase cycle perspective.
Exercises and evaluation	A full series of discussion and performance exercises should be conducted each year for cyberattacks, since there is a high probability, they will impact every jurisdiction.[42] These exercises should stand on their own and be part of various cascading incident scenarios – and cover continuity of operations/continuity of government as well. Exercises should involve the whole community in an exercise planner's invitations for both additional exercise planners/evaluators and actual players, especially since many non-local organizations (CISA, FBI, etc.) are separate legal entities from the local governments they support. In the same way, there are adversely impacted ESFs, RSFs, and CLs as noted, those elements should be exercised and evaluated so that the elements of the POETE cycle are reviewed, and revised as needed (i.e., agreed upon for change, as part of the Improvement Plan components). The AAR/IP is a critical part of this process step, and this chapter's References/Additional Reading section, as well as the footnotes throughout, contains a few examples of open-source cyberattack incident reports and research, which can be the basis for improvement planning. everywhere.

Source: Author.

NOTES

1 Ravich, S. (2022). *Strengthening the Cybersecurity of American Water Utilities.* Foundation for Defense of Democracies. See www.fdd.org/events/2022/06/08/strengthening-the-cybersecurity-of-american-water-utilities/
2 See https://tdra.gov.ae/userfiles/assets/Lw3seRUaIMd.pdf
3 CISA. (n.d.). *Water and Wastewater Systems.* USDHS. Accessed February 29, 2024. See www.cisa.gov/topics/critical-infrastructure-security-and-resilience/critical-infrastructure-sectors/water-and-wastewater-sector
4 See www.cisa.gov/stopransomware/bad-practices
5 Ballesteros, E. (2024). A Holistic Approach to Cybersecurity Risk. *Domestic Preparedness Journal.* See www.domesticpreparedness.com/articles/a-holistic-approach-to-cybersecurity-risk

6 See www.cisa.gov/sites/default/files/publications/national-infrastructure-protection-plan-2013-508.pdf
7 See www.cisa.gov/topics/risk-management/national-critical-functions
8 CISA. (n.d.). *Water and Wastewater Systems*. USDHS. Accessed February 29, 2024. See www.cisa.gov/topics/critical-infrastructure-security-and-resilience/critical-infrastructure-sectors/water-and-wastewater-sector
9 See www.publicsafety.gc.ca/cnt/ntnl-scrt/cbr-scrt/cbr-scrt-tl/index-en.aspx
10 Kitler, W. (2022). The Cybersecurity Strategy of the Republic of Poland. In K. Chałubińska-Jentkiewicz, F. Radoniewicz & T. Zieliński (Eds.), *Cybersecurity in Poland*. Springer. https://doi.org/10.1007/978-3-030-78551-2_9
11 See www.stimson.org/2023/japan-cybersecurity-policy/
12 See https://assets.publishing.service.gov.uk/media/5a81ec18ed915d74e3400c7d/water-sector-cyber-security-strategy-170322.pdf
13 See www.diplomatie.gouv.fr/en/photos-publications-and-graphics/publications/article/france-s-international-strategy-for-water-and-sanitation-2020-2030
14 Selden, J. (2024). *Shoring Up Ports to Withstand Cyberattacks*. Homeland Security News Wire. See www.homelandsecuritynewswire.com/dr20240222-shoring-up-ports-to-withstand-cyberattacks
15 See https://cyberscoop.com/cybersecurity-strategy-harmonization-critical-infrastructure/
16 See https://cyberscoop.com/epa-water-cyber-regulations/
17 See https://cyberscoop.com/water-sector-epa-rules-change-misguided/
18 Vasquez, C. (2023). *EPA Calls Off Cyber Regulations for Water Sector*. Cyberscoop. See https://cyberscoop.com/epa-calls-off-cyber-regulations-for-water-sector/
19 Marquardt, A., Leverson, E., & Tal, A. (2021). Florida Water Treatment Facility Hack Used a Dormant Remote Access Software, Sheriff Says. *CNN.com*. Accessed December 7, 2023. See www.cnn.com/2021/02/10/us/florida-water-poison-cyber/index.html
20 Teale, C. (2023). Two Recent Cyberattacks on Water Systems Highlight Vulnerability of Critical Infrastructure. *Route Fifty*. Access December 7, 2023. See www.route-fifty.com/cybersecurity/2023/12/two-recent-cyberattacks-water-systems-highlight-vulnerability-critical-infrastructure/392500/
21 Teale, C. (2022). Possible Cyber Regs Face Fragmented, Underfunded Water Sector. *Route Fifty*. Accessed December 7, 2023. See www.route-fifty.com/infrastructure/2022/08/possible-cyber-regs-face-fragmented-underfunded-water-sector/376279/
22 See www.cisa.gov/news-events/cybersecurity-advisories/aa21-259a
23 Suderman, A. (2021). Port of Houston Target of Suspected Nation-State Hack. *Associated Press*. See www.click2houston.com/business/2021/09/24/port-of-houston-target-of-suspected-nation-state-hack/
24 Texas A&M University – Galveston. (n.d.). *Texas A&M Galveston Helps Sponsor Maritime Security Conference in Galveston*. Texas A&M University at Galveston. Accessed February 29, 2024. See www.tamug.edu/newsroom/2017articles/Maritime_security_conference.html
25 Huang, F., Lin, J., & Zheng, B. (2019). Effects of Thermal Discharge from Coastal Nuclear Power Plants and Thermal Power Plants on the Thermocline Characteristics in Sea Areas with Different Tidal Dynamics. *Water*, *11*(12), 2577. https://doi.org/10.3390/w11122577
26 See www.gatech.edu/news/2023/09/27/new-resource-domestic-abuse-survivors-combines-ai-cybersecurity-and-psychology
27 See www.fema.gov/pdf/emergency/nrf/nrf-support-cikr.pdf
28 Palleti, V. R., Adepu, S., Mishra, V. K., & Mathur, A. (2021). Cascading Effects of Cyber-Attacks on Interconnected Critical Infrastructure. *Cybersecurity*, *4*, 8. https://doi.org/10.1186/s42400-021-00071-z

29 See https://hazards.colorado.edu/news/research-counts/looting-or-community-solidarity-reconciling-distorted-posthurricane-media-coverage
30 See www.fema.gov/sites/default/files/documents/fema_p-2181-fact-sheet-4-1-drinking-water-systems.pdf
31 See https://www.energy.senate.gov/hearings/2024/4/water-and-power-subcommittee-hearing-to-examine-the-federal-and-non-federal-role-of-assessing-cyber-threats-to-and-vulnerabilities-of-critical-water-infrastructure-in-our-energy-sector
32 See www.cisa.gov/news-events/news/cisa-nsa-fbi-and-japan-release-advisory-warning-blacktech-prc-linked-cyber-activity
33 See www.buckscounty.gov/CivicAlerts.aspx?AID=975
34 See www.ftc.gov/business-guidance/small-businesses/cybersecurity/cyber-insurance
35 See www.nsa.gov/portals/75/documents/what-we-do/cybersecurity/professional-resources/csi-nsas-top10-cybersecurity-mitigation-strategies.pdf
36 See www.ftc.gov/business-guidance/small-businesses/cybersecurity/cyber-insurance
37 See https://therecord.media/us-canada-water-commission-investigating-cyberattack
38 See www.thenationalnews.com/uae/2023/09/24/uae-calls-for-decisive-global-action-on-water-security/
39 See www.kdhe.ks.gov/DocumentCenter/View/22804/Cyber-Security-EOP_Annex_PRBCHS_12-23-2021?bidId=
40 See https://nsarchive.gwu.edu/sites/default/files/documents/3035768/Document-05.pdf
41 See www.esri.com/about/newsroom/arcwatch/gis-aids-in-cyberattack-recovery/
42 See www.dhs.gov/news/2021/03/31/secretary-mayorkas-outlines-his-vision-cybersecurity-resilience

REFERENCES/ADDITIONAL READING

Bello, A., Jahan, S., Farid, F., & Ahamed, F. (2023). A Systemic Review of the Cybersecurity Challenges in Australian Water Infrastructure Management. *Water, 15*(1), 168. https://doi.org/10.3390/w15010168

CISA. (2015). *Water and Wastewater Systems Sector-Specific Plan*. USDHS/USEPA. See www.cisa.gov/sites/default/files/publications/nipp-ssp-water-2015-508.pdf

CISA. (2023). *The National Cyber Incident Response Plan (NCIRP)*. USDHS. See www.cisa.gov/resources-tools/resources/national-cyber-incident-response-plan-ncirp

Fowler, B., & Maranga, K. (2022). *Cybersecurity Public Policy: SWOT Analysis Conducted on 43 Countries* (1st ed.). CRC Press. https://doi.org/10.1201/9781003259145

Kaushik, K., Dahiya, S., Bhardwaj, A., & Maleh, Y. (2022). *Internet of Things and Cyber Physical Systems: Security and Forensics*. CRC Press.

Motahhir, S., & Maleh, Y. (2022). *Security Engineering for Embedded and Cyber-Physical Systems*. CRC Press.

Teale, C. (2024). Who Should Be in Charge of Protecting Our Water Systems from Cyber Threats? *Route Fifty*. See www.route-fifty.com/cybersecurity/2024/02/who-should-be-charge-protecting-our-water-systems-cyber-threats/393927/

U.S. Environmental Protection Agency. (n.d.). *EPA Cybersecurity for the Water Sector*. EPA. Accessed February 29, 2024. See www.epa.gov/waterresilience/epa-cybersecurity-water-sector

Weed, S. A. (2017). *US Policy Response to Cyber Attack on SCADA Systems Supporting Critical National Infrastructure*. Air University Press. See https://media.defense.gov/2017/Nov/20/2001846609/-1/-1/0/CPP0007_WEED_SCADA.PDF

BIBLIOGRAPHY

Ballesteros, E. (2024). A Holistic Approach to Cybersecurity Risk. *Domestic Preparedness Journal*. See www.domesticpreparedness.com/articles/a-holistic-approach-to-cybersecurity-risk

CISA. (n.d.). *Water and Wastewater Systems*. USDHS. Accessed February 29, 2024. See www.cisa.gov/topics/critical-infrastructure-security-and-resilience/critical-infrastructure-sectors/water-and-wastewater-sector

Federal Chancellery of the Republic of Austria. (2013). *Austrian Cyber Security Strategy*. See www.enisa.europa.eu/topics/national-cyber-security-strategies/ncss-map/AT_NCSS.pdf

Huang, F., Lin, J., & Zheng, B. (2019). Effects of Thermal Discharge from Coastal Nuclear Power Plants and Thermal Power Plants on the Thermocline Characteristics in Sea Areas with Different Tidal Dynamics. *Water, 11*(12), 2577. https://doi.org/10.3390/w11122577

Kitler, W. (2022). The Cybersecurity Strategy of the Republic of Poland. In K. Chałubińska-Jentkiewicz, F. Radoniewicz & T. Zieliński (Eds.), *Cybersecurity in Poland*. Springer. https://doi.org/10.1007/978-3-030-78551-2_9

Marquardt, A., Leverson, E., & Tal, A. (2021). *Florida Water Treatment Facility Hack Used a Dormant Remote Access Software, Sheriff Says*. CNN.com. Accessed December 7, 2023. See www.cnn.com/2021/02/10/us/florida-water-poison-cyber/index.html

Palleti, V. R., Adepu, S., Mishra, V. K., & Mathur, A. (2021). Cascading Effects of Cyber-Attacks on Interconnected Critical Infrastructure. *Cybersecurity, 4*, 8. https://doi.org/10.1186/s42400-021-00071-z

Ravich, S. (2022). *Strengthening the Cybersecurity of American Water Utilities*. Foundation for Defense of Democracies. See www.fdd.org/events/2022/06/08/strengthening-the-cybersecurity-of-american-water-utilities/

Seldin, J. (2024). *Shoring Up Ports to Withstand Cyberattacks*. Homeland Security News Wire. See www.homelandsecuritynewswire.com/dr20240222-shoring-up-ports-to-withstand-cyberattacks

Suderman, A. (2021). Port of Houston Target of Suspected Nation-State Hack. *Associated Press*. See www.click2houston.com/business/2021/09/24/port-of-houston-target-of-suspected-nation-state-hack/

Teale, C. (2022). Possible Cyber Regs Face Fragmented, Underfunded Water Sector. *Route Fifty*. Accessed December 7, 2023. See www.route-fifty.com/infrastructure/2022/08/possible-cyber-regs-face-fragmented-underfunded-water-sector/376279/

Teale, C. (2023). Two Recent Cyberattacks on Water Systems Highlight Vulnerability of Critical Infrastructure. *Route Fifty*. Access December 7, 2023. See www.route-fifty.com/cybersecurity/2023/12/two-recent-cyberattacks-water-systems-highlight-vulnerability-critical-infrastructure/392500/

Texas A&M University – Galveston. (n.d.). *Texas A&M Galveston Helps Sponsor Maritime Security Conference in Galveston*. Texas A&M University at Galveston. Accessed February 29, 2024. See www.tamug.edu/newsroom/2017articles/Maritime_security_conference.html

Vasquez, C. (2023). *EPA Calls Off Cyber Regulations for Water Sector*. Cyberscoop. See https://cyberscoop.com/epa-calls-off-cyber-regulations-for-water-sector/

Part 6

Conclusion

Epilogue

This is the end of the book on water-related Emergency Management threats and hazards. It can be a never-ending story of what adverse impacts can befall a community, anywhere on the globe. The book's organizational design was heavier on the hazards than the threats – as that is where the impacts are. Emergency Management does not manage a tropical storm, but rather the preparedness, protection, prevention, response, recovery, and mitigation aligned to the adverse impacts of the tropical storm to people, places, and things.

The book was designed to be read from front to back from an overall educational and foundational knowledge basis, but also selectively as needed from a pracademic perspective. If a goal is research for a position paper or writing a grant application on a specific threat or hazard, the table of contents and index/glossary is where to quickly find what is needed. Emergency Management is a world of acronyms, this book attempted to embrace them rather than avoid them. Again, the index/glossary should be beneficial as well as Chapter 2 to help translate.

While this book was being written, in the United States, FEMA modified their list of Community Lifelines (CLs) by adding a new column for Water Systems (Potable Water Infrastructure and Wastewater Management) to separate those functions from hydration, which is under the Mass Care CL. This is a tangible sign that the U.S. Federal Government recognizes the criticality of water systems to communities nationwide. It is also a commentary that the Emergency Support Functions, Recovery Support Functions, and Community Lifelines are not static, but rather agile. Change remains the one consistent element of Emergency Management. While this book endeavored to capture the latest, most current examples of threats and hazards – there will always be something new when it comes to incidents around water.

DOI: 10.1201/9781003474685-26

Afterword

As I noted in the Preface, it was a massive tropical storm, which started my writing of a book about water-related threats and hazards. It also became quickly obvious that water was only one of the five ancient elements,[1] which need Emergency Management support and actions. It may be possible to align all of the possible threats and hazards to the elements. Water is just the proverbial tip of the iceberg (sorry).

The other ancient foundational elements are Air, Earth, Fire, and Spirit (sometimes described as "aether") (see following figure). While writing this book, I would often come across a threat or hazard which did not fit directly with water, such as an earthquake or a landslide, and instead of ignoring it, I would append it to another elemental document. Effectively, I have started outlining the next three books, and envision the one on Spirit will include elemental aspects of the following:

- Chemical and biological threats and hazards (for those who know the acronym CBRNE, those are the first two letters. Next is radiological which I can put into the Earth book, and both nuclear and explosive I can add to the Fire book).
- Electricity-related threats and hazards (this will most likely be the longest chapter).
- Human-spirit threats and hazards, including cyberattacks in more detail, plus a chapter on general evilness of humans.

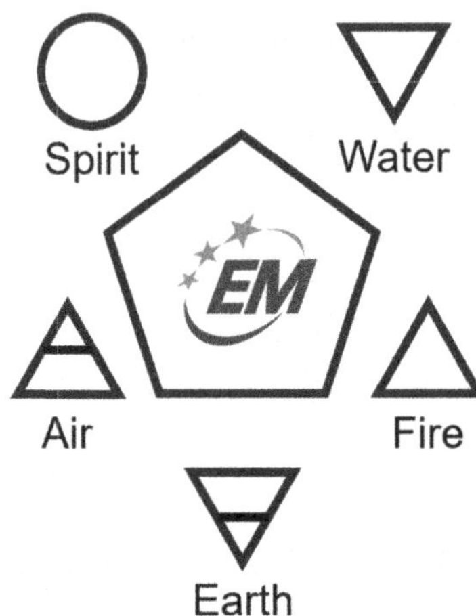

THE FIVE ELEMENTS FOR EMERGENCY MANAGEMENT THREATS AND HAZARDS

Source: Barton Dunant.

DOI: 10.1201/9781003474685-27

It is my goal that enough copies of the book – in print and/or electronic format – are purchased, so the publisher considers adding more elements to the series, for a total of five books. And certainly, if I missed the mark on anything in this book, I will correct it in the electronic version and/or a future edition. If you have any suggestions or ideas for any of the elemental topics – including this one on water – please reach out to me directly.

NOTE

1 See www.learnreligions.com/elemental-symbols-4122788

Index

Note: Unless an organization's reference has a county's name in it, organizations listed are located in the United States.

T

U

V

W

Y

Printed in the United States
by Baker & Taylor Publisher Services

Printed in the United States
by Baker & Taylor Publisher Services